Estimating Impact

Alexander Kott · Gary Citrenbaum
Editors

Estimating Impact

A Handbook of Computational Methods
and Models for Anticipating Economic,
Social, Political and Security Effects
in International Interventions

 Springer

Editors
Alexander Kott
Army Research Laboratory
2800 Powder Mill Rd
Adelphi, MD 20783
USA
alexander.kott1@us.army.mil

Gary Citrenbaum
System of Systems Analytics, Inc.
3877 Fairfax Ridge Rd, Suite 201C
Fairfax, VA 22030
USA
gcitrenbaum@sosacorp.com

ISBN 978-1-4419-6234-8   e-ISBN 978-1-4419-6235-5
DOI 10.1007/978-1-4419-6235-5
Springer New York Dordrecht Heidelberg London

Library of Congress Control Number: 2010936359

Printed on acid-free paper

Springer is part of Springer Science+Business Media (www.springer.com)

Acknowledgments

This book, like most of its genre, represents a collective effort in more ways than one. Not only does it draw from the explicit efforts of its authors, but it also benefited substantially from a large support team – our teachers, advisors, coworkers, and families. We are delighted to recognize, and thank all of these individuals, who are, fortunately for us, too numerous to call out separately.

We are particularly happy to acknowledge a small set of individuals whom our authors, as individuals, have suggested warrant special recognition. Specifically:

Alex Kott, Steve Morse, and Gary Citrenbaum thank Len Hawley, former National Security Council (NSC) official, for introducing them to the ideas of Next State Planning and for sharing with them the insights he gained as senior mentor for the Integrated Battle Command (IBC) program (latter referred to as the Conflict Modeling, Planning, and Outcome Experimentation [COMPOEX] program).

Gary Citrenbaum is pleased to recognize the invaluable support of his coworkers in assembling and fleshing out this book. These include: Sara Citrenbaum, who drew, or redrew all the graphics, Gina Mansell, who assembled and proofread the manuscript, and Jordan Willcox, who developed many of the chapter summaries.

Mark Abdollahian, Jacek Kugler, Brice Nicholson, and Hana Oh acknowledge the continuous support of Sandra Seymour, Gwen Williams and Lynda Marquez at the School of Politics and Economics, Claremont Graduate University.

Ravi Bhavnani, Dan Miodownik, and Rick Riolo jointly thank Petra Hendrickson and Roman Kareev for research assistance. Ravi Bhavnani acknowledges the support of the Department of Political Science, Michigan State University. Rick Riolo acknowledges the support of the Center for the Study of Complex Systems, University of Michigan. Dan Miodownik acknowledges support from the Israel Science Foundation (Grant #33007).

Bruce Skarin acknowledges the financial support of the Air Force Research Lab (AFRL) for portions of his research.

Alan Graham acknowledges the support of PA Consulting Group's Federal and Defense Services Practice, where most of his chapter was written, and where he received much of the valuable professional experiences behind it. Without PA's generosity and professional excellence, his work would not have been possible. In particular, the collaboration with Elizabeth Farrely, Tamara Greenlaw, Donna

Mayo, Tom Mullen, Joshua Park, Whitney Pickels, Ritu Sharma, Erik Waldron, and many others was both highly productive and enjoyable.

David Jonker and William Wright wish to acknowledge the senior subject matter experts they had the privilege of working with on the DARPA COMPOEX program. These Generals, Admirals, Ambassadors and U.S. Aid leaders provided invaluable insight and visualization guidance into how they "saw" complex systems.

William Bennett thanks Ed Waltz and Pat Allen for introducing him to the challenge of modeling media effects. He also thanks Jonathan Goldstein and Howie Zhu for their support in developing data interfaces and computational modules.

On a bittersweet note, Ed Waltz, Gary Citrenbaum, and others gratefully acknowledge the pioneering vision of Dr. John G. Allen, the first Program Manager of the DARPA Integrated Battle Command (IBC) program [later referred to as the Conflict Modeling, Planning and Option Exploration (COMPOEX) program]. John's passion and direction led to the development of the first computational framework for rapidly and practically integrating PMESII models in support of analysis and planning. Without his vision and encouragement, several of the technologies discussed in this book would not be available today. Although John did not live to see this book or many of the other fruits of his labors, he and his contributions will not be forgotten.

Contents

About the Authors

Dr. Mark Abdollahian is cofounder and Chief Operating Office of Sentia Group, Inc. a Washington DC based consulting and software services company that utilizes social science modeling to forecast and shape political outcomes. Dr. Abdollahian is in charge of research, development and engineering efforts, including SENTURION™ political modeling software. He has been a consultant to the World Bank, the United Nations, Department of Defense and the private sector, including Arthur Andersen, Motorola, McKinsey, Raytheon, British Aerospace, Chevron, and DeBeers. His research focuses on fusing micro- and macro-social behavioral theories into business intelligence and analytic solutions. He is coauthor of Power Transitions and a number of scholarly articles on strategic decision-making and conflict. Dr. Abdollahian earned his Ph.D. in political economy and mathematical modeling from Claremont Graduate University. Dr. Abdollahian is a coauthor of Chap. 3, Politics and Power.

Dr. William Bennett has over 30 years experience modeling and analyzing information and decision systems with application to military intelligence and operations planning. He earned both a Master of Science and Ph.D. in Electrical Engineering from the University of Maryland at College Park. He has held positions in technology development and management at ALPHATECH, Atlantic Aerospace Electronics, Techno-Sciences, Inc. and US Naval Research Laboratory, and has served as adjunct faculty and advisor to chairman of the Electrical and Computer Engineering Department of University of Maryland. Dr. Bennett is a senior member of IEEE, the American Institute of Aeronautics and Astronautics and Society for Industrial and Applied Mathematics and Military Operations Research Society, Dr. Bennett currently serves as Principal Research Engineer at BAE Systems, Advanced Information Technologies. His current research addresses complex systems analysis and decision aids for sociotechnical systems, and modeling methods to support strategic communication for the US Department of Defense. Dr. Bennett is the author of Chap. 5, Modeling of Media Influence.

Dr. Ravi Bhavnani earned his Ph.D. in political science from the University of Michigan, Ann Arbor with a degree certificate from the UM's Center for the Study of Complex Systems. His thesis research, conducted under the supervision of Dr. Robert Axelrod, explored the link between intraethnic and interethnic violence.

Dr. Bhavnani is currently an Assistant Professor in the Department of Political Science at Michigan State University, where his current research focuses on the microfoundations of violence in ethnic and nonethnic civil wars using agent-based computational modeling. Recent publications have appeared in the *Journal of Politics* (2009), *Journal of Conflict Resolution* (2009), *Complexity* (2008), and *Journal of Artificial Societies and Social Simulation* (2008). Dr. Bhavnani is a coauthor of Chap. 7, Groups and Violence.

Dr. Gary Citrenbaum serves as the President and Chief Scientist of System of Systems Analytics, Inc. (SoSACorp), headquartered in Fairfax, VA. He has been designing and consulting on complex systems for the United States Government for over 35 years. His specialty is concept development, as well as system design, development, utility assessment, and deployment. Dr. Citrenbaum's work has spanned a wide range of real-world problems: treaty monitoring, operations planning, intelligence collection and analysis, approaches for analyzing and mitigating international crises and to the development of tools for counterterrorism problems. His research interests include interagency planning approaches, large-scale optimization, modeling using agent-based, system dynamic, Markov approaches, system and process utility assessment, structured argumentation, and social network analysis. Dr. Citrenbaum served as coeditor of this book.

Dr. Deborah Duong earned her Ph.D. in Computational Science and Informatics from George Mason, majoring in Computational Social Science, in 2004. She currently researches modeling and simulation of irregular warfare that has created several modeling and simulation programs for the United States Government, including the Nexus Cognitive Agent-Based Simulations and the Oz Wargame Integration Toolkit; and the MICCE and Indra Natural Language Systems to extract Role ontologies from text. Her specialty is the use of AI techniques to simulate the phenomena of Interpretive Social Science. Her academic works include her dissertation work SISTER, the Symbolic Interactionist Simulation of Trade and Emergent Roles, published in JASSS. Dr. Duong is a coauthor of Chap. 9, Crime and Corruption.

Dr. Alan K. Graham earned his Ph.D. in Control Theory and Decision Science in Electrical Engineering from MIT. As faculty and staff at MIT, he was Research Director for the System Dynamics National Economic Model Project. Dr. Graham's experience includes 13 years at PA Consulting (an international management and technology consultancy) where he led teams using system dynamics models of market competition to address corporate strategy (in marketing, market forecasting, technology, and capital investment) in several industries (wireless and landline telcos, telecoms equipment, specialty and commodity chemicals, oil, pharmaceuticals, and shipping). For the government clients, Dr. Graham has led analyses for government regulation (electric power, wireless and landline telco), and government policy (counterinsurgency, strategic deterrence, economic impacts in multiple regions, and economic warfare). Coauthor of two books and numerous articles, he was on the Editorial Board of the highly cited *System Dynamics Review* for 23 years.

He is a Senior Member of the IEEE. Dr. Graham is the author of Chap. 4, Economics and Markets. Since 2009, he has been Chief Knowledge Officer for Greenwood Strategic Advisors.

Dr. Dean Hartley III is the Principal of Hartley Consulting. Previously, he was a Senior Member of the Research Staff at the Department of Energy Oak Ridge Facilities (Oak Ridge National Laboratory, Y12 Site and East Tennessee Technology Park). He is a past Vice President of the Institute for Operations Research and Management Science (INFORMS), a past Director of the Military Operations Research Society (MORS), past President of the Military Applications Society (MAS), and a member of the College on Simulation of INFORMS. Dr. Hartley is a Senior Fellow with the George Mason University School of Public Policy, a consultant for the Naval Postgraduate School (NPS), Modeling, Virtual Environments and Simulation (MOVES) Institute, and a Research Fellow with the University of Alabama in Huntsville, Center for the Management of Science and Technology (CMOST). His expertise includes modeling of combat, operations other than war (OOTW), stability and support operations (SASO), and stability, security, transition, and reconstruction (SSTR) operations, verification and validation of models, psychopharmacology modeling, and simulation. Dr. Hartley has published *Predicting Combat Effects*, contributed chapters to three other books, and written more than 150 articles and technical documents. Dr. Hartley is a coauthor of Chap. 11, Verification and Validation.

David Jonker is a Partner and cofounder at Oculus Info Inc. He is a senior system architect and senior visualization designer with over 15 years experience in visualization system research, design, and development. Mr. Jonker served as the human computer interface and visualization lead designer and technical architect on a major research program dedicated to modeling complex societal effects. He led the design and implementation of a system that helps senior decision-makers to visualize large complex social, political, and economic behaviors, to explore alternative actions in those domains, and to understand effects. Mr. Jonker is chief technical architect for Oculus main products. His research interests include high performance information visualization technical architectures, as well as visualization semiotics. He has authored papers on information visualization. Mr. Jonker earned a B.E.S. and B.Arch. from the University of Waterloo. He is coauthor of Chap. 10, Visualization and Comprehension.

Dr. G. Jiyun Kim earned his doctoral degree in Political Science, together with a graduate certificate in Complex Systems, from the University of Michigan, where he specialized in political transitions and conflict resolution. He was a postdoctoral researcher at the University of Pennsylvania in the Department of Electrical and Systems Engineering. His research involved: 1) building and empirically validating virtual countries using a multi-resolution, agent-based model; and 2) building an international crisis simulator to monitor and assess the North Korean nuclear crisis. He is currently an Adjunct Professor in the Department of Politics at New York University. His current research and teaching focus on the economic and financial ramifications of political instability in emerging markets. Dr. Kim is a coauthor of Chap. 1, Introduction, Judgment and Computation.

Dr. Alexander Kott currently serves as a Division Chief at the United States Army Research Laboratory (ARL). Prior to joining ARL he worked at the Defense Advanced Research Projects Agency (DARPA) – the central R&D organization of the United States Department of Defense. At DARPA, he initiated and managed several large-scale research programs, including the Conflict Modeling, Planning and Outcomes Exploration (COMPOEX) program. His research interests include modeling of societal systems, phenomena in social and information networks, dynamic planning in rapidly changing, uncertain, and adversarial environments, and instabilities in decision-making systems. Dr. Kott has published over 70 technical papers and has served as the editor and coauthor of several books, including *Information Warfare and Organizational Decision Process*, *Adversarial Reasoning*, and *The Battle of Cognition*. Dr. Kott served as senior editor of this book and coauthored Chap. 1, Introduction: Judgment and Computation, Chap. 8, Insurgency and Security, and Chap. 12, Conclusions: Anticipation and Action.

Dr. Jacek Kugler is the Elisabeth Helms Rosecrans Professor of International Relations at the School of Politics and Economics, Claremont Graduate University, where he has also served as Chair. He is a cofounder of Sentia Group Incorporated. Dr. Kugler is the editor of *International Interactions*, a past president of the International Studies Association and the Peace Science Society. His numerous publications in world politics and political economy are widely available in scholarly journals. He is the coauthor of *The War Ledger*, *Births, Deaths, and Taxes*, and *Power Transitions* as well as the coeditor of *Parity and War*, the *Long Term Stability of Deterrence*, and *Political Capacity and Economic Behavior*. Dr. Kugler earned his Ph.D. in world politics from the University of Michigan. Dr. Kugler is a coauthor of Chap. 3, Politics and Power.

Dr. Corey Lofdahl is a Consulting Engineer at IGEN Corporation in Carlisle, Massachusetts where he uses advanced simulation and statistical techniques to analyze hard policy and strategy problems and develops supporting software. He was the conflict economics modeler for DARPA's Integrated Battle Command (IBC) project. Dr. Lofdahl earned degrees in electrical engineering, computer science, and international relations from the University of Colorado at Boulder, Brown University, and MIT, and is the author of *Environmental Impacts of Globalization and Trade: A systems study* (MIT Press, 2002). Dr. Lofdahl is the author of Chap. 6, Governance and Society.

Dr. Dan Miodownik earned his Ph.D at University of Pennsylvania, Political Science Department. He currently serves as an assistant professor in the Departments of Political Science and International Relations at the Hebrew University of Jerusalem, where he studies the emergence, unfolding and regulation of antiregime mobilization, protest behavior, ethnic polarization, and civil wars. Dr. Miodownik also has significant interest in the development of computer simulations – Agent-Based Modeling in particular – to assist comparative political analysis of these and other complex social phenomena. His work has been published in journals such as the *American Political Science Review*; *Journal of Conflict Resolution*; *Comparative Politics*; *Studies in Comparative International Development*; *Nationalism & Ethnic Politics*; *Social Science Computer Review*; and *Journal of Artificial Societies and Social Simulations*. Dr. Miodownik teaches courses on ethnic conflict and mobilization, civil wars, the study of politics and research methods. Dr. Miodownik is a coauthor of Chap. 7, Groups and Violence.

Dr. Stephen Morse has over 30 years of experience as a technical consultant to various institutions of the U.S. Government. His contribution to the current volume is based on his research in use of model predictive control for optimized decision-making in complex dynamic environments. His other relevant efforts include: PMESII modeling; rapid optimization algorithms for adaptive replanning; remote sensing; high performance computing and parallel processing; and data mining on very large data warehouses. Dr. Morse is the author of two books and numerous technical articles and presentations. Dr. Morse is a coauthor of Chap. 12, Conclusions: Anticipation and Action.

Brice Nicholson is a Ph.D. student at Claremont Graduate University, where his research interests include systems dynamics applied to security, stability, transition, and reconstruction operations (SSTRO), and integration. His most recent project was the development of a formal model of stabilization and reconstruction operations while working at the National Defense University. Mr. Nicholson earned his B.A. in Political Science and History from California State University Northridge and his M.A. in International Political Economy from Claremont Graduate University. Mr. Nicholson is a coauthor of Chap. 3, Politics and Power.

Dr. Hana Oh earned her Ph.D. in political economy and comparative politics from Claremont Graduate University. She has worked with the World Bank, the United Nations, and the private sector, and is currently a Senior Analyst at Sentia Group, where she is responsible for political risk analysis and mitigation strategies. Her research spans computational modeling, social network analysis, and behavioral game theory applied to real world policy problems, and she has published articles on social and dynamic modeling of human behavior. Dr. Oh is a coauthor of Chap. 3, Politics and Power.

Dr. Rick Riolo earned his Ph.D. in Computer Science at the University of Michigan (UM) for research focused on genetic algorithms and classifier systems. He currently works at the UM's Center for the Study of Complex Systems, where he provides agent-based modeling expertise in support of interdisciplinary research projects involving topics such as urban sprawl and its ecological impacts; decision making in closed political regimes; the evolution of cooperation, supply chain dynamics, coordination of agents competing for limited resources; the relationship between phenotype plasticity and the structure and dynamics of food webs; and the spread of antibiotic resistance in nursing homes and other nosocomial settings. Dr. Riolo is a member of the Swarm Board of Directors (SDG) and the Repast Organization for Architecture and Development (ROAD), groups responsible for the development and distribution of Swarm (see http://swarm.org) and Repast (http://repast.sourceforge.net), software packages for creating agent (individual)-based models (ABM/IBM). Dr. Riolo is a coauthor of Chap. 7, Groups and Violence.

Karl Selke is currently a Senior Systems Scientist for the Center for Complexity Analysis (CCA), a second year Ph.D. student at George Mason University's Center for Social Complexity, and works as an irregular warfare analyst for a U.S. government organization. Before joining CCA in August 2008, he spent 2½ years as a Systems Engineering Analyst at Evidence-Based Research, Inc. on several projects involving numerous clients such as the Defense Advanced Research Projects Agency, the Defense Information Systems Agency, and the Navy's Second Fleet. Mr. Selke has a BS in political science from Lake Superior State University and a MS in systems engineering focusing on operations research and management science from George Washington University. His area of research is focused on the processes and techniques for integrating computational social science models into strategic-operational wargaming. Mr. Selke is a coauthor of Chap. 9, Crime and Corruption.

Bruce Skarin is a Simulation Scientist leading projects that utilize computational models to aid decision making, collaboration, and knowledge management at Aptima, Inc. Mr. Skarin is also a part-time MS student at Wocester Polytechnic Institute where he earned a BS in System Dynamics. His experience includes system dynamics, agent-based, and discrete event modeling with a focus on simulating sociocultural behavior, social networks, and organizational dynamics. While at Aptima he has designed and built an agent-based model that forecasts changes in local populations that is used to assist ongoing strategic planning. He is a coauthor of Chap. 8, Insurgency and Security.

Dr. Stuart H. Starr is a Distinguished Research Fellow at the Center for Technology and National Security Policy (CTNSP), National Defense University (NDU). Concurrently, he serves as President of Barcroft Research Institute (BRI), where he consults on Command and Control (C2) and Modeling and Simulation (M&S) issues (e.g., The Institute for Defense Analyses (IDA)), lectures to audiences world-wide on C2 and M&S issues, and participates on Blue Ribbon panels (e.g., member of the Army Science Board). Prior to founding BRI, Dr. Starr was Director of Plans, The MITRE Corporation; Assistant Vice President for C3I Systems, M/A-COM Government Systems (currently a unit of SAIC); Director of Long Range Planning and Systems Evaluation, OASD(C3I), Office of the Secretary of Defense (where he was member of the Senior Executive Service); and Senior Project Leader, IDA. Dr. Starr earned his Ph.D. and MS in Electrical Engineering from the University of Illinois, a BSEE from Columbia University, and a B.A. from Queens College. He has received the Clayton Thomas medal (2004) and the Vance Wanner medal (2009) from the Military Operations Research Society (MORS) for lifetime accomplishments in operations analysis. Dr. Starr is a coauthor of Chap. 11, Verification and Validation.

Robert Turner leads the systems analysis efforts at IMAG in the Office of the Secretary of Defense. He has over 30 years of experience in systems analysis, including the creation and management of innovative analysis tools, such as FSAT, PET, the CAST software suite, and the Oz Wargame Integration Toolkit. He specializes in the analysis of both conventional and Irregular DoD simulation models, including JAS, TACAIR, JICM, and the Nexus Agent-Based Model Suite. Mr. Turner is a coauthor of Chap. 9, Crime and Corruption.

Ed Waltz is the Chief Scientist, Intelligence Innovation Division of BAE Systems Advanced Information Technologies, where he leads intelligence analysis and information operations planning research for government organizations. For the past decade, his research has focused on modeling human systems, including foreign leadership, organizations, and social populations. He earned a BSEE from the Case Institute of Technology and an MS in Computer, Information and Control Engineering from the University of Michigan. He has over 35 years of experience in developing and deploying signal processing, data fusion, and intelligence analysis capabilities. He is the author of *Knowledge Management in the Intelligence Enterprise, Information and Warfare Principles and Operations*, the coauthor of *Counterdeception: Principles and Applications for National Security* and *Multisensor Data Fusion*. Mr. Waltz is the author of Chap. 2, Emerging Techniques and Tools.

William Wright is a Senior Partner at Oculus Info Inc. A researcher and practitioner in information visualization since the late 1980s, he founded Visible Decisions in 1992 and cofounded Oculus Info Inc. in 2001. He works directly with major organizations in New York, Washington and elsewhere to create and apply visual analytic methods for competitive business advantage. Mr. Wright has served as Principle Investigator on major research programs sponsored by the U.S. government. Results of his research include several widely used software systems. His current research interests include intelligent, mixed initiative, and visual analytic systems that enhance human understanding and decision-making. He has written over 25 papers on information visualization and visual analytics. Mr. Wright earned a B.A.Sc. and M.A.Sc. in Systems Design Engineering from the University of Waterloo. Mr. Wright is coauthor of Chap. 10, Visualization and Comprehension.

Contributors

Mark Abdollahian
Sentia Group, Inc, 1066 31st Street, NW, Washington D.C. 20007, USA
maa@sentiagroup.com

William H. Bennett
BAE Systems, Inc., Technology Solutions, 6 New England Executive Park,
Burlington, MA 01803
william.bennett@baesystems.com

Ravi Bhavnani
Department of Political Science, East Lansing, Board of Trustees, MI 48824, USA
rvibhav@gmail.com

Gary Citrenbaum
System of Systems Analytics, Inc., 3877 Fairfax Ridge Road Suite 201C, Fairfax,
VA, 22030-7425, USA
gcitrenbaum@sosacorp.com

Deborah Duong
NPS, 555 Dyer Road, Quarters "B" on Stone Road, Monterey, CA 93943, USA
dduong@aciedge.com

Alan K. Graham
Greenwood Strategic Advisors AG, Unteraegeri, Switzerland
alan.graham@greenwood-ag.com

Dean Hartley
Hartley Consulting, 106 Windsong Ln, Oak Ridge, TN, 37830, USA
dshartley3@comcast.net

David Jonker
Oculus Info Inc, 2 Berkeley Street, Suite 600, Toronto ON, M5A 4J5, Canada
david.jonker@oculusinfo.com

G. Jiyun Kim
New York University, 19 West 4th Street 2nd Floor, New York, NY 10012, USA
gjiyunkim@gmail.com

Alexander Kott
Army Research Laboratory, 2800 Powder Mill Road, Adelphi, MD 20783, USA
alexander.kott1@us.army.mil

Jacek Kugler
Claremont Graduate University, 150 E. 10th Street, Claremont, CA, 91711, USA

Corey Lofdahl
IGEN Corporation, 84 Hobblebush Lane, Carlisle, MA 01741, USA
clofdahl@igencorp.com

Dan Miodownik
Department of Political Science and International Relations,
Hebrew University of Jerusalem, Mount Scopus, Jerusalem, 91905, Israel

Stephen Morse
Cobham Analytic Solutions, 5875 Trinity Parkway Ste 300 Centreville, VA 20120

Brice Nicholson
Claremont Graduate University, 150 E. 10th Street, Claremont, CA, 91711, USA

Hana Oh
Sentia Group, Inc, 1066 31st Street NW, Washington, DC, 20007, USA

Rick Riolo
Center for the Study of Complex Systems, University of Michigan,
515 E. Jefferson Street, Ann Arbor, MI, 48109-1316, USA

Karl Selke
Center for Social Complexity, Krasnow Institute for Advanced Study, George Mason
University, Rms. 373–384 Research 1 Building, MS 6B2, Fairfax, VA, 22030, USA

Bruce Skarin
Aptima Inc, 12 Gill St., Ste. 1400, Woburn, MA, 01801-1753, USA

Stuart Starr
Center for Technology and National Security Policy, National Defense University,
300 5th Avenue SW, Fort Lesley J. McNair, Washington, DC, 20319, USA

Robert Turner
IMAG, 6707 Democracy Blvd, Suite 200, Bethesda, MD, 20892, USA

Ed Waltz
Intelligence Innovation Division, BAE Systems Advanced Information
Technologies, Ann Arbor, MI, USA
ed.waltz@baesystems.com

William Wright
Oculus Info Inc, 2 Berkeley Street, Suite 600, Toronto ON, M5A 4J5, Canada

Chapter 1
Introduction: Judgment and Computation

Alexander Kott, Gary Citrenbaum, and G. Jiyun Kim

Regardless of political views, the consensus of observers is that international interventions in Afghanistan (beginning in 2001) and Iraq (beginning in 2003) could have been planned better. In particular, objective observers suggest that the undesired effects of the plans could have been anticipated more accurately, and that, in turn, many of the resultant difficulties and tragedies could have been avoided. However, is this true? Do there exist rigorous and practical techniques for anticipating the effects of complex interventions, and for planning in accordance with such anticipations? The last few years have, in fact, seen the emergence of quantitative, computational tools and techniques that attempt to answer this question. Inspired by such efforts, this book aims to offer the reader a broad and practical introduction to computational approaches for anticipating the effects of interventions.

The book is written for practitioners, by practitioners. We envision that readers of this volume will include developers of computational tools and models, as well as professionals who acquire, integrate, and adapt such tools. This book should also benefit analysts, who use such tools, and decision-makers who wish to learn about the opportunities and limitations that the new computational approaches offer.

As will become apparent, we use the term intervention as a generalization of a broad spectrum of actions and operations – diplomatic, informational, military, economic, and others – by which a country or coalition may attempt to influence the events in, and the actions of a region. Although the word intervention suggests to many a military invasion, historically most interventions have had little or no military component. In many cases, they have been purely diplomatic, or purely economic.

Interventions may focus on different areas: e.g., crisis prevention or crisis management; peace building or peacekeeping; or security, stability, and recon-

A. Kott (✉)
Army Research Laboratory, 2800 Powder Mill Road,
Adelphi, MD 20783, USA
e-mail: alexander.kott1@us.army.mil

A. Kott and G. Citrenbaum (eds.), *Estimating Impact*,
DOI 10.1007/978-1-4419-6235-5_1, © Springer Science+Business Media, LLC 2010

struction operations. However, in other cases, fortunately relatively rare, they involve full-scale war or extensive counter-insurgency operations. Furthermore, many interventions may entail acts not of sovereign governments or coalitions of states, but of Non-Governmental Organizations (NGOs) such as famine relief organization. These interventions, as well as those of international business, can have multiple, often unanticipated effects and far reaching international consequences.

International business is a form of intervention in its own right, but it is also highly dependent on other forms of intervention. Governmental and NGO intervention in a region can, for instance, impact the value of a corporation's investments in a specific country, change the business and labor climate, affect expenses on security, and increase or reduce corruption, among many other effects. Thus, an ability to estimate the effects of such interventions is of critical importance to business analysts and planners.

Intervention is inherently international and transnational. In today's smaller, flatter, and interdependent world, interventions of all sorts – from military to economic sanctions or aid, to natural-disaster relief, to various coalition and civil-military engagements – are likely to persist. They are also likely to become yet more complex and difficult as the world changes.

In current practice, intervention-related decisions, planning, and effects estimation relies on historical analogies, on qualitative theories, and on expert opinion, experience, and intuition. Without minimizing the importance of soft theories, experience, and intuition, this book argues for a broader, more balanced view. It describes how emerging computational techniques can support anticipation and planning in interventions. Our challenge is to maintain a balanced and critical coverage of technological aspects of the intervention problem, without dismissing those aspects that may not be amenable to quantitative approaches.

Emergence of quantitative, computational methods in a previously qualitative discipline is a universal trend. To employ an analogy (admittedly imperfect, as is any analogy) one can consider an ancient and complex human endeavor – construction of bridges. Today, building a bridge without prior rigorous computational analyses of stresses in the proposed design would constitute an irresponsible and possibly criminally negligent act. Yet, the first book on computing stresses in bridge frames appeared only relatively recently, in 1847. "Before that time bridge members were proportioned according to the judgment of experienced builders, which was often defective" (Tyrell 1911).

Similarly, some years from now it may be considered irresponsible and perhaps even criminally negligent to undertake an intervention (or to decide not to intervene) without employing a computational analysis – along with expert assessments – of the effects of candidate intervention. Few will want to return to the days when decisions on international interventions relied exclusively on qualitative opinions and judgments.

1   Common Recent Practices

One finds few examples prior to about 2005 of rigorous quantitative analyses aimed at predicting or anticipating the effects of interventions. To be sure, there are exceptions. Financial interventions and economic aid are and have for many years been subject to rigorous analysis of potential effects (Taylor 2007), and military aid was sometimes preceded by an analysis of its effects on the regional military power balance. Other types of interventions, however, particularly those involving a complex set of multiple interacting social phenomena, rarely benefited from a systematic predictive analysis of effects and outcomes.

Observers of interventions frequently note and lament this lack of rigorous planning and analysis. Consider the case of international sanctions imposed on Iraq beginning in 1991. This complex intervention combined economic pressure (embargo), military intervention (maintenance of no-fly zones), a food-for-oil program, and humanitarian aid. To magnify this complexity, the application of the intervention mechanisms differed between Iraqi Kurdistan and the rest of Iraq. And, the effects of the intervention were just as complex and multifaceted. Impacts on the economy included inflation, changes in food availability and distribution control, and growth of the illicit economy. Equally extensive were the social impacts, which included effects to family and tribal relations, crime, and emigration and even to the nature of the society's reliance on the state and to the state's approach to the manipulation of social groups.

The planning of this intervention, however, did not appear to include a predictive analysis of such effects. The history of the international community's decisions on the extent and nature of the sanction mechanisms is one of confusion, bickering, and internal partisanship. Conspicuously absent is any mention of a quantitative or systemic predictive analysis of the sanction's effects. Particularly with humanitarian aid, reliance on short-term measures and lack of planning and predictive analysis resulted in unanticipated negative effects, such as manipulation of the aid by a few political and business figures, underappreciated difficulties in resettling the population from refugee camps, and a self-perpetuating dependence of the population on food aid (Graham-Brown 1999, pp. 303–325).

1.1   Planning Guides

In response to concerns about inadequate planning of interventions, a large number of publications have emerged: planning guides, doctrinal manuals, descriptions of best practices, checks lists, and so forth. These tend to recognize that planning does not have to include – and in practice rarely does include – a prediction of effects. However, they also suggest that there is an important connection between planning and effect prediction, and make the implicit assumption that a plan based on good guidance has the best chance of succeeding.

The Guide for Participants in Peace, Stability, and Relief Operations (Perito 2007), for example, offers readers a "Framework for Success for Societies Emerging from Conflict." The guide includes a table in which columns represent desired end-states – security, rule of law, stable democracy, sustainable economy, and social well-being – and rows contain tasks and key objectives for leadership – e.g., "protect human rights," "promote sound fiscal policy," etc. In effect, the guide links suggested tasks to desired end-states, and in this way implies the anticipated effects of the tasks.

A very similar table appears in the Essential Tasks Matrix of the Office of the Coordinator for Reconstruction and Stabilization (2005). Here, detailed lists of tasks are organized by the goals that they are intended to accomplish. The authors of the matrix view it as a tool for planners, and are careful to stress that timing, appropriateness, and priority of tasks will vary from case to case, depending on the circumstances underpinning a particular intervention. The US Army (2008) field manual on stability operations refers to the same matrix, discusses the tasks in detail, and offers guidance regarding detailed planning and execution of these tasks.

The weakness of such planning aids – as their authors generally recognize and acknowledge – is that intervention scenarios are very complex, and that similar intervention activities can produce very different effects in different situations. The upshot is that a standard procedure that works well in some interventions may, in fact, prove quite counterproductive in others. Ashraf Ghani, the finance minister of Afghanistan in 2002–2004, alludes to this point in describing the debilitating effects of the standard procedures applied by NGOs in providing food aid to Afghanistan. In a vicious chain of events, international food aid depressed grain prices and induced farmers to grow poppies, which led to a poppy eradication campaign, which caused resentful farmers to support the Taliban insurgency, which undermined the state's capacity to support economic development, and so forth (Ghani and Lockhart 2008).

To some extent, this weakness can be mitigated by providing more specific guidance on the applicability and effects of actions. A certain guide on famine aid, for instance, explicitly differentiates recommendations based on the situation. It describes conditions suitable to such approaches as income-support projects, food ration distributions, market interventions, direct and indirect monetization, and payment-in-kind programs. Moreover, special methods are suggested for operating in conflict zones, such as border enclaves. The guide even offers qualitative rules for anticipating effects of intervention actions when, for example, free food delivery may cause systemic or psychological dependency, or destruction of local agriculture (Cuny and Hill 1999). Such guides provide planners a collection of triples – situation, actions, and effects – and a planner's challenge devolves to that of matching the situation under consideration with the ones available in the guide.

1.2 Historical Analogy

Drawing analogy between a given situation and a historic case is another way to predict potential effects of proposed actions. With references to "another Munich" or "another Vietnam," analysts and decision-makers often make a claim that the situation under consideration is similar to a certain historic case, and argue that similar actions or inaction will lead to similar outcomes.

For example, debates about the need for an international intervention in Darfur frequently referred to the Rwanda's genocide of 1994. Decision-makers felt uncommon urgency specifically because the experience of Rwanda was fresh in their collective memory. It was unthinkable to permit a repetition of Rwanda's grim outcomes by repeating the inadequate responses to the early stages of Rwanda's crisis. Discussing ethnic cleansing in Darfur, a key UN official insisted on drawing a direct analogy with Rwanda (Slim 2004). And, relying on the same analogy, a number of experts agreed that it was important to openly call genocide by its name, and thereby mobilize the will of the international community (Straus 2006).

Yet, finding a suitable historical analogy and interpreting it in application to a given situation is by no means easy. History does not repeat itself. There are usually significant differences between historical situations. And, even when differences are small, there is still the possibility that they will, in nonlinear fashion, produce significant differences in the final effects. Consequently, experts rarely agree on whether differences between two situations are small enough to allow a suitable analogy to be drawn (Siegel 2007).

It is possible that the importance of historical analogies is seriously overrated. Some argue that historical analogy is not a major source of decisions, and that intervention decisions (presumably including assessments of potential outcomes of the decisions) arise largely from the ideologies and partisan interests of decision-makers. Analogies, they claim, serve merely as post hoc justifications (Taylor and Rourke 1995) for decisions based on other factors.

Ideally, one would like to use a rigorous, objective methodology for deriving decision-relevant information from historic cases. Social scientific analysis could help analysts and decision-makers diagnose intervention situations and infer the likely effects of a particular intervention under consideration. Multivariate quantitative approaches could contribute to the analysis of multiple features in multiple cases, as opposed to a few analogies that a human mind can draw at any given time.

Unfortunately, although there is an abundance of multivariate quantitative approaches, the approaches do not always yield consistent and conclusive results concerning the impact of interventions, and a significant amount of research will be required before they meet the emerging analysis needs. There are, for instance, needs for taxonomies of interventions and their outcomes, for theories of causal mechanisms that explain workings of interventions, and for hypotheses of the links between characteristics of interventions and their outcomes. All of these remain subjects of research (Stern and Druckman 2000) and, for now, the drawing and interpretation of analogies remain matters of judgment.

1.3 Expert Judgment

Historical analogy is merely one of the tools that analysts employ. Expert judgment – informed opinions that experts offer in response to a given problem – is also widely used in many technical and social disciplines (Meyer and Booker 2001). The cognitive mechanisms of expert judgment are complex and to a large degree intuitive (Hammond et al. 1987), however, the resulting judgments are thought to be useful, and expert judgment is today probably the most common approach to intervention outcome prediction.

There are a number of advantages to using expert judgments for predicting effects. For instance, such judgments can make use of a diverse body of experience, education, knowledge, and skills, and thus can be very robust. Moreover, the judgments can be produced rapidly, without technical tools, and with only a modest investment of labor. And importantly, when experts reach a consensus, it is particularly compelling to decision-makers. Expert judgment is particularly useful when a phenomenon is poorly understood and information is uncertain and open to conflicting interpretations – as is often the case in planning an intervention.

On the other hand, expert judgment has its limitations. The solicitation method, i.e., the process of gathering opinions, has a significant impact on expert judgment. The expert's personal agenda, biases, ideology, and desire to conform may all influence the judgment as well. It is often difficult to trace the logic of the judgment, and even when the expert provides a line of reasoning in support of the judgment, it may be merely a post hoc justification. Different experts may offer well-argued yet entirely conflicting judgments, further complicating the task of the decision-maker.

Consider how the strengths and weaknesses of expert judgment manifested themselves in the case of the surge: the 2007 decision by the U.S. President George W. Bush to quell the growing violence in Iraq by injecting additional U.S. troops. Insurgency and sectarian violence in Iraq had been growing steadily since the 2003 invasion by the U.S.-led coalition. By 2006, the rate of U.S. casualties and of civilian deaths in Iraq had reached such an alarming level that the President felt strong public pressure to make significant changes in the conduct of the war. Most experts, including the Joint Chiefs of Staff and top commanders of U.S. troops in Iraq, advised a strategy based on maintaining the existing level of U.S. troops in Iraq, increased focus on development of the Iraqi army, and gradual disengagement of U.S. troops from security duties in Iraqi cities. They envisioned long-term positive effects of these actions but not a rapid reduction in violence (Woodward 2008).

A much smaller group of experts, arguably of significantly lesser stature and without formal positions in the U.S. military or government, consisted largely of the retired general John Keane and the think-tank scholar Frederick Kagan. This group of experts recommended executing a large and sustained surge of U.S. forces to secure and protect critical areas of Baghdad. They anticipated a major reduction in violence within 18–24 months as a key effect of the proposed intervention (Kagan 2006).

President Bush, therefore, faced two drastically different recommendations for action and different estimates of the resulting effects. It is difficult to see what rational analysis could lead the President to choose between the two conflicting expert judgments. Both judgments were well-articulated and offered careful, compelling arguments. The antisurge judgment originated with a much larger and stronger-credentialed group of experts. Yet, the President elected to follow the pro-surge judgment, possibly for ideological and emotional reasons. In January of 2007, he announced his decision and the corresponding plan. As the surge of troops in Iraq proceeded, in June–July of 2007 the level of violence in Iraq started to drop drastically, reaching a 3-year low 6 months later.

After the events, although the drop in violence was undisputable, expert judgments on the nature and cause of the drop varied widely. A number of experts argued that the drop in violence had nothing to do with the surge and that the timing was merely a coincidence. Instead, they identified a range of alternative mechanisms as causes of the drop. Some argued that by the time the surge started, the ethnic cleansing had reached its completion, and few targets for intercommunal violence remained living in mixed neighborhoods (Agnew et al. 2008). Others pointed to the Sunni tribal revolt against anti-American insurgents (Petraeus 2007; Simon 2008). Yet other experts credited the new, more efficient techniques employed by the U.S. military to find and kill insurgents (Woodward 2008).

Even after the events occurred, even when pertinent facts can be collected and examined, experts' judgments may remain far apart. The differences and uncertainty are even greater in judgments made a priori. Clearly, a decision-maker faces major challenges in choosing between conflicting expert judgments and lacks a rational, systematic methodology for doing so.

2 Emerging Computational Approaches

Recent major international interventions – e.g., the invasion of Afghanistan in 2001 and the invasion of Iraq in 2003 – sparked an explosive growth of research on methods and tools, particularly computational tools for estimating the effects of interventions.

A distinguishing feature of this research and development was its focus on the tight coupling of and interaction between phenomena normally studied in different disciplines: e.g., economics, social and political sciences, and military science. Interventions, it seems, produce complex, multifaceted ramifications that do not lend themselves well to techniques and models confined to a single discipline.

The focus on intertwined effects of such interactions led to the coining terms like Diplomatic, Informational, Military, Economic, Financial, Intelligence and Law enforcement, referred to as DIMEFIL or just DIME (JFCOM 2004), to capture the totality of interrelated means by which intervening parties can influence a subject of intervention. A related term – Political, Military, Economic, Social, Infrastructure and Information, or PMESII (JFCOM 2004) – was coined to denote the totality of an intervention's effects. To oversimplify, in this school of thought an intervention is a process in which DIME is the vector of inputs and PMESII is a vector of outputs, or effects of the inputs. In 1998, one could find virtually no modeling or simulation tools, computational or manual, that took the intertwined DIME-PMESII perspective, yet by 2008 there were dozens of such tools (Hartley 2008).

The roots of many computer-based tools lie in manual wargaming, an analytical role-playing technique widely used in military and business decisions

(Gilad and Stitzer 2008). Consider a wargame where an international corporation analyzes its plans to initiate business operations in a country where it has not operated before. The wargame involves establishing several teams: one representing the corporation itself, another representing a competitor, another, the primary market segment, yet another, the country's government, and so forth. The teams gather in an office space where each team has its own conference room, and a larger room where all teams meet periodically. There is also a neutral team responsible for managing the wargame and adjudicating the results of each round. Before the wargame, each team receives a detailed brief on the purpose and rules of the game, and on key data pertinent to the country situation, competition, and corporate plans.

The wargame proceeds in rounds, 5–10, perhaps, and the role-playing teams execute each round in a few hours, taking into account the fact that each round may represent days, weeks, or even months of real world time. At the beginning of a round, each team discusses plans and role-plays its actions; e.g., the team playing the corporation begins executing various business and legal actions necessary to initiate the operations in the country; the team playing the country's government responds with legal or taxation actions, etc. After each round, the participants present their actions to an adjudication team, which determines the outcomes of the round – changes in the attitude of consumers or changes in prices and market share, for example. The process continues until the game's objectives are met or the prepared scenario is exhausted.

The products and lessons of the wargame are diverse. Participants are able to explore the likely outcomes and often unanticipated effects of a plan, to identify possible reactions and counteractions of competitors and government bodies, to characterize the evolution of market attitudes, and to identify alternative implementations of a plan (Kurtz 2003).

Although frequently used for business decision-making, wargaming is even more common in military practice. There, a wargame simulates a planned military operation in order to explore possible actions, reactions, and counteractions of the opposing forces, and to estimate the likely outcomes of the operation. Traditionally, the focus of wargames was rather narrowly circumscribed; military wargames focused on forces, weapons, movements, and fires; while business wargames focused on marketing, investments, revenues, and profits.

However, by about 2005, spurred by difficulties in then-ongoing international intervention in Iraq, many organizations began conducting wargames of new kinds. Due to concerns about the effects of interconnected multidisciplinary phenomena, these wargames started to encompass a far broader range of phenomena than traditional wargames. Wargamers considered how political developments might impact media and population perceptions; how these in turn might impact the economy of a country; how economy could affect employment and food production; how these would reflect on resources available to the country's government and antigovernment movements; and so on (Walters 2008; Colaizzi 2005). Thus, it is not surprising that some of the earlier tools for computational analysis of intervention-related phenomena are based directly on the manual wargaming paradigm.

One example of a wargaming-based approach to planning is the Peace Support Operation's Model (PSOM, Marlin 2009). PSOM is a computerized, time-stepped wargame in which human players decide the actions of insurgent and counterinsurgent forces. The PSOM's geographic area of operations (the wargame board) consists of 50 km squares, each assigned attributes such as degree of urbanization, nature of terrain, population density, quality of infrastructure, cultural values, populace's perception of security, and populace's degree of support to the government.

In PSOM, human players operate the insurgency and counterinsurgency forces. At the beginning of the wargame, the players allocate their respective force units to selected squares of the wargame board. Players assign particular missions to these force units: enforce, stabilize, disrupt, and others. During each time-step, the computer determines the outcome of each force unit's mission based on the current condition in the square, and on actions and strengths of the opponents' forces in the square. The outcome then leads to changes in the square's attributes values. For example, if a counterinsurgent force unit is deemed successful in its security-enhancement mission, the value of the security attribute in the square increases. The game then proceeds to the next time-step, and so on.

In other tools, computer programs replace human players. This permits dramatically faster and less expensive wargames, and analysts are able to explore numerous cases, reflecting alternative scenarios and assumptions.

The representation of humans or groups of humans by software is a forte of what is referred to as agent-based modeling, a paradigm in which actors are represented by so-called agents. An agent possesses at least three components: (1) attributes, e.g., a political party may be characterized by its influence in its native country and its attitudes toward foreign corporations; (2) the set of actions it can take, e.g., a party can introduce legislation that opposes the operations of foreign corporations in the country, or demand additional taxes on foreign corporations; and (3) a computational means, e.g., rules or other reasoning algorithm by which an agent can make decisions about the actions it will take.

An example of an agent-based tool is Senturion (Abdollahian et al. 2006). This tool focuses on the political dynamics of a country struggling to deal with the diverse influences and changes brought about by, for example, an intervention. Senturion's agents possess a complex decision-making mechanism that employs algorithms drawn from game theory, decision theory, spatial bargaining, and microeconomics. The tool has successfully predicted several important real-world intervention-related situations, including those in Iraq, Palestinian Territories, and Darfur in 2004–2009 (Sentia 2008).

Other models make use of the so-called system dynamics approach. This paradigm entails defining equations that relate each significant variable to the variables that influence it; a foreign product's market share, for instance, might be specified in terms of the product's price and of the extent of negative publicity associated with the invasion of foreign products. The resulting system of equations (typically a system of coupled nonlinear differential equations) is typically solved by numerical simulation, and the solution indicates how each variable evolves over time. System dynamics model development packages provide tools aimed at simplifying the specification and solution of such systems of equations. In particular, they make it easy to create variables representing "stocks" of goods, inflows and outflows representing temporal changes to variable, and flow "valves" that open and close as a function of other variables (Sterman 2001).

Some computational tools combine multiple models and multiple modeling paradigms. For example, the Conflict Modeling, Planning, and Outcome Exploration system (COMPOEX) uses a combination of models, some agent-based and others based on system dynamics. COMPOEX (Kott and Corpac 2007) was designed to aid decision-makers in planning, visualizing, and executing major operations such as interventions. It includes a collection of PMESII models (we use terms PMESII model and societal model interchangeably) which modelers compose into an integrated multiresolution model. The integrated model can simulate the future behavior of a country (e.g., stagnant economic growth, increasing corruption, expanded terrorist influence and unrest) as well as the effects of intervention actions (Waltz 2008).

The planning tool of COMPOEX allows planners to schedule coordinated DIME actions along multiple lines of effort (e.g., economic, governance, strategic communications, etc.) in synchronization matrix format. The planner enters the attributes unique to each discrete action (e.g., time of economic action start, action duration, rate of investment, source of investment, targeted economic sectors, targeted geographic region or population, etc.) and the resources required (e.g., financial resources, personnel, etc.). Despite the tools' relative immaturity, a U.S. government agency recently used COMPOEX in a major study to formulate a plan of potential international actions in support of an unstable country (Messer 2009).

Development of such tools, and of the underlying science, is a continuing and growing effort, and a broad range of organizations sponsor and perform such research and development efforts. One example is the US government's Human Social and Cultural Behavior program (HSCB 2009). This extensive, multiyear effort explores ways to model how individuals, organizations, groups, tribes, and whole nations behave under various conditions. The program pays special attention to diversity of cultures and societies, and to the profound influence that cultural and societal differences can have on attitudes, perceptions, and behaviors.

To be sure, the usefulness of such computational techniques and tools remains to be proven at this time. The history of developing these tools is very short, and even the better-known ones are far from mature. Nevertheless, already there is evidence of significant value in using PMESII modeling tools. Whether they "predict" the future is, perhaps, the least important issue. Arguably, human experts cannot predict the future either, at least not in a way that would allow us to know a priori, which of multiple experts' predictions is correct. Still, neither experts nor tools are useless. Computational tools help analysts, planners, and decision-makers in a number of ways other than predicting the future: e.g., by highlighting unwarranted assumptions, generating alternative approaches, elucidating details, and uncovering the potential for unanticipated effects (more on this in Chap. 12).

Regardless of limited and unproven capabilities of the current, first-generation tools, the trend is unmistakable. Their use is rapidly growing, and their value is beginning to be appreciated. There is a rising recognition that quantitative, computational methods are indispensable elements – although by no means a panacea – for producing prudent decisions about an international intervention.

3 The Roadmap of the Book

Following this introductory chapter, we offer readers a broad review of emerging computational techniques and tools (Chap. 2). Next, we proceed with a sequence of chapters that discuss specific techniques for intervention-related phenomena and effects (Chaps. 3–9). Then (in Chaps. 10 and 11), we pay attention to two common challenges – validation of models and visualization of results. In the concluding chapter (Chap. 12), we explore how to apply the tools presented in this book.

Readers need not necessarily approach this book in a linear fashion. Except for Chaps. 1 and 2, chapters are largely independent, thus readers can select chapters in any order, based on their specific interests. To aid readers, we offer below some guidance in making their selections. Table 1 shows where to find discussions of various mechanisms of intervention, such as Diplomatic, Information, Military, and Economic, i.e., the DIME dimensions of interventions.

Table 2 shows where to find discussion of intervention's effects, such as Political, Military, Economic, Social and other areas subsumed under the term PMESII. Some authors prefer to classify effects under the rubrics of Governance, Security, Rule of Law, Social Well-being, and Economics. Of these, Governance and Rule of Law are included in Table 2.

Because agent-based and system-dynamics techniques are currently the most popular computational approaches to intervention-related phenomena, they appear in this book most commonly. However, a number of other modeling techniques are also mentioned. This is reflected in Table 3.

Finally, examples and case studies illustrate the techniques reviewed in this book. The list below helps the reader locate examples that might be of special interest:

State stability	Chap. 3
Government reconstruction	Chap. 6
Political decision-making	Chap. 1
Elections	Chaps. 3 and 8
Embargo	Chap. 1
Insurgency	Chaps. 1, 3, 7 and 8
Tribal conflict	Chap. 7
Sectarian conflict	Chap. 8
Growth of crime	Chaps. 7 and 9
Law enforcement	Chaps. 4 and 9
Transportation	Chap. 4
Unemployment	Chap. 4
Information campaign	Chap. 5

Table 1 DIME aspects in the book's chapters

Mechanisms of intervention	Chapter								
	1	2	3	4	5	6	7	8	9
Diplomatic			X		x	x			
Information			x		X	x	x	x	x
Military	x		X				x	X	
Economic			x	X		x	x		x
Financial				X		x	x		x
Law enforcement			x			x	x	x	X

Table 2 PMESII effects discussed in the book's chapters

Effects of intervention	Chapter								
	1	2	3	4	5	6	7	8	9
Political	x		X			x			
Military	x		X				x	X	
Economic			x	X		x	x		x
Social			x	x	x	x	X		x
Information					X		x	x	x
Infrastructure				x		x			
Governance			x	x		X	x		x
Rule of law				x		x	x		x

Table 3 Modeling techniques discussed in the book's chapters

Modeling techniques	Chapter								
	1	2	3	4	5	6	7	8	9
System dynamics	x	x	x	X		x	x	x	x
Agent-based	x	x	x		x	x	X	X	
Influence diagrams		x		x				X	x
Statistical						x	x	x	
Game theory			x				x		x
Wargaming	x							x	
Bayesian		x							X
Equilibrium			x	x					x

A few disclaimers are in order here. First, because this book is the work of multiple authors, the terminology is not, despite the editors' valiant attempts, always consistent. Second, not every author agrees with the opinions of every other author. Thus, the reader may occasionally encounter contradictory advice. Third, all opinions belong to the respective authors and not to their employers or clients.

As we conclude this introduction, it should be clear to the reader that we and this book do not intend to advise decision-makers on whether to, or how to conduct an intervention. Rather, we limit ourselves to explaining the strengths and weaknesses of technologies discussed. On the other hand, we stress that, despite the still disappointing immaturity of these technologies, they are already demonstrating significant potential and useful, albeit limited capabilities. Thus, we believe that decision-makers are no longer justified in dismissing the insights the technologies offer.

At the same time, we wish to warn the reader against over-reliance on these or any other technologies. It would be an imprudent abrogation of a decision-maker's responsibility to rely on technology as a substitute to human ingenuity, experience, and leadership.

References

Abdollahian, M., Baranick, M., Efird, B. & Kugler, J. (2006). *Senturion. A Predictive Political Simulation Model.* Center for Technology and National Security Policy National Defense University. http://www.ndu.edu/ctnsp/Def_Tech/DTP%2032%20Senturion.pdf.

Agnew, J., Gillespie, TW., Gonzalez, J. & Min, B. (2008). Baghdad nights: evaluating the US military 'surge' using nighttime light signatures. *Environment and Planning A*, 40, 2285–2295.

Colaizzi, J. (2005). *Joint Urban Warrior 05 Provides Catalyst for Coordination in Urban Environments.* http://www.jfcom.mil/newslink/storyarchive/2005/pa052505.htm.

Cuny, F. C. & Hill, R. B. (1999). *Famine, Conflict and Response: a Basic Guide.* West Hartford, Connecticut: Kumarian Press.

Ghani, A. & Lockhart, C. (2008). *Fixing Failed States.* New York, NY: Oxford University Press.

Gilad, B. & Stitzer, T. (2008). *Business War Games: How Large, Small, and New Companies Can Vastly Improve Their Strategies and Outmaneuver the Competition.* Franklin Lakes, NJ: Career Press.

Graham-Brown, S. (1999). *Sanctioning Saddam: The Politics of Intervention in Iraq.* London: I.B. Tauris.

Hammond, K. R., Hamm, R. M., Grassia, J. & Pearson, T. (1987). Direct comparison of the efficacy of intuitive and analytic cognition in expert judgment. *IEEE Transactions on Systems, Man, and Cybernetics, SMC*, 17(5), 753–770.

Hartley, D. (2008). *DIME/PMESII Tools: Past, Present and Future, Workshop on Analyzing the Impact of Emerging Societies on National Security.* Argonne National Laboratory, Argonne, IL, 14–18 April 2008. http://www.mors.org/events/08es.aspx.

HSCB Newsletter (2009). http://www.dod.mil/ddre/doc/Spring2009_HSCB_Newsletter_Issue1.pdf.

JFCOM (2004). *Operational Net Assessment: a Concept Paper for Joint Experimentation.* Norfolk, VA: US Joint Forces Command.

Kagan, F. W. (2006). *Choosing Victory: A Plan for Success in Iraq.* A report by the American Enterprise Institute. http://www.realclearpolitics.com/RCP_PDF/ChoosingVictory.pdf.

Kott, A. & Corpac, P. S. (2007). COMPOEX Technology to Assist Leaders in Planning and Executing Campaigns in Complex Operational Environments, *12th International Command and Control Research and Technology Symposium*, Newport, Rhode Island, June 19–21.

Kurtz, J. (2003). Business wargaming': simulations guide crucial strategy decisions. *Strategy & Leadership*, 31(6), 12–21.

Marlin, B. (2009). *Ascertaining Validity in the Abstract Realm of PMESII Simulation Models: an Analysis of the Peace Support Operations Model (PSOM).* Masters Thesis, Naval Postgraduate School, Monterey, CA.

Messer, K. (2009). The Africa Study. *Presented at the HSCB Focus 2010 Conference*, August 5–7, 2009, Chantilly, VA.

Meyer, M. A. & Booker, J. M. (2001). *Eliciting and Analyzing Expert Judgment: a Practical Guide.* Philadelphia, PA: SIAM.

Office of the Coordinator for Reconstruction and Stabilization, Post-Conflict Reconstruction Essential tasks, United States Department of State (2005). http://www.crs.state.gov/index. cfm?fuseaction=public.display&id=10234c2e-a5fc-4333-bd82-037d1d42b725.

Perito, R. M. (2007). *Guide for Participants in Peace, Stability, and Relief Operations.* Washington DC: United States Institute of Peace Press.

Petraeus, D. H. (2007). *Report to Congress on the Situation in Iraq.* Testimony Presented Before the House Subcommittee on 11 September 2007. http://www.dtic.mil/cgibin/GetTRDoc?AD= ADA473579&Location=U2&doc=GetTRDoc.pdf.

Sentia Group (2008). *Implications of a U.S. Drawdown in Iraq.* http://www.sentiagroup.com/pdf/ SentiaInsightMonthly-USDrawdownInIraq-July2008.pdf.

Siegel, R. (2007). *Scanning History for Analogies to Iraq War,* Transcript of NPR Radio Broadcast, August 22, 2007. http://www.npr.org/templates/story/story.php?storyId=13872684.

Simon, S. (2008). The price of the surge. *Foreign Affairs,* 87(3), 57–76.

Slim, H. (2004). Dithering over Darfur? A preliminary review of the international response. *International Affairs* 80(5), 811–828.

Sterman, J. D. (2001). System dynamics modeling: tools for learning in a complex world. *California Management Review,* 43(1), 8–25.

Stern, P. C. & Druckman, D. (2000). Evaluating interventions in history: the case of international conflict resolution. *International Studies Review,* 2(1), 33–63.

Straus, S. (2006). Rwanda and Darfur: a comparative analysis. *Genocide Studies and Prevention,* 1(1), 45–46.

Taylor, A. & Rourke, J. (1995). Historical analogies in congressional foreign policy process. *The Journal of Politics,* 57(2), 460–468.

Taylor, J. B. (2007). *Global Financial Warriors: The Untold Story of International Finance in the Post-9/11 World.* New York, NY: W.W. Norton & Co.

Tyrell, H. G. (1911). *History of Bridge Engineering.* Chicago: The G.B. Williams Co.

US Army (2008). *Field Manual FM 3-07, Stability Operations.* Headquarters Department of the Army. http://usacac.army.mil/CAC2/Repository/FM307/FM3-07.pdf.

Walters, E. (2008). *Problems and Prospects of Defense COIN Wargaming. Small Wars Journal.* http://smallwarsjournal.com/blog/2008/04/problems-and-prospects-of-defe-1/.

Waltz, E. (2008). *Situation Analysis and Collaborative Planning for Complex Operations, 13th ICCRTS.* Bellevue, WA, June 17–19.

Woodward, B. (2008). *The War Within: a Secret White House History 2006–2008.* New York, NY: Simon & Schuster.

Chapter 2
Emerging Techniques and Tools

Ed Waltz*

International interventions require unconventional approaches to modeling and analysis. According to Alberts et al. (2007, p. 5), the characteristics of intervention problems include:

1. The number and diversity of the participants is such that

    (a) There are multiple interdependent lines of management and control,
    (b) The objective functions of the participants conflict with one another or their components have significantly different weights, or
    (c) The participants' perceptions of the situation differ in important ways; and

2. The effects space spans multiple domains and there is

    (a) A lack of understanding of networked cause-and-effect relationships, and
    (b) An inability to predict effects that are likely to arise from alternative plans of action.

In such situations, analysts and planners need to follow a set of principles that are very different from those of situations with unified management structure, a clear objective and a situation understood by all, and an environment that has little adaptation and whose behavior is reliably predicted. First, they must be aware of the numerous arenas and domains involved in complex adaptive systems. Second, because of the lack of predictability in complex systems, planners must take steps to produce agile plans.

Effective computational models hold the promise of enabling planners to explore the deep dynamics of complex situations, and to explore effects across a wider range of candidate policies or plans. By evaluating a wide range of plans across a

---

*All statements of fact, opinion, or analysis expressed are those of the author and do not reflect the official positions or views of the Office of the Director of National Intelligence (ODNI) or any other U.S. government agency. Nothing in the content should be construed as asserting or implying the U.S. government authentication of information or ODNI endorsement of the author's views. This material has been reviewed by the ODNI to prevent disclosure of classified information.

E. Waltz (✉)
Intelligence Innovation Division, BAE Systems Advanced Information Technologies,
Ann Arbor, MI, USA
e-mail: ed.waltz@baesystems.com

A. Kott and G. Citrenbaum (eds.), *Estimating Impact*,
DOI 10.1007/978-1-4419-6235-5_2, © Springer Science+Business Media, LLC 2010

variety of situations, analysts and planners can achieve more agile and robust strategies that account for the uncertainties in knowledge of the actual situation. Bankes has described this goal of plan robustness as follows:

> A level set provides much more information than does a single optimal policy. Combining this idea with that of policy landscapes, the computer can be used to discover policies that are robust across multiple scenarios or alternative models, and to identify and graphically depict sets of policies with satisfactory robustness (Bankes 2002).

Computational models and simulations will specifically allow analysis and planning teams to address the two principles by encouraging the rigorous analysis of complex international actions by explicit modeling in the following ways:

> First by *structural analysis*, the process of decomposing the situation into fundamental components and their interactions, and quantifying the relationships between components,

> Second, by the *dynamic analysis* of the behavior of the system of interconnected systems, using simulation to gain familiarity with the interaction between systems; this includes the *analysis of sensitivity* of the systems to key factors.

> Next, by *exploratory analysis* of the effects (anticipated and unanticipated) of a range of potential actions by a variety of parties and groups using computational simulations.

This chapter introduces the modeling and simulation technologies available to represent complex situations: the political, military, economic, social, information, and infrastructure (PMESII) states of systems and the effects of diplomatic, information, military and economic (DIME) actions on those systems. First, the methods of explicitly representing complex situations are described, illustrating the methods to translate the tacit knowledge of subject matter experts' (SMEs) mental models to explicit conceptual models and then to computational models. Second,[1] We introduce the means by which these models can be applied to represent systems in which physical elements dominate (e.g., infrastructure, etc.) and systems in which nonphysical elements dominate (e.g., the human, social and cultural factors that dominate political, social and economic systems). Then, we explain the uses of these models to estimate the system state, hidden relationships, variables, uncertainties, and dependencies.

The alternative methods of implementing static computational models and dynamic simulations are described, introducing the application and appropriate roles for discrete event, system dynamic, Markov, Bayesian and agent-based modeling implementations. The composition of integrated simulations using these alternative modeling approaches is described. Next, the chapter describes the use of modeling technology to perform exploratory analysis to determine the effects of our actions, to develop effective and acceptable courses of action, and to perform assessments of the models in situ. The chapter concludes with an overview of the issues of

---

[1] We use the term "tacit" in the general sense defined by Michael Polanyi (1891–1976) in *The Tacit Dimension*, "*we can know more than we can tell*": a prelogical phase of knowing that has not been articulated. Therefore, it is not explicit knowledge that has been codified. Tacit knowledge includes sensory information, perceptions, and the higher-level conceptions that attempt to make sense of them.

model validations, describing approaches to verification, validation, and accreditation in the context of uncertainty and exploratory analysis.

1   Representing Situations

Analysts have long sought efficient methods to describe, with precision, the makeup of complex economic, political, and military situations. *Situation Assessment* (also called political analysis) is the process used by analysts to identify the key actors (individual leaders, organizations, or aggregate population segments) involved in political competition or conflict over resources, policy, or other aspects of power. The actor's interests and goals, roles, constraints, abilities, behaviors, and lines of influence to other actors are identified. In addition, the context of the situation (e.g., economic environment, political landscape, cultural-social setting) is described. The process is generally a static enumeration of the situation at a point in time and is generally reported in narrative form with supporting tabular data, where appropriate. (For a detailed list of the elements of a comprehensive situation assessment, see Covey et al. 2005, p. 45). Consider, for example, the typical narrative situation assessments of Iraq in three documents with significant influence on the U.S. policy:

- *Prospects for Iraq's Stability: A Challenging Road Ahead, National Intelligence Estimate*, ODNI, *January 2007*. The unclassified judgments in this national intelligence estimate include three pages of narrative assessment, followed by a half-page judgment on three "security paths" or adverse trajectories that could occur if violence does not subside (ODNI 2007).
- James A. Baker, III, and Lee H. Hamilton, *The Iraq Study Group Report*, NY: Vintage, 2006. This report includes a 30-page narrative assessment, followed by a 3-page projection of the consequences of a continued decline in security in Iraq. This assessment preceded a 60-page analysis of alternative courses of action and recommendations (Baker and Hamilton 2006).
- *Stabilizing Iraq: An Assessment of the Security Situation*, Statement for the Record by David M. Walker Comptroller General of the United States, GAO-06-1094T, Sept. 11, 2006. This document, supporting congressional testimony, includes an 18-page narrative assessment, supported by two graphs of violent incident trends and two tables of data on Iraqi security force readiness (Walker et al. 2006).

In each case, the analysis enumerates the major factors and the interrelations between the factors and provides a narration of possible scenarios (i.e., the dynamics of alternative outcomes). This process of *decomposition* (breaking apart, or factoring) identifies the component parts (or subsystems) and their interconnections to allow the situation to be more easily understood, analyzed, and described. It also allows individual factors (e.g., politics, economics) to be described in greater detail.

Modeling technology is now allowing us to go beyond these static enumerations and narrative descriptions of potential scenarios and effects, even as analysts (who seek to understand a situation) and planners (who develop approaches to change the situation to achieve an objective) are seeking more breadth of enumeration and depth in dynamic analysis. A planner identified the need for more effective means of analysis and planning:

> Analytical tools have improved dramatically. Unfortunately, questions over effects-based [approaches to] operations persist: the adequacy of intelligence, the lack of cultural sensitivity, the risk of studying inputs rather than outputs, and the need for models to account for cognitive, cultural, political, and social factors. These are serious questions, and their solutions are not obvious (Meilinger 2004, p. 122).

Solutions would include a rigorous process to decompose and represent the behavior of a real-world system, S (e.g., a regional political-military competition between states, a single nation-state, a provincial insurgency, or the stabilization of a major urban area), which comprise interdependent physical and nonphysical (e.g., social, economic) contributing elements with behaviors in component models, m, and interactions, i, in a composite model, M, such that:

> (Completeness) The decomposed set of m and i, once composed into M, can be shown to achieve a measurable level of coverage, C, of the elements and behaviors of S to represent a specified level of causal granularity, G.

> (Behavioral Specificity) The component models and interactions between models in M can be specified to achieve an aggregate level of G, and the dynamic behavior of M can achieve a specified degree of the behavior of S.

> (Descriptive causality) The level of G can be related to specific causes and effects achievable in M and observable in S.

> Note that this challenge *presumes* decomposability of *S to some degree*; if all elements at the finest granularity are independent in some significant degree, then decomposition at a higher level is not possible (Table 1).

A representative process of decomposing a situation into PMESII elements and then interacting computational models proceeds by *decomposing the situation* into key elements (or systems) and their interactions and then *composing models* that represent the situation by these elements and their interactions. The process proceeds in the following steps (Fig. 1):

1. *Describe the situation* informally by discussion with SMEs who can enumerate the key elements (actors and systems), their relationships and interactions, the critical factors of influence, and the behaviors of these elements. These discussions are often in narrative form (stories), and the quantification process requires careful translation of the SMEs' qualitative narratives into conceptual models. In this step, it is also critical to recognize if there are multiple concepts (hypotheses) of how a situation operates. For example, one group of SMEs may believe that a nation is driven by its underground economy and external influence, but another group may believe that it is driven by the official economy and internal cultural factors. In these cases, it may be necessary to maintain *both* models: two hypotheses that may be used to evaluate plans, in order to develop robust plans that can address either of the two views of how the country operates.
2. *Identify the Elements, Relationships, and Systems*. From the initial discussions, guided by the typical PMESII factors, develop conceptual representations

Table 1 Representative PMESII elements and aspects

Elements	Qualitative aspects	Quantitative aspects
Political	Intent; public opinions of political leadership (via polls)	Leadership power, ability, stability, coherence, external support, diplomatic strength
	Leadership strength	
	Organizations, parties, groups, factions and relationships	Power structure; national, provincial, city governments
	Values, motivations, goals, activities	Regulations, policies
Military	Will; intent, resolve	Traditional military order of battle; units of force. Physical assets
	Cohesion; readiness	
		Physical networks, lines of communications
Economic	Public confidence	GDP; GDP growth
	Financial outlook	Inflation
	Government ownership, participation; forms of commercial activity	Trade balance (import, export, capital inflow)
	Wealth distribution, relationships with factions	Construction; public finance; debt
	Illegal economic activities	Economic status of population elements, shortages, subsidies
Social	Culture: languages, religions; social, ethnic/tribal, backgrounds and relationships	Security/law and order (includes crime and criminal organizations)
	Demographics of attitudes and perceptions; historical context, customs	Public health; mortality rates, disease rates
	Culturally based perceptions, temperaments	Demographics presence, distribution in city and environs
	Social outlook	
Information	Messages; time of dissemination, location if relevant	Broadcasting/publishing/website organizations
	Medium (includes electronic, print, speeches/harangues, person-to-person); authority/legitimacy of source (from outsider point-of-view); intended audience(s), perceived legitimacy of source	Local, foreign (including US) media channels
		Transmission sites locations, media traffic; political orientation, role
	Message contents; events, activities	Media, volume, bandwidth, coverage
	Assertions, declarations, threats, directions/imperatives; opinions, stated or implied perceptions	Content originators (political/social groups, writers, producers)
Infrastructure	Public utility service satisfaction; heat, light, water, sewer	Electrical power production
		Water, sewer
	Public transportation efficiency, availability	Transportation efficiency factors
		Manufacturing production
	Manufacturing production	Gas, petroleum production, flow rates, efficiency
	Manufacturing transportation efficiency (rail)	Telecommunications bandwidth, coverage

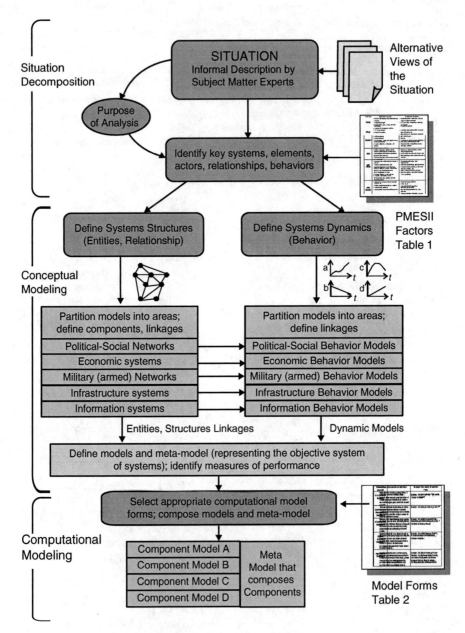

Fig. 1 Situation decomposition and model composition

(generally, tabular lists of elements and graphical depictions of relationships) of the major elements; review these with SMEs and refine until the SMEs agree that these conceptual models represent the *structure* of the situation. Also represent the major dynamics of the situation (e.g., "legitimate economy will go up as the illegitimate economy goes down and corruption is reduced") and confirm these major behavioral factors with SMEs.

3. *Develop Component Models of System Structure That Will Produce Expected Behaviors.* The component models of PMESII subsystems (e.g., the legitimate and illegitimate, or underground economy subsystem models) are created and tested to produce the behavior expected by the SMEs. The models are evaluated for a range of behaviors, using historical data when available, and by the SMEs to compare model behavior to SME-expected behaviors.

4. *Compose the Component Models into a Metamodel.* Finally, the models are *composed* (integrated) into a unified model of models; the interconnections between models (e.g., the agricultural impact on the legitimate economy and the drug-crop interconnections with the illegitimate economy, and their interactions) make up, in fact, a model in its own right. This metamodel of interconnections will produce large-scale system behaviors that are not inherent in the independent models, producing effects that *emerge* from the interaction of the models. At this stage, the measures of systems performance (metrics) that characterize the situation must be defined and checked to verify that the model can be compared to the real situation; this will also aid in the identification of means in the real world that can be used to compare model results to real-world situation dynamics.

2   Conceptual Modeling

In the preceding section, we used the general term *model* to refer to any abstract representation of a system, but we now distinguish among:

- *Mental models* of systems or phenomena that are understood (or believed to be understood to some degree) by the SMEs.
- *Conceptual model* representations of mental models that may be presented in a variety of narrative or graphical means to explicitly represent elements, relationships, and causal functions of a system or phenomena.
- *Computational models* that implement the structure and causal behavior of a conceptual model in a computational form that allows the dynamic behavior of the modeled system to be simulated.

The process for representing a given situation, using the PMESII categories, proceeds from the tacit mental models of experts to computational models that allow analysts to explore the dynamics of interacting PMESII systems (Fig. 2). The process of *abstraction* – representing the real (concrete) world in qualitative structures and quantitative functional relationships – requires the capture of SMEs' tacit knowledge of the particular PMESII systems of an area in explicit conceptual models. These models are first captured in narrative form, and in lists of enumerated actors, systems, and dependent interrelationships. From these lists, graphical structures

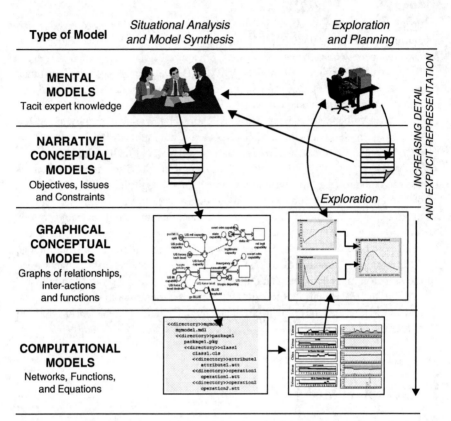

Fig. 2 The relationships between mental, conceptual, and computational models

are created at a common level of granularity. An appropriate computational modeling paradigm (approach to modeling causality, e.g., Bayesian model of influence, Petri net model of a sequential process, system-dynamics model of process flows, agent-based model of social behavior) is selected. The computational model is constructed and operated over a range of environments and perturbations to evaluate and refine its behavior relative to known behaviors in the real world.

The upward process in the figure illustrates how the results of computational experiments flow upward from computational model results to conceptual displays of the dynamics of conceptual model variables, in a form understandable to the modeler and analyst. The results of experimental simulation refine the analysts' and SMEs' mental models as results are questioned and are used to refine the models until confidence is built in the results and models become useful for exploring large-scale dynamics. The quantitative results can again be translated into narrative "stories" that describe potential outcomes of candidate actions.

It is important to distinguish between empirical modeling and the kinds of causal models that we apply in this chapter (Fig. 3). *Empirical modeling* refers to those

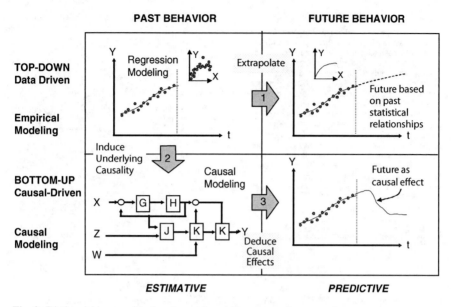

Fig. 3 Distinguishing empirical and causal modeling approaches

methods that represent a system or phenomenon based on the data produced from prior experience or experimentation. Generally, this involves *quantitative* methods of regression that seek to establish the functional relation between selected values of x and observed values of y (from which the most probable value of y can be predicted for any value of x). In the figure, the use of direct regression (path 1) can produce a functional model that can extrapolate future values as a function of the past. *Causal modeling* (paths 2 and 3 in the figure) seeks to induce the underlying causality (functional processes) of phenomena or systems and represent them explicitly. Such a causal model can then create representations of future behavior deductively from input variables, and the internal system behavior can be explicitly observed and compared to the real world.

Causal modeling often proceeds from the narrative model of an SME to a corresponding graphical causal model (example, Fig. 4) that represents the major elements (actors, systems, processes, etc.) and the structure (relationships between elements) of the model. The process for causal modeling often proceeds:

1. SME is interviewed to describe the system and its major elements, the factors that influence its behavior, and the key relationships between elements. A narrative description is developed, with a list of elements and relationships for review by the SME.
2. The underlying empirical data is sought to develop the empirical basis for the current situation; the accepted theoretical basis is also sought to develop the relevant causal model.

Narrative Model

Rule of law is enabled by the governance process that establishes policy to fund the building of appropriate capacity to maintain policing and the criminal justice pipeline, comprised holding jails, courts, and prisons. The capacity building must include policies to maintain funding, reduce corruption and, increase competence in all functions. The policing policy, should fund a robust security sector, training new police to build the operational police units. The security sector includes coordinated activities of police, the military, and supporting intelligence. The criminal organizations maintain a criminal network that provides goods and services for the black market, while the market funds the operation of the organization. The civil population interacts with the black market (indirectly) as a purchaser, and (directly) as a victim of crime, extortion, and other predatory crimes.

Corresponding Conceptual Model

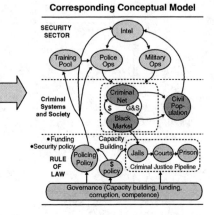

Fig. 4 Translating a narrative conceptual model to a corresponding graphical conceptual model

3. A corresponding graphical (functional) model is developed and again described to the SME to refine the modeler's understanding of the system. This may be a much different perspective of the same system the SME knows well, and the discussion may reveal more insight as the SME is asked to detail more explicitly the causal behavior, thus refining the conceptual model. (This process is the beginning of internal model validation: building confidence in the underlying theory on which the model is based.)

4. A computational model of the system is developed and the *behavior* of the model over a range of conditions is recorded and presented to the SME to perform a comparison to known real-world behavior. Discrepancies between the model output and empirical data must be examined, explained, and the model refined until the model behavior compares to the real world sufficiently for its intended use. (This process is the beginning of external model validation: building confidence by comparison of the model behavior to the SME's empirical understanding of real-world behavior and, if appropriate, historical cases.)

Of course, the preceding method is nothing more than an instance of the general procedure of the scientific method, which is based on hypothesis (model) building, prediction of behavior, and testing against empirical data.

Consider, for example, three approaches to decomposition of the primary systems that represent the competitive structure of an insurgency and counterinsurgency (COIN) and their representation in high-level (of abstraction) conceptual models (Fig. 5). The first decomposition is Manwaring and Fishel's SWORD model that decomposes the competition into principal actors and their interrelationships using seven dimensions: (1) military actions of the intervening power, (2) support actions of the intervening power, (3) host-government legitimacy, (4) degree of outside support to insurgents, (5) actions against subversion, (6) host-country military actions, and (7) unity of effort (Manwaring and Fishel 1992, pp. 272–305).

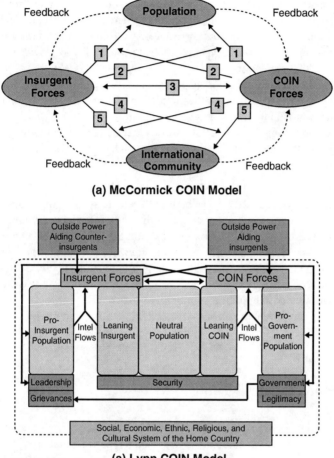

(a) McCormick COIN Model

(a) Lynn COIN Model

Fig. 5 Two representative conceptual models of counterinsurgency

While the SWORD model identifies key static indicators, further insight into the dynamic modeling requirements can be found in conceptual insurgency-COIN models developed by McCormick and Lynn (McCormick 1999; Lynn 2005) that describe the elements (entities) and relationships between insurgent and COIN forces (Fig. 5b). The essential elements (entities or actors) of both models include:

- Insurgent Force(s): The leadership, combatants (guerillas), financers, and supporting population that carries on the insurgent political message and a coordinated campaign of violence to undermine the legitimate government and demonstrate its inability to provide security and services.

- Government and COIN Force (s): The leadership, combatants (military), and supporting population that endorses the current government, its political message and legitimacy. The government carries out a COIN campaign of information and action to support its legitimacy and demonstrate its ability to provide security and services to the population.
- Civil Population includes elements that support the legitimate government, those that support the insurgents, and the population in the middle for which both sides compete to prove legitimacy and gain support.
- External supporting powers include those external parties (states, organizations, etc.) that supply ideological, financial, material, or human resources to either side of the conflict.

Both models also represent the basic relationships between these entities by simple arrows that describe the interactions between key actors. The graphical representations of the conceptual models distinguish the actors (leaders or elites, organizations or institutions, and population groups) and the relations between the actors. In both models, government and insurgents compete for population support, and the competition is conducted across the many relationships that exist between the parties (political, military, economic, etc.). The U.S. Army's Field Manual for COIN acknowledges the value of narrative insurgency models of history and a conceptual modelmaking process for understanding the COIN environment (US Army 2006, p. 1–4, para. 1–76 and p. 4–3, para. 4–9).

3   Computational Modeling

Computational models include a wide range of models that compute output functions as a result of inputs. Computational *models* include the computation of complex yet static functions (such as a computer spreadsheet) or dynamic *simulations* that implement models as they operate over time. Simulation tools provide a means for analysts and planners to be immersed in the modeled structure, dynamic behavior and responses (effects) to courses of action; simulation provides a tool for experimentation and exploration of behavior.

The modeler may choose from a variety of computational approaches to implement the component models and to compose them into an integrated model. The primary computational modeling approaches to simulate processes over time (Table 2) are distinguished by three characteristics:

The method used to move the model through time: Time-continuous functions may be represented in time-discrete steps (time-sampled), and the simulation proceeds to compute all functions and interactions in a time-discrete (incremental step-by-step) fashion; in this case, the unit of simulation progress is a time clock, and all models apply a uniform time constant to represent processes that occur within the time step. Alternatively, the unit of progress in the simulation may be chosen to be discrete events; event-based simulations increment from event to event by event triggers that represent the causal propagation from any given event to any other event-producing processes.

Table 2 Representative modeling and simulation tools (Waltz 2006)

Simulation approach	Description and characteristics	Example commercial simulation tools
General causal modeling	Static Bayes networks represent chains of actions to effect nodes and resulting effects; Dynamic Bayes nets add a representation of complex states, and transitions at nodes to represent the aggregate dynamics of a causal networks	Example tools: Netica™(Norsys); Bayes Net Toolbox for Matlab™(MathWorks)
Discrete-event simulation	Simulate event-based systems using queuing models of queue-servers, Petri nets, Markov, and other models that define nodes, links, and resources to simulate process interactions, synchronization, and scheduling of discrete events	Example tools: Matlab® SimuLink® and SimEvents® (MathWorks); (Ptolemy) University of California at Berkeley; FlexSim Software (FlexSim); SIMAN (Systems SIMAN Modeling Corp.), ProModel (ProModel Corp.), and GPSS/H (Wolverine Software)
Discrete-time simulation	Time-based simulation of continuous or time-discrete processes defined by differential equations; represent continuous processes by state-machine simulation of all processes for each discrete-time increment	Example tools: ExtendSim™ (Imagine That Inc.)
System dynamics simulation	System dynamics flow models are based on the principle of stock accumulation and depletion, representing the flow of resources to accumulate "stock" variables. System dynamics causal models account for positive and negative feedback across processes and represent nonlinear behavior	Example tools: iThink™(ISEE Systems); PowerSim Studio (Powersim Software); Vensim® (Ventana)
Agent-based simulation	Agents represent interacting autonomous rational cognitive actors, their goals, beliefs and autonomous behavior to study social behavior of individuals, groups or populations. Goal-seeking adaptation produces realistic emergent behavior not predictable from the underlying models	Example tools: Power Structure Toolkit (Soar Technology); DyNet (Carnegie Melon University); RePast (University of Chicago's Social Science Research); SWARM (Santa Fe Institute), SOAR (University of Michigan)

The approach to deal with process functions and functional interrelationships: The number of functions (N), functional complexity (C), their interrelationships (R), and the relative autonomy of functions (A) that characterize a model distinguish the models that are relatively compact system-process models (e.g., system-dynamics models in which N, C, R, A attributes are low, high, low, low, respectively) and highly interactive models (e.g., agent-based models in which N, C, R, A attributes are high, low, high, high, respectively).

The approach to deal with uncertainty: The uncertainty in inputs to the model and uncertainty in the internal model functions themselves influence the uncertainty in simulation outputs. The models may directly represent and propagate uncertainty throughout the model (e.g., Bayesian networks use probabilities to propagate uncertainty) or a deterministic model may be used to represent uncertainty by varying input or internal variables in a controlled manner (e.g., by being sampled from a distribution) across many simulations to assess the effects of uncertainty. This approach is called Monte Carlo simulation.

For each model category, there exist commercially available model-building tools that allow the modeler to develop, test, and validate models with available empirical data (see Table 2 for representative commercial tools). The characteristics of the major computational modeling approaches are summarized in the following paragraphs.

General Causal Models: A fundamental interest in modeling is the representation of the causal relationships between entities or events. (One event, *the cause A*, must be prior to or simultaneous with another event, *the effect B*.) Models of causality are represented by directed acyclic graphs that represent events as nodes and edges (or links) as the causal relationship. From this simple representation, a number of sophisticated model implementations can be created:

Influence Models: Directed graphs that represent functional relationships (influences) between variables are called influence diagrams and are computed as influence models. Such models are often used in decision modeling, and the graph proceeds from decision nodes (alternatives that can be chosen by a decision-maker) and independent variable nodes (deterministic or probabilistic variables) to the dependent objective node: the function influence by decisions and variables that is to be optimized. In simulations, these models may be used to represent the effects of the decisions of rational human decision-makers seeking to optimize an objective.

Bayesian Network Models: These models can represent *probabilistic causation*, allowing an effect to be probabilistically related to a cause. The models permit effects to be represented by conditional probabilities; for example, $P(B|A)$ represents the probability of the occurrence of the effect B, under the condition that A occurred. Bayesian networks allow the calculation of the propagation of probabilities across complex directed acyclic graphs to compute posterior probabilities, as a function of prior and computed conditional probabilities (Pearl 2000).

General Discrete Event Models: In these causal models, a system's behavior is represented as a sequence of events in which the triggering of each event can be determined by external conditions or conditioned by the state of other events. These models are implemented as simulation tools that represent systems in which the state evolves at discrete points in time (events) rather than continuously as in time-discrete models. Discrete-event models readily represent processes with transaction, flows, delays, and queues; they are well suited to traffic, production, inventories, and movement of commodities (Banks et al. 2004).

Markov Models: These models represent the dynamic of systems by their states, the state-transition probabilities that move the system from state to state, and the available information to observe the state. A Markov model represents a system with the state directly observable, and a Hidden Markov Model represents a system in which the state is not directly observable (although it may be partially observable or inferred from variables related to the state). These models are used to model systems that move from discrete state to state in their operation.

Petri Network Models: These network models represent interactive and distributed processes, modeling the communication, synchronization of messages and processes, and sharing of resources across distributed processes. Models are represented in a directed graphical notation that represents communication flow and the flow of process activities, represented by tokens being passed across the network. (When the tokens take on value, the nets are referred to as Colored Petri nets.) These models are applied to representing political policy processes that follow legislative sequences, logistic processes, communication distributed computing (network) processes, and commodity delivery processes.

Discrete Time Models: Many models are implemented to operate in discrete time steps, in which the simulation unit is a fixed time interval (as opposed to discrete event simulations in which the simulation units are events that may have varying time intervals). Processes represented in differential equations are well suited to direct implementation as time-discrete models that are updated at the Δt interval. The general causal models previously described are often implemented as discrete-time simulations, as well as the models that follow.

System Dynamics Models: The fundamental principle in representing systems in this model form is the dynamic flow of critical "stocks" in the system modeled; stocks are accumulated or depleted over time (the "flow" of capital or stock). Stocks can refer to material entities (e.g., crops harvested, children born, steel produced) or more immaterial entities (shares of securities owned, financial capital invested, human or intellectual capital, etc.). In this modeling paradigm, the modeler must identify the key stocks that represent the fundamental flow dynamics of the system. For an insurgency organization model, for example, the stocks may be financial capital, insurgent fighters, and weapons; the flows are a function of donations-expenditures for weapons, recruitment and attrition of insurgent fighters, and weapons purchased-weapons consumed or expended, respectively. Once the fundamental stock and flows are defined, the functions that influence the flows are modeled to "throttle" the accumulation and depletion of stocks. These functional relationships allow the modeler to represent critical time delays, queues, and feedback loops that provide positive reinforcement (growth) or negative reinforcement (balancing) behaviors. The completed model provides a simulation of the time-dynamic behavior of the system, the changing level of the critical stocks that describe the system, and the effects of initial conditions and the time delay and feedback functions. The models can readily simulate nonlinear systems and can simulate general equilibrium behavior that is exhibited by economic, production and social systems, as well as the conditions that disturb such stability. Numerous modeling tools allow the model to be created graphically and simulated rapidly, using a standard system dynamics graphical formalism. The graphical symbols are compiled onto ordinary differential equations that represent the flows and conditioning parameters that represent the time delays and feedback loops that couple the differential equations. For a comprehensive overview of system-dynamics models, see Sterman (2000).

Agent-Based Models: These models represent the interaction of a network of autonomous actors, interacting with an awareness of their environment and individually operating by an internal behavior (goal-directed, able to cooperate or compete with other actors). The actors in real life may represent leaders, organizations or the aggregate behavior of population groups, and they are represented by software agents that perceive their environment (e.g., sociopolitical, economic, security, or other aspects), reason about the situation compared to their interests and goals, perform decision-making, and then act in the environment to respond to the situation. Agent-based simulations are described as *generative* because they autonomously generate behavior as a result of the interaction between agents, generating equilibrium as well as the emergence of higher order (complex) behaviors, not predictable in the behaviors of the individual actors. These simulations uniquely allow the modeler to represent individual and group decision-making to simulate the effects of interactions between large numbers of actors in a dynamic environment. In particular, models that employ agents with relatively modest rules can produce relatively complex behaviors, due to the high level of interactions within the network of agents. The models are most often applied to political and social modeling (e.g., political power struggles over policy positions and social interactions between groups or groups and elites), as well as modeling economies, logistics, and transportation behaviors, and the spread of disease. Because they explicitly represent the decision-making of individuals or groups, they are well suited for the study of organizations. Tools for creating agent-based models include NetLogo (Northwestern University), Swarm (Santa Fe Institute), and Power Structure Toolkit (Soar Technology). For an overview of agent-based models, see Epstein and Axtell (1996) and Axelrod (1997).

Hybrid simulations: Because each modeling approach has a particular strength, it may be appropriate to implement a simulation that integrates (or composes) different types of models to apply the advantages of each. This is often the case when modeling situations in which the interactions of political and social systems, economic systems, and physical systems (computer systems, production, infrastructure, transportation) must be represented. The next paragraphs describe the approaches to model composition to develop hybrid simulations.

4   Model Composition

Once a major system (e.g., an unstable or failing state) has been decomposed into component subsystem models (e.g., PMESII), the analysis of effects across multiple models requires that the individual models be composed into an overall system for analysis. This issue of integrating diverse component models into a composite model has long been a challenge to the modeling community (Davis and Anderson 2004). Consider the two alternative approaches to composition of multiple models.

- An *analytic composition* process runs models independently in time but considers the interaction effects by running model excursions to describe the effects of interdependencies. The results are composed by an external analysis of the independent simulation dynamics and basing one model's inputs on the results of others, but the models do not directly interact.
- A *computational composition* process integrates multiple component computer models of individual PMESII systems into a single *metamodel* (or *metasimulation*) that describes a larger situation than any one component and synchronizes their interacting operations at the same run-time. When the components are of varying resolutions (or causal granularities), the metamodel is a multiresolution model (MRM). The composed MRM structure may represent a hierarchy of fine-granularity submodels that contribute upward to a lower-granularity model that integrates the results, or lower-granularity models may provide contextual information downward for finer-granularity models.

A computational composition of models has been performed in the system called COMPOEX (COnflict Modeling, Planning, and Outcomes EXperimentation). It is an example of a large-scale simulation framework that composes a diverse set of modeling paradigms into a single run-time metamodel (Fig. 6). The COMPOEX tool architecture (Fig. 6) includes:

- A planning tool that organizes and schedules the injection of actions to models along the simulation time sequence, and
- An option exploration tool that hosts the integrated model and runs simulations of the synchronized sets of composed models.

All models are plugged onto a "backplane" that represents the state vector of PMESII state variables. The models are stepped in time-discrete manner, generally in 1-week increments, simulating behavior over a 2–3 year period of time. Characterizing the integrated simulation as a finite state machine, the state vector is the memory that stores current state; the sequence of states for any given variable over 156 weeks of a 3-year simulation represents the behavior of the variable. A typical COMPOEX model may include well over 10,000 such state variables. The visualization service allows users to customize views of any of the variables and their relationships; it also detects and displays discrete effects that should be brought to the attention of the planner. It furthermore allows the user to trace causality within the simulation, allowing the user to trace the (upstream) variables on which an effect is dependent and the (downstream) variables that are dependent on the effect variable (Waltz 2008).

The model of power actors and relationships is at the core of the COMPOEX simulation, providing the major abstract dynamic within a virtual world of economic, material services, media and sources of information exchange, physical

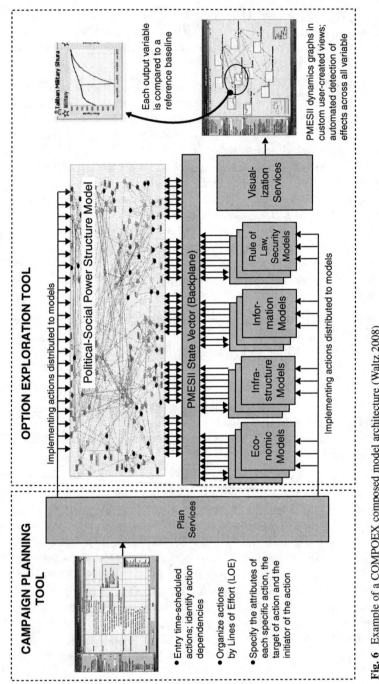

Fig. 6 Example of a COMPOEX composed model architecture (Waltz 2008)

violence, and infrastructure. Power struggle behavior is included across the many composed models within the simulation environment. The COMPOEX approach to abstraction is based on two major partitions of the model:

- Power Influence Network: Competing actors for power are represented in agent-based models in which autonomous agents compete for power, represented as the abstract capital commodity in four dimensions (political, social, economic, and armed military). This network represents all human decision-making, influence, and action. The operation of the agent-based actor simulation is described in more detail in Taylor et al. (2006).
- Virtual World: The context within which the actors compete (or cooperate) for power is represented by a set of interconnected process models, implemented by a variety of modeling paradigms (e.g., system dynamics, discrete time models, Bayesian networks). These models may represent aggregate human behavior (e.g., aggregate economics, production, large-scale population behavior), but do not represent the core competition for political power.

The structure of the composed power network and virtual world models (Fig. 7) illustrates the interaction between the actor net and the virtual world. The agent-based actors perform goal-directed behavior to compete in the power struggle; each actor behaves to achieve political, social, economic, and armed power (capital) objectives relative to all other actors in the simulation (Waltz 2008).

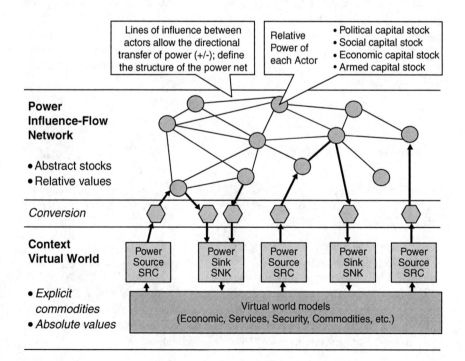

Fig. 7 Integrating the power structure competition and the virtual physical world

A composed metamodel may be organized into multiple levels of causal granularity, such that lower (finer) levels of granularity produce results that influence higher (coarser) levels; the higher levels may also set contextual variables for the lower level models. Consider three typical levels of granularity in such simulations:

- National or regional level: A top-level country-level model sets the context for the lower-level models, representing country-level political policy, national power struggles and economic base.
- Province: The overall behavior of individual provinces – political, social, economy – representing the dynamics of the political power struggle, behavior of the social populations, and relations between provinces.
- City: Major urban areas may be modeled individually (local political struggles, economic powers, civil health services, infrastructure, etc.) and are aggregated upward to the province city level.

All models interact by exchanging variables at common time increments, across a common state vector of variables that represents the PMESII state of the system at any time increment; the MRM operates as a time-discrete state machine allowing models of various modeling paradigms (e.g., agent-based, Bayesian, Petri net, system dynamics) to plug and play on the PMESII state vector.

Large-scale computational PMESII models are excellent candidates for high-performance computing implementation. Such highly-interactive models must be interpreted in the context of the uncertainty inherent in model parameters and system interactions, requiring behavioral uncertainty to be observed over large ensembles of runs (using Monte Carlo methods) that may be distributed across computing nodes on cluster before (1) analyzing the statistics of results to understand aggregate behavioral dynamics and (2) mining all results to discover emergent properties of the complex interactions interaction.

5 Exploring with Models

We are careful to distinguish two desired capabilities when we apply computational models of intervention situations to conduct analyses of internal dynamics and the effects of a potential plan of action:

- *Prediction*: the ability to foresee a specific, individual future event or scenario; generally, prediction refers to a high degree of accuracy of outcomes for specified model fidelity, resolution, and granularity.
- *Anticipation*: the ability to foresee a "landscape" of feasible futures, an "envelope" or "range" of many point predictions. This allows us to explore the range of dynamic behaviors, feasible events, and consequences, providing awareness of emergent situations that would surprise us if we had not simulated them.

Indeed, the results of computational experiments are, in fact, *exploratory*: lacking the specificity expected from physics-based model prediction of well-established physical processes (e.g., prediction of the trajectory of the bullet). The PMESII modeling processes track currently available information to drive causal simulations to create an *envelope* or *landscape* of many feasible outcomes. The simulations create an envelope of other parties' decisions, actions, and effects, then estimate our courses of action and effects. Simulation tools allow analysts to explore predictive envelopes, not point predictions. The models of decision-making and physical activities are refined over time, and the accuracies of the predictive envelopes can be tracked over time to estimate their predictive performance.

The challenge of intervention modeling, then, is to create explicit models that, by exploration, will reveal assumptions, explicitly show interactions, and simulate complex dynamics of PMESII systems to help the user understand the critical instabilities, potential domains of and emergent chaotic behaviors not expected by the tacit intuition of SMEs. Anticipating PMESII system of systems behavior requires a description of the behavior of humans with free will, organized in social networks, with varying beliefs, desires, motivations, perceptions, and goals. A realizable PMESII prediction methodology confronts the challenge of explaining social systems that exhibit unknowable causality. Jervis has pointed out how the high degree of interaction between policymaking actors in such situations confounds analysis and causal prediction: (1) results of the system cannot be predicted from separate actions of individuals, (2) strategies of any actor depend upon the strategies of others, and (3) the behaviors of interacting actors even change the environment in which they interact (Jervis 1997a, b).

Complexity is the emergent property of social system behavior, caused primarily by the interactions of its independent actors, rather than on the properties of the actors. This behavior cannot be predicted by models of the properties of the actor or by a linear combination of them. Some such linear systems exhibit responses that may have a predictable range of responses (to some degree) or not; other deterministic nonlinear systems exhibit such sensitivity to initial conditions that they exhibit behavior described as chaotic (Gleick 1988). The approach to studying such problems is not analytic (decomposition to reduce to a closed form solution); rather, it requires a synthetic approach, whereby *representative* models synthesize (simulate) behavior that may be compared to the observed world and refined to understand behavior in a more holistic manner.

6   Building Confidence in Models

PMESII models are not excluded from the necessity to provide a means for users to develop confidence in their validity in order to provide analytic value. Without user confidence in their faithful representation of reality and credible simulation of system behaviors, such models will fail to gain user acceptance, and users will return to the "tried and true" methods of situation analysis: oral discussion and the traditional enumeration of factors and narrative description of plausible scenarios.

Model developers apply a process of confidence-building in the credibility of a model by evaluating the model against two criteria to determine how faithfully it represents reality, for the intended application:

- *Internal criteria*: Behavior of the model is consistent with a theory or understanding of phenomena or causality (i.e., the model is internally consistent with a coherent explanation of a system and its phenomena; this theory should be an accepted general theory of structure and behavior).
- *External criteria*: Behavior of the model output is consistent with observed real-world behavior (i.e., the model is consistent with at least one particular instance of such a system observed in the real world; it is preferable, of course, if the model can be shown to be consistent over a wide range of conditions, if such data are available).

The formal *validation* process determines the degree to which a model or simulation is an *accurate representation* of the real world from *the perspective of the intended uses* of the model or simulation (emphasis added). This definition focuses on the external criteria (DoD 1994). (The process is distinguished from *verification*, the process that precedes validation to evaluate the correctness of a model with respect to a certain formal specification of a theory, using the formal methods of testing, inspection, and reviewing.)

When considering validation of PMESII models, it is important to distinguish between those models that are used as a *substitute* for thinking and those that serve the purpose of stimulating deep thinking (Table 3). A fire-control computer, for example, uses physical kinematic models to compute ballistic trajectories in support of an artillery officer by *eliminating* the need for thinking about trigonometry. In contrast, the models described in this chapter are for analysts and planners and serve the purpose of aiding them to think deeply and broadly about the structure and dynamics of a situation and the effects of alternate actions. In this case, validation is not performed once and trusted thereafter. The very phenomena of social situations remain in flux, and the validation process must often be performed in situ, on a day-to-day basis. In the earlier case, gravity, ordnance mass, and the influence of other physical factors remain constant; in the case of models of human, social and cultural systems, the entities have free will and the modeler cannot count on a constant human behavior.

A recent RAND study described the basis of validation in such models, in which uncertainty in the model (e.g., application of a particular theory of social behavior and response to media appeals) and in the source data inputs (e.g., uncertainty in demographic data on tribal affiliations) is deep:

> [Our conclusions] apply when the models or their data are more afflicted with uncertainty. For example, no one has a "correct" model of war with all its notorious complications, and, even if such a model existed, it would have large numbers of uncertain inputs. ... In such cases, we believe that model validation should be construed quite differently than might be suggested by the usual definition of validity. A validation process might reasonably conclude by assessing the model and its associated databases as "valid for exploratory analysis" or "valid, subject to the principal assumptions underlying the model, for exploratory analysis" (Bigelow and Davis 2003).

Table 3 The roles of validation in modeling

Approach to the use of a model	Conventional modeling as a substitute for thinking	Unconventional modeling as a stimulus for thinking
Operational need	Act quickly, react and respond (trust accuracy and automation)	Think hard and deep, reason, explore, discover (Insight; understanding)
Metaphor	Black box: model as trustworthy tool to provide answers	Guide: model as a tool used to learn and plan for complex endeavors
User's central value	Accuracy of the model	Usefulness (utility) of the modeling process
Validation	Validation before use: trust in authority that reviewed validation and certified the model for a given use	Validation during use: construction-comparison-refinement builds trust in representation
Ownership	User is not the owner of the models	User is the owner of the model (user is creator, modifier, explorer)
Basis of validity	Confidence in the model based on authority (Approved prior accreditation of validation process)	Confidence in the model developed and refined on a daily basis during use and refinement of models

Similarly, the National Research Council (NRC) has recognized that the techniques used to validate models in the physical sciences are not appropriate for modeling the behavior of individual, organizational, and societal (IOS) systems:

> Verification, validation, and accreditation: These important functions often are made more difficult by expectations that verification, validation, and accreditation (V&V) – as it has been defined for the validation of models of physical systems – can be usefully applied to IOS models. ... Current V&V concepts and practices were developed for the physical sciences, and we argue that different approaches are needed for IOS (individuals, organizations, and societies) models (Zacharias et al. 2008).

The RAND report by Bigelow and Davis (2003) on validation of multiresolution models concluded that *comprehensibility*, *explainability,* and *uncertainty representation* are the critical elements for such models:

> The authors believe that when working within this troubled but common domain, it is particularly important for two criteria to be met in assessing a model (and its associated data):

- The model should be *comprehensible* and *explainable*, often in a way conducive to explaining its workings with a credible and suitable "story."
- The model and its data should deal effectively with uncertainty, possibly *massive uncertainty*.

Referring to the use of societal models (note that in this book the terms PMESII model and societal model are used interchangeably) to increase our understanding of complex system behavior, a pioneer of social modeling wisely noted: "The moral of the story is that models that aim to explore fundamental processes *should be judged by their fruitfulness, not by their accuracy. For this purpose, realistic representation of many details is unnecessary and even counterproductive. ... the intention is to explore fundamental social processes … the interactions of adaptive agents typically lead to nonlinear effects that are not amenable to the deductive tools of formal mathematics."* (Axelrod 1997, p. 6).

7 Summary

Emerging analytic and planning tools allow analysts and planners to capture models of interventions and help anticipate their effects. We distinguish between mental models of systems or phenomena; conceptual model representations of elements, relationships, and causal functions; and computational models that implement a conceptual model and simulate the time-dynamic behavior of the modeled system. Empirical modeling represents relations between variables of a system or phenomenon based on the data from experience or experimentation, e.g., via regression methods. Causal modeling explicitly represents underlying causality (functional processes) of phenomena and derives future behavior deductively from input variables. Computational simulations of interventions are developed by decomposing the political, military, economic, social, information and infrastructure (PMESII) elements of a situation, and then representing them in component models. Modeling techniques for representing human-social systems include agent-based models, Bayesian network models, time-discrete and event-discrete models, system dynamics models, and Markov and Petri models. A model composition framework is required to integrate diverse models. For example, the DARPA COMPOEX (COnflict Modeling, Planning, and Outcome EXperimentation) program developed a large-scale simulation framework and an associated PMESII model component library, and demonstrated an ability to compose a diverse set of modeling paradigms into a single run-time metamodel. Large PMESII models face the validation challenge of demonstrating that they represent the real world well enough to support their intended uses. The uses may include prediction, i.e., the ability to foresee a specific, individual future event or scenario; or anticipation, i.e., the ability to foresee a "landscape" of feasible futures. In particular, such tools can offer awareness of emergent situations that would surprise us if we had not simulated them. The tools can aid the validation process by permitting analysts to compare modeled behavior to situation data and to refine both their models and their understanding of the systems and phenomena they represent.

8 Resources

1. *US DOD M&S Organizations*

   DoD Modeling and Simulation Coordination Office (M&SCO)
   http://www.msco.mil/

   Information Analysis Center (MSIAC) Modeling & Simulation
   http://www.dod-msiac.org/

DoD Modeling and Simulation Resource Repository (MSRR)
http://www.dod-msiac.org/

DoD Standards Vetting Tool (DSVT)
http://140.32.24.71/

DoD VV&A Documentation Tool (DVDT)
http://dvdt.nmso.navy.mil

2. *US Military Services M&S Organizations*

Army Modeling & Simulation Directorate
http://www.ms.army.mil/

Army Program Executive Office for Simulation, Training and Instrumentation
(PEO STRI)
http://www.peostri.army.mil/

Navy Modeling & Simulation Office (NMSO)
https://nmso.navy.mil/

Air Force Agency for Modeling & Simulation (AFAMS) (Public)
http://www.afams.af.mil/
Air Force Environment Scenario Generator (ESG)
https://esg.afccc.af.mil/index.php,
https://ine.aer.com/esgsite/

ESG Operational Test & Evaluation
https://ine.aer.com/

Marine Corps M&S Management Office (MCMSMO)
https://www.mccdc.usmc.mil/MCMSMO/index.htm

3. *NATO M&S Organization*

NATO Modeling and Simulation Group (NMSG)
http://www.rta.nato.int/panel.asp?panel=MSG

Technical Cooperation Program (TTCP) – Joint Australia, Canada, New Zealand,
the United Kingdom, and the United States
http://www.dtic.mil/ttcp/

4. *Modeling and Simulation Society and its Journals*

JDMS: The Journal of Defense Modeling and Simulation: Applications,
Methodology, Technology
Simulation: Transactions of The Society for Modeling and Simulation
International

5. *Modeling and Simulation Conferences*

European Simulation Conference
http://www.itec.co.uk/

Flight Simulator Engineering & Maintenance Conference
http://www.aviation-ia.com/fsemc/

Winter Simulation Conference (WSC)
http://www.wintersim.org/

SISO Spring and Fall Simulation Interoperability Workshop
http://www.sisostds.org/

MODSIM Modeling and Simulation World
http://www.modsimworld2008.com/

References

Alberts, D., & Hayes, R., DoD Command and Control Research Program. (2007). *Planning: Complex Endeavors*. Washington, DC: CCRP Publications.

Axelrod, R. (1997). *The Complexity of Cooperation: Agent-Based Models of Competition and Collaboration*. Princeton, NJ: Princeton University Press.

Baker, J. A. III, & Hamilton, L. H. (2006). *The Iraq Study Group Report*. New York, NY: Vintage Publications.

Banks, J., Carson, J., Nelson, B., & Nicol, D. (2004). *Discrete-Event System Simulation (4th ed.)*. New York, NY: Prentice-Hall.

Bankes, S. C. (2002). Tools and techniques for developing policies for complex and uncertain systems. *Proceedings of the National Academy of Sciences*, 99, 7263–7266.

Bigelow, J., & Davis, P. K., RAND Corporation. (2003). *Implications for Model Validation of Multiresolution Modeling*. Santa Monica, CA: RAND.

Covey, J., Dziedzic, M., & Hawley, L. (eds.) (2005). *The Quest for Viable Peace: International Intervention and Strategies for Conflict Transformation*, Washington, DC: U.S. Institute of Peace Press.

Davis, P. K, & Anderson, R. A., RAND Corporation. (2004). *Improving the Composability of Department of Defense Models and Simulations* (MG-101-OSD). Santa Monica, CA: RAND Distribution Services.

Department of Defense. (1994). *Modeling and Simulation (M&S) Management* (DoD Directive 5000.59).

Epstein, J. M., & Axtell, R. (1996). *Growing Artificial Societies: Social Science from the Bottom Up*. Cambridge, MA: MIT Press/Brookings Institution.

Gleick, J. (1988). *Chaos: Making a New Science*. New York, NY: Penguin Books.

Jervis, R. (1997a). *System Effects: Complexity in Political and Social Life*. Princeton, NJ: Princeton University Press.

Jervis, R. (1997b). Complex Systems: The Role of Interactions. In: Alberts, D. S. & Czerwinski, T. J. (eds.), *Complexity, Global Politics and National Security*. Washington, DC: National Defense University.

Lynn, J. A. (2005). Patterns of insurgency and counterinsurgency. *Military Review*, 85(4), 22–27.

Manwaring, M. G., & Fishel, J. T. (1992). Insurgency and counterinsurgency: towards a new analytical approach. *Small Wars & Insurgencies*, 3(3), 272–305.

McCormick, G. H. (1999). People's War. In: Ciment, J. (ed.), *Encyclopedia of Conflicts Since World War II, Vol. I: Afghanistan Through Burundi*. New York, NY: Schocken Press.

Meilinger, P. (2004). The origins of effects based operations, *Joint Forces Quarterly*, 35, 116–122.

Office of the Director of National Intelligence (2007). *Prospects for Iraq's Stability: A Challenging Road Ahead*. (National Intelligence Estimate). Available at www.dni.gov/press_releases/20070202_release.pdf.

Pearl, J. (2000). *Causality: Models, Reasoning, and Inference*. Cambridge, MA: Cambridge University Press.

Sterman, J. D. (2000). *Business Dynamics: Systems Thinking and Modeling for a Complex World*, New York, NY: McGraw Hill Publishing.

Taylor, G., Bechtel, R., Morgan, G., & Waltz, E. (2006). *A Framework for Modeling Social Power Structures*. Proceedings of Conference of the North American Association for Computational Social and Organizational Sciences. Notre Dame, IN.

U.S. Army. (2006). *Counterinsurgency* (Field Manual FM 3-24, MCWP 3-33.5). Washington, DC: Department of the Army.

Walker, D. M., Comptroller General of the United States, Government Accounting Office (GAO). (2006). *Stabilizing Iraq: An Assessment of the Security Situation*. (Statement for the Record, GAO-06-1094T). Author.

Waltz, E. (2006). Means and Ways: Practical Approaches to Impact Adversary Decision-Making Processes. In: Kott, A. (ed.), *Information Warfare and Organizational Decision-Making*. Boston, MA: Artech Press.

Waltz, E. (2008). Situation Analysis and Collaborative Planning for Complex Operations. *Proceedings of the 13th International Command and Control Research Symposium*. Bellevue, WA.

Zacharias, G., MacMillan, J., & Van Hemel, S. B. (eds.) (2008). National Research Council. *Behavioral Modeling and Simulation: From Individuals to Societies*. Washington, DC: National Academies Press.

Chapter 3
Politics and Power

Mark Abdollahian, Jacek Kugler, Brice Nicholson, and Hana Oh

P in PMESII stands for Political. Perhaps P was placed first merely to make the acronym easier to pronounce. However, more likely P's position of prominence was intended to signify its relative importance in international affairs.

Analysts, in particular, count redistribution of political power as one of the most notable effects of an international intervention. There is strong motivation for this; for example, a diplomatic intervention or international information campaign may shift power toward political groups which support the position favored by intervening groups; an economic or humanitarian intervention can strengthen political groups that control or, at least, take credit for aid; or a military intervention (a blockade, weapons provisioning, or invasion) can affect the military power of various political parties in a region, for better or worse.

Furthermore, the importance of politics does not end here. Political developments are frequently the very cause (vis a vis the result) of intervention. In response to economic or environmental shortfalls, for example, politics, through its decision-making processes, attempts to ameliorate competition for scarce resources and, in doing so, often produces conflict – ranging from the trivial (a local school board divided over the location of a new schoolhouse) to the catastrophic (a war between superpowers). Conflict, in turn, exerts an influence on the political and economic decisionmaking of intervention planners by creating uncertainty (e.g., regarding elections, economic trends, and overall stability) – often sufficient uncertainty to blur the boundary between profit and loss, or between victory and defeat scenarios.

Clearly, it is in the interest of intervention analysts to be able to understand and rigorously model political dynamics, but, aside from anecdotal instances, political scientists have received little support from the modeling community. However, this is beginning to change as the field blossoms. This chapter discusses recent work in this area, assesses the challenges faced, and provides a flavor of what is on the horizon. As with the rest of this book, it is hoped that this introduction stimulates further research as well as interest by intervention practitioners.

M. Abdollahian (✉)
Sentia Group, Inc, 1066 31st Street, NW, Washington D.C. 20007, USA
e-mail: maa@sentiagroup.com

A. Kott and G. Citrenbaum (eds.), *Estimating Impact*,
DOI 10.1007/978-1-4419-6235-5_3, © Springer Science+Business Media, LLC 2010

1   The Challenge of Political Modeling

As with many young disciplines, the world of political models is labyrinthine and balkanized; it comprises a plethora of different methodologies to analyze different aspects of similar phenomena. War, for example, is one of the primary foci of political science, as it is the most dramatic and destructive event that occurs in the political arena. As such, political scientists have long attempted to model different aspects of intra and interstate hostilities, ranging from the probability of war being initiated to the breadth, severity, and duration of a conflict, and to predicting the victor (Midlarsky 2000; Kugler and Lemke 1996; Mearsheimer 2001; Small and Singer 1982; Bueno de Mesquita and Lalman 1992; Gilpin 1981; Waltz 1979; Kahneman and Tversky 1979; Gilbert and Troitzsch 1999; Mearsheimer 2001; Bahaug and Gates 2002; Bahaug and Lujala 2004; Abdollahian and Kang 2008; Arbetman and Johnson 2008; Abdollahian et al. 2009; Levy and Thompson 2010). These inquiries result in a variety of models for different aspects of conflict and cooperation from probability through planning to reconstruction, but as yet, political scientists have no *general, integrated* model of when individuals, organizations, states, or collections of states will cooperate or fight.

The chief impediment to the construction of general, integrated political models is the broad theoretical question of determining which influences are most crucial in affecting political outcomes. Do individuals shape events or does history constrain individuals? This issue is known as the "levels-of-analysis problem"; i.e., whether the analyst should examine the individual decision-maker, organizational and interest group mobilization, national preferences, or the structure of the international system. Former House Speaker Tip O'Neil famously noted that "all politics are local" (O'Neill 1993); a useful (but less eloquent) corollary may be that *all politics are the aggregation of preferences and power at each level of analysis.*

Analytical questions regarding political modeling generally fall into three categories: micro-level dynamics (the expected actions and interactions of individuals, groups, or governments), intranational structural dynamics (the subnational, structural factors that politically propel a nation, such as economic prosperity, democratization, or other national indicators), and international structural dynamics (the cross-national comparisons of national factors). Each of these categories is best assessed using a particular methodology and the appropriate theoretical assumptions.

As Table 1 indicates, agent-based modeling is best applied to the near-term, micro-level dynamics of how individuals, groups, or nation-states interact. This bottom-up approach allows for detailed granularity in understanding how individuals interact in a given political environment. A vast literature (Schelling 1960; Axelrod 1986) on rational choice and microeconomic theories explains the drive behind individual-level behaviors and interactions (discussed later in this chapter). For example, how do specific insurgent groups gain support of likely sympathetic target populations: through coercion, influence, or the distribution of public goods? (Fearon and Laitin 2003).

Intra- and international dynamics have traditionally been captured at the structural level (Organski 1958; Waltz 1979; Goldstein 1988; Rasler and Thompson 1994;

Table 1 Political Methodology & Applications

Time Horison	Application Space	Methods	Available Tools
Near Team	Micro Level	Agent Based Models Neural Networks	Senturion
		Genetic Algorithms Game Theory	SEAS
		Expert Systems Bayesian Updating	POFED
Mid to Long Term	Intra-State	Social Network Analysis Artificial Intelligence	PERICLES
		Statistical Models Dynamic Modeling	COMPOEX
	Inter-State		Power Transitions

Tammen et al. 2000; Mearsheimer 2001; Lemke 2002; Doran 2003). Using equation-based dynamic modeling, most political theories (Richardson 1960; Intriligator and Brito 1984; Muncaster and Zinnes 1988; Saperstein 1994) at this level focus on indicators of national attributes and how those indicators interrelate. Here, a multitude of theoretical and empirical research exists (Zinnes and Gillespie 1976; Nicholson 1998; Brown 1995; Kadera 1995; Abdollahian 1996, 2008, 2009). Game-theoretical approaches (Schelling 1960; Powell 1987; Fearon 1994; Zagare and Kilgour 2000; Abdollahian and Alsharabati 2003) can be applied to any of the aforementioned, albeit, generally, with lower levels of fidelity unless highly tailored – and thus less generalizable – to a larger variety of political circumstances.

Of course, macro inputs, such as international events and government action, influence micro, or individual and group outcomes. Individuals, groups, and nations interact embedded in an environment that is defined, shaped, and constrained by macro dynamics of our political milieu. Changing national attributes, such as decreasing economic production or highly unstable political environments, can significantly alter micro-level interactions and even individuals' decision calculus. For example, during domestic political disturbances, decision time horizons of individuals as well as companies become shorter in the face of increasing uncertainty, driving more selfish behavior and eroding trust (Axelrod 1986).

A political modeler should account for the nexus between intra- and interstate dynamics that influence the decision calculus of the individuals and groups at the micro level and vice versa. While local political interactions of individuals are influenced by national and international conditions, the sum of those interactions can shape national and international conditions as well. Currently, there are good political models at the micro level (Bueno de Mesquita 1985; Kugler and Feng 1997; Rasler and Thompson 1994) and at the macro level (Grossman 1991; Fearon and Laitin 2002; Collier and Hoeffler 2004), but very few bridge the gap. For that reason, this chapter surveys political modeling theories from several subdisciplines in political science, spanning macrostructural theories of conflict, deterrence, war, and political economy to micro-level theories of political motivations and decision-making. While each of these literatures defines a portion of political interactions, together they outline the phenomenology of conflict and the boundaries of our current knowledge.

2   Theoretical Building Blocks

The first step in the construction of a political model is to select an appropriate
theoretical foundation to inform and validate the underlying assumptions about the
political behaviors to be modeled. A classification system to assess the applicability of various political theories is outlined to aid the reader. This system includes
unit of analysis, model type, assumptions, key variables, structure of the environment, and the core logic of how the variables are related, in addition to main
implications, empirical support, and shortcomings of these theories. Tables 2 and 3
describe several micro-, intra-, and international political theories and their relevant discriminating attributes.

Political phenomena occur in a multidisciplinary environment, including not
only the specific political factors but also the economic, sociological, psychological,
and even technological factors that can motivate political behavior. To explain terrorism or failed states, for example, one *single* political theory will not suffice.
In the absence of any grand, unified field theory, analysts must combine best-in-breed
theories. Theories are building blocks that researchers combine in various ways to
model different political phenomena. In order to do so, the inputs and outputs of the
theoretical blocks must be consistent and interlocking so that they can be combined
in meaningful ways.

Once the foundation of a political model has been laid with the building blocks
of theory, however, an analyst must determine what will be constructed upon that
foundation. What is the artificial environment in which individuals, groups, or
nations will interact? For example, the methodological engines that model individual behavior can be game theoretical (Intriligator and Brito 1984; Zagare and
Kilgour 2000; Powell 1987), microeconomic (Schelling 1960; Fearon 1994), or
rule-based expert systems (Bennett and Stam 2000; Abdollahian and Alsharabati
2003; Gilbert and Troitzsch 1999). If using a conceptual representation of political
bargaining space, then theories such as the median voter theorem (Kim and Morrow
1992; Bueno de Mesquita 1980), subjective expected utility comparisons (Edwards
1996; Camerer and Lowenstein 2003), and Arrow-Pratt risk aversion (Pratt 1964)
or Prospect Theory (Battalio et al. 1990; Cacey 1994; Kahneman and Tversky
1992; Levy and Levy 2002) are among many that are commonly used by political
modelers. The next section examines some of the typical approaches for creating
artificial environments or models in which political phenomena at the micro-,
intra-, and interstate levels may be simulated and tested.

3   Key Approaches to Conflict and Cooperation

Below are surveyed a few of the typical best-in-breed political science approaches,
detailed above in Tables 2 and 3. We first explore the macro analysis of conflict and
cooperation by using nation-states as the unit of analysis to understand conflict
behavior among and between nations. We then turn to a few main theories that

Table 2 Macro Theories and their attributes

	Classical Deterrence	Game Theoretic Deterrence	Dynamic Deterrence	Arms Race	Power Transitions	Endogenous Growth	POFED
Unit of Analysis	Nation State	Nation State	Nation State	Nation State	Nation State	Nation State	Nation State
Model Type	Structural	Microeconomics Individual Decision making Rational Choice	Microeconomics Individual Decision Making Rational Choice	Coupled Differential Equations	Structural Differential Equations	Formal Modeling Dynamic Equilibrium	Structural at the domestic level Dynamic general equilibrium model
Assumptions	Every response should be credible Every actor is rational	Nuclear war is so costly that only an irrational leader could consider it a means of conflict resolution	Every response should be credible Every actor is rational Risk averse individual is one who always prefers the expected value of a monetary gamble to the gamble itself while a risk loving individual has an opposite preference.	Deterministic Dynamic Rationality Utility Maximization	Hierarchy in the international system States are unitary & rational actors Status Quo & international hierarchy is based on relative power of state State power is based on domestic factors	Utility maximization Economic growth derived endogenously "Precisely" constant returns to scale in the production process	Individuals seek to maximize their lifetime utility by choosing how much to consume, save, and how many children to have Policy makers choose tax rate, amount of public investment, and military spending to maximize their chances or remaining in power

(continued)

Table 2 (continued)

	Classical Deterrence	Game Theoretic Deterrence	Dynamic Deterrence	Arms Race	Power Transitions	Endogenous Growth	POFED
Key Variables	Absolute cost of war	Individuals' preferences Individuals' choices in determining interstate conflict behavior	Missile stocks, Casualties, Rate of firing its missiles, Proportion of the counterforce attack, Proportion of counter value attack Effectiveness of missiles against enemy missiles Effectiveness of missiles against enemy cities	Missiles for two countries A and B and rates of change in missiles. Decision to launch missiles for country A and B Counterforce effectiveness: the ability to destroy the other nation's missiles. Counter-value effectiveness: the ability to kill the other nation's population.5. Casualties	Relative Power = $\dfrac{GDP*}{Population*}$ Political Capacity Status Quo Evaluations (satisfaction or dissatisfaction)	Capital Labor Innovation Human Capital Patent Laws (some)	Income: GDP per capita Fertility Human capital. Relative instability: Number of deaths in a given year relative to the maximum number of deaths in the country Political capacity: Ratio of revenues a government extracts compared to extract compared to predicted levels a society could attain based on economic endowment
Structural Environment	The cost of war is the key difference between the nuclear and pre-nuclear war. Nuclear arms races ensure peace.	Realism In a non-cooperative game: anarchy condition In a 2-persons game: bipolar international system	The cost of war is the key difference between the nuclear and pre-nuclear war. Nuclear arms races ensure peace.	Anarchy—no overarching authority to guide (or constrain) the individual collection of sovereign states.	Dominant nation at the top of international hierarchy Other nations under the dominant power Dominant power structures system to maximize its security	Interplay of economic and innovation factors	Interplay of political, economic, and demographic indicators anticipate the impact of interventions in fragile states

Core Logic						
Nuclear arms races ensure peace\ Ultra stable solution can be reached through mutually assured destruction by proliferating nuclear weapons	Contrary to classical deterrence theory which finds key to interstate stability in the structure and distribution of power, game theoretical deterrence focuses on the interplay of preferences, choices in determining interstate conflict behavior, and outcomes.	A country chooses both a rate of fire and a targeting strategy either city or arsenal The 4 stages of nuclear deterrence strategies are: a. Conventional Balance of Power b. Massive retaliation c. Nuclear capabilities for two countries are equal but arsenal is not enough to assure retaliation. d. Mutually Assured Destruction	Arms races will lead to the "cone of mutual deterrence" based on the parameter estimates and nation's current weapons stock.	Power of a dominant state is overtaken by a challenging state Rise of populous, less developed state is inevitable Wars are most likely when nations are in parity and challenger is dissatisfied with status quo	Innovative firms generate knowledge: Knowledge is a public good which spills over into the economy as a whole resulting in increase in productivity. Technological Progress Formulation: Ensures the profitability of knowledge development to fuel innovation leading to increases in productivity.	Domestic factors such as fertility, income, and political effectiveness can lead to domestic instability leading to state failure

(continued)

Table 2 (continued)

	Classical Deterrence	Game Theoretic Deterrence	Dynamic Deterrence	Arms Race	Power Transitions	Endogenous Growth	POFED
Implications	Nuclear proliferation will stabilize the world politics.	Irrational and accidental wars are possible. Nuclear powers are unlikely to fight when they are of roughly same size but disparity provides incentives to initiate a nuclear strike	Possibility of war exists only when costs are acceptable. Path from Balance of Terror to Mutually Assured Destruction is unstable as one nation acquire nuclear weapons unilaterally	Arms races can have deterrent effects if nations have "run away" armaments that propel them into the "cone of mutual deterrence" or MAD.	Preponderance = peace Impact of relative power on the likelihood of war is conditioned by satisfaction or dissatisfaction within the international system	Differences in growth are maintained due to innovation Innovation is a function of production externalities Growth is determined by national differences	Economic fragility is self reinforcing Growth leads to stability creating further growth. Political capacity accelerates growth High rising capacity = stability Low & declining capacity = instability
Empirical Support	Brodie (1959): urged to turn our attentions from "Win-the-War Strategies" to "Deterrence Strategy" facing the era of nuclear warfare that is "too large-scale, too menacing to all our hopes. Waltz (1979) and Mearsheimer (2001): argues that if Mutual Assured Destruction is stable, expanding the scope of nuclear deterrence should dramatically reduce the likelihood of war whether contenders are large or small.	Schelling (1960): Chicken game captures realism and classical deterrence theory Fearson (1992): Crisis bargaining model shows that nations in adversarial relationships violate deterrence when one nation suggest possible use of threat Powell (1987): Mutual Assured Destruction is a cure all while failure of deterrence comes from random moves of nature Zagare & Kilgour (2000): Prisoner's dilemma better explains reality than chicken game.	Intriligator and Brito (1976, 1984): Path from the balance of terror to Mutually Assured Destruction is unstable as one nation acquires nuclear weapons unilaterally	Gilbert, Rider, and Hutchison (2005): The "arms race" variable has statistically significant, positive effect of the likelihood of military dispute. Also an arms race between strategic rivals does not facilitate deterrence, but instead facilitates war.	Organski & Kulger (1980): Nations equal in power are more likely to fight than nations with unequal power Lemke & Werner (1996): Study power and war of dyads including dominant nations Kim (1996): Status quo evaluations may be important in determining war than parity Abdollahian & Kang (2008)	Jones (1995): These theories have largely failed to perform empirically. Cavusoglu and Tebaldi (2006): The empirical work to date suggests that the "conditional convergence" hypothesis of the revised classical model seems to provide a more accurate empirical description of growth.	Feng, Kugler and Zak (2000): Provides overview of international dynamics that ferment domestic instability and lead to sate failure. Crossnationally validated for 78 countries and sub national level for 6 countries

Weaknesses	In order to achieve the "ultra-stable" nuclear world, proliferation optimists such as Mearsheimer and Waltz advocate a virtue of spreading nuclear weapons in unstable regions (Mearsheimer, 1990; Waltz, 1990).Waltz (1979) and Bueno de Mesquite and Riker (1982) suggest that, since nuclear weapons guarantee peace, more nations should acquire them.	Same as the weaknesses in classical and dynamic deterrence theories	Cost is the only critical factor in decision to engage in war Individuals' preferences are assumed to be the same Nations with or without nuclear weapon all have the same attitude of risk aversion	Little empirical support for this presentation of the theory outside of military buildups.	The logic of powerful therefore satisfied raises the question, If rising states are successful, why are they dissatisfied with the Status Quo? Prevention problem rises as myopic states forego a preemptive attack. There are critiques on when a war is most likely to occur during the transition process.	Lack of comprehensive empirical support	The approach cannot be applied effectively to states which have experienced total governmental collapse

Table 3 Micro Theories and their attributes

	Mediau Voter	Von Neumann & Morgenstern Utility Functions	Risk Propensity	Subjective Perceptions & Utility
Unit of Analysis	Individual Decision Maker	Individual Decision Maker	Individual Decision Maker	Individual Decision Maker
Model Type	Microeconomic	Microeconomic	Microeconomic	Microeconomic
	Rational Choice	Rational Choice	Rational Choice	Rational Choice
Assumptions	Agent has a unique ideal point	State of nature: individual utility is dependent on underlying uncertainty & payoffs	A risk averse individual in one who always prefers the expected value of a monetary gamble to the gamble itself while a risk loving individual has an opposite preference.	Contrary to von Neumann-Morgenstern theory, probabilities are assumed as "subjective" numerical frequencies based upon individuals perceptions
	Single / multi Peaked Preferences	Gamble: decision making under uncertainty		
	1-demensional choice space	Lottery: individual choices denoted by $p*x+(1-p)y$		
	Sincere preference expression			
	All agents are utility maximizers			
Key Variables	Constituents' preference	Probability	Probability	Probability
	Politicians	Utility	Utility	Utility
	Political parties	Risk Propensity	Risk Propensity	Risk Propensity
	Election rules			Perceptions
	Electoral institution			

Structural Environment	Democracy or any majority rule voting system	Investigates the motives of an agent decision-making under uncertainty & risk	Neumann-Morgenstern & Friedman and Savage (1948) constructed the concepts of utility and univariate risk propensity in seeking the ways of analyzing economic issues with expected utility framework.	Investigates the motives of an agent decision-making under uncertainty & risk according to subjective probabilities
Core Logic	In a majority election, where two politicians commits to policy position closest to his own preference, if the politicians want to maximize the number of votes, they race to the median to win If either candidate deviates to a different policy position, the deviating candidate receives less than half the vote.	The individual's expected utility of a gamble is represented as $V=E(u(w))=pu(w1)+(1-p)u(w2)$ The expected value of gamble is represented as $u(E(w))=u(pw1+(1-p)w2)$	The individual's expected utility of a gamble is represented as $V=E(u(w))=pu(w1)+(1-p)u(w2)$ The expected value of gamble is represented as $u(E(w))=u(pw1+(1-p)w2)$ If the former is smaller than the latter, such behavior is called risk aversion. If the former is larger than the latter, such behavior is called risk acceptant.	Similar to Utility & Risk but perceptions drive subjective estimates

(continued)

Table 3 (continued)

	Mediau Voter	Von Neumann & Morgenstern Utility Functions	Risk Propensity	Subjective Perceptions & Utility
Implications	Majority rule voting system is imperfect and failures exist on many levels Each and every election mechanism produces its losers and inefficiencies so its only a trade off of inefficiencies Democracy has a danger of producing a leader without much support Politicians and political parties gather to the median to maximize their votes	The expected utility property says that the utility of a lottery is the expectation of the utility from its prizes. We can compute the utility of any lottery by taking the utility that would result from each outcome, multiplying that utility times the probability of occurrence of that outcome, and then summing over the outcomes. Utility is additively separable over the outcomes and linear in the probabilities.)	The utility of the expected value is higher than the expected utility of the gamble, so the agent is risk averse. A level of certain wealth provides the same utility as does participating in his gamble. The individual will be willing to pay anything up to some value to avoid participating in the gamble. Even when costs are paid, this riskaverse person is as well off as he would be if forced to face the world uninsured.	If an individual's behavior satisfies restrictions sufficient to ensure the existence of subjective probabilities, those probabilities must satisfy Bayes' law. Bayes' law is important since it shows how a rational individual should update his probabilities in the light of evidence, and hence serve as the basis for most models of perceptual and rational learning behavior. Misperceptions between individuals' subjective estimates can drive different behavior.

| **Empirical Support** | Black (1948): Stated that political parties will pursue policies that appeal the most to median voters
Downs (1957): Introduced left-right axis to economic theory. He claimed that as voters do not have perfect information regarding the candidates, voters will resort to economic issues | von Neumann and Morgenstern (1944): Investigated the motives of an agent making a decision under risk. By defining a real-valued preference function over the set of cumulative functions, enables us to model agent preference over alternative probability distributions in a manner completely analogous to non-stochastic decision making theory. | Bueno de Mesquita (1985): Computed hypothetical risk scores by sampling 30,000 configurations per year and selecting the global minimum and maximum from this search as the hypothetical minimum and minimum for all states.
Kim and Morrow (1992): Showed rising state's risk attitude is increasing while declining state's risk attitude is decreasing Tversky & Kanaheman (1979) show opposite results in Prospect Theory. | Knight (1921): Proposed distinguishing between risk and uncertainty according to whether the probabilities are given to us objectively or not.
In a sense, the theory of subjective probability nullifies this distinction by reducing all uncertainty to risk through the use of beliefs expressible as probabilities.
Theory is therefore a far-reaching generalization of expected utility theory. |

(continued)

Table 3 (continued)

	Mediau Voter	Von Neumann & Morgenstern Utility Functions	Risk Propensity	Subjective Perceptions & Utility
Weaknesses	Majority rule decision mechanism reveals a failure of at least one type of inefficiency	Violations of interpersonal comparisons of utility	There is no measurement to capture the degree of agent commitment with its direction.	Violations of interpersonal comparisons of utility
	No one majority rule mechanism fully reveals preference of the voters	Must estimate objective probabilities	The game structure is strictly non cooperative. Decision makers do not make binding, strictly enforceable commitment to the status quo.	Must estimate subjective v. objective probabilities
	The non-verification of single peaked preference can lead to majority cycle trap where agenda maker has the power to choose any outcome by manipulating the order of propositions	No possibility of misperceptions	The incidence of conflict between agents is over-predicted	

explain the nation-state itself and how its economic factors, demographics, and other national indicators can lead a nation to war or peace. Finally, several key theories that drive micro-, individual-level behavioral dynamics will be mentioned to explain how preference, behavior, and perception of individual leaders in a nation can be combined to anticipate politics, peace, or conflict.

3.1 Macro Interstate Approaches

One of the earliest models of relations between states was the so-called arms race model originally introduced by Richardson (1960). Such models investigated the dynamics of nations' armament buildups by using coupled differential equations. Richardson's equations posit a simple deterministic relationship between two states based on action and reaction, in which a small buildup by one side would lead to a larger counter by the opponent. Thus, an arms race could produce a wide gap in capabilities; this gap in capabilities was expected to prompt conflict. The equations define a precise movement of armaments through time, in which the pace of armament or disarmament is a function of how far one nation is away from its long-run equilibrium point. Depending on the initial conditions, the equilibrium can be peace or war. An "equilibrium point" in a dynamic system is a solution for the equations that does not change with time. The "initial condition" is the value of the variables at the onset of the simulation. Thus, the rate of armament buildup is expressed as:

$$dx / dt = ay - bx + g$$

$$dy / dt = cx - dy + h$$

where

1. dx/dt (for nation x) and dy/dt (for nation y) are the rates of armament
2. x and y are the amount of armaments
3. a and c are "threat" parameters
4. b and d are "fatigue" parameters
5. g and h are "grievance" parameters

Based on the value of the model's parameters, the nations experience either "runaway" (i.e., unchecked) armament or disarmament (based on the model's initial conditions) or convergence at an equilibrium point. Stability in the system is determined by whether states place relatively more emphasis on the threat of the other nation's arms or on fatigue from armament buildup and expenditure. Although the Richardson model represents only an early attempt to simulate and predict political behavior, his work influenced later scholars.

A major extension of this perspective led to the evolution of deterrence theory at the macro interstate level. One of the cornerstones of deterrence theory was laid early in the Cold War by Brodie (1959), who urged a focus on deterrence rather than victory, as nuclear warfare is "too large-scale, too menacing to all our hopes." The expectation that nuclear terror can credibly compel potential opponents to avoid confrontations is rooted in the high cost of nuclear war; thus, the implication of deterrence theory is that nuclear arms races ensure peace. This classic notion behind Mutually Assured Destruction (Huth and Russett 1990; Waltz and Sagan 1995) – that nuclear proliferation leads to highly stable international conditions – was refined by deterrence scholars such as Intriligator and Brito (1981). They assume that when nations anticipate that the costs of war will exceed a threshold above which said nations are not willing to initiate conflict, nations will fight only in self-defense. When a second threshold is exceeded, a nation is no longer willing to confront the opponent, and that nation will be deterred from war or yield to the aggressor's demands. Therefore, the possibility of war exists only when costs are "acceptable." Unstable conditions occur when contending actors have only conventional capabilities and cannot impose sufficient costs to deter opponents. In their research, this scenario is divided into four stages. The first is called the "Cone of War," in which nuclear parity stabilizes world politics. Second is "Massive Retaliation," in which one side initiates a nuclear buildup. Third is the "Balance of Terror," in which both nations have nuclear capabilities, but their arsenals are not large enough to assure retaliation if the other side attacks preemptively. This stage is tenuously unstable, until a credible second-strike capability is developed. The fourth and final stage is "Mutually Assured Destruction," in which equality of nuclear capabilities and secure second-strike capabilities on both sides minimizes the likelihood of war because the costs become unacceptably high. The basic model is as follows:

$$dMa \ / \ dt = -\alpha Ma - \beta'\beta \, Mb \times fb$$

$$dMb \ / \ dt = -\beta \, Mb - \alpha'\alpha \, Ma \times fa$$

$$dCa \ / \ dt = (1 - \beta')\beta \, Mb \times vb$$

$$dCb \ / \ dt = (1 - \alpha')\alpha \, Ma \times va$$

where:
   Country: a, b
   Initial time: $t=0$
   $M(t)$: missile Stocks
   $C(t)$: casualties
   α, β: the rate of firing a country's missiles
   α', β': proportion of counterforce (against enemy missiles) attack
   $(1-\alpha')$, $(1-\beta')$: proportion of counter value (against enemy cities) attack
   f: effectiveness of missiles against enemy missiles
   v: effectiveness of missiles against enemy cities

The implication of dynamic deterrence is that during the development of nuclear capabilities, one nation cannot fully deter the other and war is possible. It is in the last stage at which mutual destruction is assured that the cost of war prevents both nations from initiating war. As one nation acquires weapons unilaterally, however, the path to the last stage is very unstable. Deterrence is therefore not entirely stable; additionally, because terrorists and other violent nonstate actors do not have a "return postal addresses," deterrence is not credible toward such threats. Deterrence is the ability to prevent attack by a credible threat of unacceptable retaliation. The calculation of the cost of war is the main rationale in deterrence; thus, the theory is difficult to apply to violent nonstate actors. One nonobvious insight from the Intriligator and Brito theory is that a strategy that assures retaliation and minimizes communication among contenders may produce conflict.

Organski (1958) proposed that nations would fight when they are dissatisfied with international norms and hold equal capabilities. This is based on the assumption that hierarchy exists in the international system. Here, hierarchy is defined as a system in which a dominant nation (the "defender" or "dominant power") is at the top of an international power hierarchy, with "great powers," "middle powers," and "small powers" under the dominant power. In this power transition theory (PTT), power is measured relatively (in comparison to other states) based on demographic and industrial indicators, where Relative Power = GDP × Population × Political Capacity where Political Capacity measures the state's domestic control, a ratio between anticipated and actual tax receipts. Additionally, the dominant power enforces the status quo of the international system, while lesser powers are either satisfied or dissatisfied with the status quo.

PTT anticipates interstate dynamics by analyzing this relative power distribution across the international system and the member's satisfaction with the status quo. Under conditions of parity and dissatisfaction, the theory predicts the highest probability of international conflict; when nations dissatisfied with the status quo accrue enough power to challenge the dominant nation, PTT postulates that war is most likely. For example, in the middle of the Cold War, a PTT-based analysis (Organski and Kugler 1980) concluded that the conflict in Europe would not be repeated because of integration, that the USSR would fall from the rank of competitors by 2000, that China would emerge as the leading challenger to the United States, and that the political center would shift from the West to Asia by the end of this century. The dominant power is committed to defending the international treaties and norms that constitute the status quo, which reflects the dominant power's preferences (as it is the most powerful nation within the international hierarchy).

Using a system of symmetric, coupled nonlinear differential equations, Abdollahian and Kang (2008) formalized and tested a system-dynamic model to identify to what extent and degree policymakers can maintain stability in rival dyads, such as the U.S.–China case. Their model explores some of the structural conditions of how conflict or cooperation affects the growth and transition from the PTT literature. The work suggests specific, strategic policy prescriptions for managing conflict or cooperation and highlights the nonlinear and nonmonotonic effects of foreign policy actions.

The entire power parity model system of nonlinear ODEs is the combination of the following equations:

$$\frac{dP_{\mathrm{D}}}{dt} = B_{\mathrm{D}} P_{\mathrm{D}} \left(1 - \left(P_{\mathrm{D}} + P_{\mathrm{C}} \right) \right) - H_{\mathrm{D}} C_{\mathrm{C}}$$

$$\frac{dC_D}{dt} = -S_D P_D C_C \frac{1}{\sqrt{2\pi\sigma}} \exp{-\frac{1}{2}\left(\frac{P_D / P_C - 1}{\sigma}\right)^2} \frac{1}{2.50599}$$

$$\frac{dP_C}{dt} = B_C P_C \left(1 - \left(P_D + P_C\right)\right) - H_C C_D$$

$$\frac{dC_C}{dt} = -S_C P_C C_D \frac{1}{\sqrt{2\pi\sigma}} \exp{-\frac{1}{2}\left(\frac{P_C / P_D - 1}{\sigma}\right)^2} \frac{1}{2.50599}$$

where:

P_D is the systemic power level of the dominant nation.
P_C is the systemic power level of the challenger.
B_D is the national growth rate coefficient of the dominant nation.
B_C is the national growth rate coefficient of the challenger.
H_D is the dominant nation's cost coefficient for competition.
H_C is the challenger's cost coefficient for competition.
C_C is the conflict level that the challenger targets toward the dominant nation.
C_D is the conflict level that the dominant nation targets toward the challenger.
S_D is the foreign policy stance of the dominant nation toward the challenger.
S_C is the foreign policy stance of the challenger toward the dominant nation.
σ is the parity variance condition coefficient.

The variables in the power parity model include systemic power levels, conflict levels, foreign policy stances, and the value of the parity ratios for a rival dyad. The parameters in the power parity model are the national growth rates, the cost of competition, and the parity variance condition. By varying the parameter values and initial conditions of variables for rival dyads, an analyst can explore the performance of the dynamic model under various circumstances, not only reconstructing historical relationships between dyads but also forecasting simulations. Figure 1 demonstrates the policy results using U.S.–China data (Abdollahian and Kang 2008).

Figure 1 depicts a scenario in which China adopts a highly hostile foreign policy stance ($s_c = -0.9$), and the U.S. policy response is allowed to vary. Notice that at aggressive U.S. foreign policy response values, the effects of competition on systemic power levels produce a significant, detrimental impact on both countries. As the United States begins to question the rise of China, small changes in the firming of the American policy stance produce sharp increases in dyadic conflict. Hence, the structural stage is set for prompting early conflict initiation and war escalation. At the other extreme, an acquiescent foreign policy stance toward China produces sustainable levels of systemic power for a while, although a Chinese overtaking is guaranteed within about 15 years. In this case, after China surpasses the United States in systemic capabilities, possible minor conflicts or hostile incidents are still expected between the two countries. At a neutral foreign policy stance, levels of U.S. conflict remain low throughout the transition period as a result of small changes in the U.S. conflict equation. Under these simulation conditions, only a neutral U.S. policy stance can secure the window of opportunity for peace and stability.

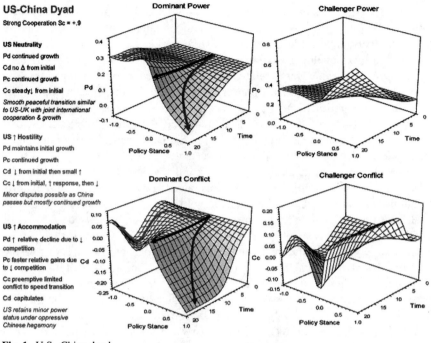

Fig. 1   U.S.–China dyad

3.2   Macro Intrastate Approaches

The systematic empirical research on *intrastate* and *nonstate* conflict has a very long and distinguished record based on innumerable case studies and a vast array of alternative propositions accounting for the rise of nation-states and the emergence of institutions and dissatisfaction in the polity (Brinton 1952; Crenshaw 1995; Huntington 1968; Gurr 1974; Diamond 1992; Inglehart 1997; Welzel et al. 2003). These contributions span the fields of not only political science and economics but also sociology and cultural anthropology. Tilly (1975) in a classic assessment and Poggi (1990) summarize systematically the process of state formation. Barnett (2004) links the motivations of international terrorism to economic modernization of states. Lemke (2009) shows that when a nation emerges from a cooperative aggregation of states – such as the unification of the Italian states in the nineteenth century or expansion of the United States – that legacy leads to relatively stable and evolving governance. On the other hand, when the birth of a nation or its reconstruction is associated with serious conflict, insurgencies and developmental lags are introduced. Nation-building propositions can profit from such long-term assessments, but understanding the political motivations and mechanisms, let alone modeling causation, remains submerged in such summaries. For a recent general review, see Midlarsky (2009).

Davies (1962) was among the first to systematically relate insurgency with an inverse U curve of development. He demonstrated that internal instability was not likely among the least and most developed societies but maximized among the less developed societies, particularly those undergoing fast economic development. A large related literature subsequently developed relating opportunity to the likelihood of insurgency to predict conflicts based on the economic and political incentives or constraints they face (Grossman 1991; Collier and Hoeffler 2002, 2004; Elbadawi and Sambanis 2002; Fearon and Laitin 2003; Fearon 2004; Barnett 2004; Hegre and Sambanis 2004; Abdollahian et al. 2009). A number of explanatory variables, including ethnicity, culture, absolute deprivation, language, and race are discarded, while economic well-being, the strength of political institutions, and reliance on commodity exports – mainly oil – are consistently associated with conflict. Thus, affluent societies that have institutionalized effective governance, and do not rely on exports of commodities such as oil, are least likely to experience insurgencies. Unfortunately, this literature is, however, confounded by the lack of reliable and consistent historical data for most countries. The revival of interest in insurgency studies in the early 2000s refocused researchers on how nonstate actors can generate intrastate instability. Previous studies had demonstrated a very weak link between civil war and the initiation of international conflict (Tanter 1966; Rosenau 1964), and for that reason much of the earlier research focused on intrastate conflict at the state-society level. Demands to link intrastate with substate actors and interstate conflict forced most researchers to rely on national rather than subnational data for their exploration. Collier and Hoeffler (2004), for example, show that "greed" rather than "grievance" is associated with the initiation of intrastate conflicts. The causal relation from "grievance" and "dissatisfaction" to domestic instability is established but not directly related to the source of conflict. The concern here is that substate actors and their representative populations that have "grievances" or are "dissatisfied" with fiscal performance or political governance are not directly identified. Rather, differences across nations help to determine the likelihood of intranational instability. Buhaug and Gates (2002) among others challenge such results showing that applying aggregate measures says little about whether conflicts are located in these areas. This leaves a void where integration of political models across levels of analysis can help.

A second major contribution to the emerging understanding of intrastate instability is driven by the contribution of Fearon and Laitin (2003) that found, contrary to most case study results, little relation between conflict and ethnic, religious, linguistic, or cultural differences. Instead, they show that rough conditions identified by Guevara (1968) are cross-national correlates of insurgency and guerrilla activity. Based on these aggregate results, geography and the flow of populations are used to explain intrastate conflict (Hendrix and Glaser 2007; Salehyan and Gleditsch 2004). The standard argument is that rough terrain confers tactical advantages on insurgents by mitigating the advantages enjoyed by state armies, which can mobilize disenfranchised groups to rebel. Outnumbered and outgunned insurgents can avoid direct engagement and gain access to safe havens where they can recruit and replenish supplies. Collier and Hoeffler (2002) find that mountainous, forested terrain aids in insurgencies. Likewise, Fearon and Laitin (2003) find a significant positive relationship between rough terrain and the onset of conflict. Most likely, climate and terrain provide the preconditions for effective insurgencies but are not and cannot be the variables that *cause* intrastate conflict as geography does not significantly vary over time. Raleigh (2004) is right on point when he challenges the aggregate approaches, arguing that modern insurgencies are as likely to be based in urban areas as rural ones that provide excellent safe havens. Moreover, safe havens in neighboring states where porous borders exist would make rough terrain irrelevant. He argues that weak states, defined as those with low GDP per capita and weak political institutions, are limited in their ability to project political authority regardless of terrain. Fragile states rather than geography place governments at risk, and this is shown in more recent work that suggests economic, demographic, natural resources, and political factors trump geographic variables (Humphreys 2005). Controlling for these foreign safe havens and economic and political development shows that rough terrain, as defined by Fearon and Laitin (2003), is not a significant predictor of conflict onset (Bahaug 2002, 2004; Rodrik 2004; Engerman and Sokoloff 2002).

Recent work focuses on the causes of domestic instability as based on insights from the long case study record and on intrastate and stakeholder data that emphasize *differences within* a national unit. Cederman (2004) links ethnic groups that inhabit mountainous terrain suggesting national formation. Using agent-based modeling techniques, he finds that violent separatist movements are much more likely to occur in mountainous terrain and tropical climates that provide shelter for guerrilla activities when ethnic groups are hierarchically organized and not otherwise. This type of approach to identify the pathways of a nation moving down the road to intrastate conflict requires a detailed, subnational level analysis at the provincial, district, or individual level. Our approach starts with the respecification of successful models that account for international conflict at the intrastate level.

3.2.1 Applications: Relative Political Capacity

For a detailed example, we first examine a key political indicator of relative political capacity and then explore one structural model of domestic political economy called the Politics, Fertility, and Economic Development model [POFED] (Feng et al. 2000). Relative political capacity (RPC) is the ability of a government to extract resources from its population as evidenced by relative performance of actual versus expected tax collection efforts for a given level of economic development (Arbetman and Kugler 1997). Recent work (Arbetman and Johnson 2008) on the dynamic effects of changes in political capacity suggests that as a government loses its ability to extract resources and advance its goals, the potential for competitors willing to fill that gap rises. Unexpectedly, as the political capacity of the challenger rises, a competitor usually replaces the government. If the competitor gains footing, the political capacity rises anticipating the lowering in instability. When the new government establishes control and achieves normal levels of political performance, this cycle of instability comes to a close. The pattern suggests a relationship between a government's level of political capacity, changes in the level of political capacity, and intrastate instability. Not only do the levels of political capacity matter but also the rates of change as shown in Fig. 2.

Here, we see political capacity changing from positive to negative, in both level and rate, from 1996 to 2005 for a particular nation with the associated size of participants in demonstrations and those killed or wounded in such (Kugler et al. 2008). Disaggregating political capacity to the provincial level within nations shows even more clearly the areas from which a national government will be challenged. Arbetman and Johnson (2008) show that without a strong central government presence, provincial governments face a political challenge from groups that are themselves capable. Such information is essential in assessing to what degree providing economic assistance to an area – such as Darfur in Sudan – would limit casualties without destroying the central political foundations required for continual stability.

RPC & Political Stability

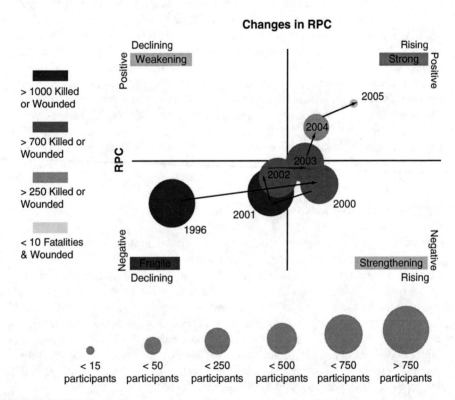

Fig. 2 Political capacity

Clearly, instability results from the interaction between economic growth and political capacity, so the linkages between political capacity and domestic political economy are crucial. Here, we look at an example of a structural model of detailed domestic intrastate politics, POFED which highlights the effects of growing political resources, economic constraints, and demographic pressures on the promulgation of conflict (Feng et al. 2000; Kugler et al. 2005). The model grew out of extant literature on modernization, human capital formation, institutional capacity, and economic development as a means of tracing the dynamic interrelationships between productivity, fertility, political effectiveness, and social stability. By using a statistically validated system-dynamics approach for the behavior of individuals and policymakers in a dynamic world focusing on antecedents for state failure and insurgency, POFED accounts for the political and economic structural environments that cause a country's living standards and political position to grow or decline.

The interplay of political, economic, and demographic indicators is modeled in POFED to anticipate the impact of interventions in fragile states and can thus be used to identify direct policy levers to mitigate state fragility. In addition to identifying policy, investment, and business actions that impact structural conditions to decrease state fragility, POFED can provide detailed tactical leverage points when applied at higher geospatial resolutions, such as provincial or district level analysis. As shown in Fig. 3, POFED has five major components that capture factors of state fragility and the effects of potential intervention: Income (y), Fertility (b), Human Capital (h), Instability (S), and Political Capacity (X). POFED is a dynamic general-equilibrium model based on the intersection of political and economic maximization function. The models show maximizing behavior of individuals seeking to maximize their lifetime utility by choosing how much to consume and save and how many children to have, while policymakers choose the tax rate and the amount of public investment and military spending to maximize their chances of remaining in power. In equilibrium, prices move endogenously so that supply balances demand for all goods and investments in the economy, while policymakers set fiscal policies. Further, the political-economic market equilibrium is a dynamic curve in the phase space tracing the evolution of the political-economic system. The model is specified as:

$$b_t = By_t^{-1}$$

$$y_{t+1} = Ay_t^{\eta}(1 - S_t)^{\alpha} \chi_t^{1-\alpha} h_t^{1-\alpha}$$

$$h_{t+1} = \omega h_t / b_t^{\theta}$$

$$S_{t+1} = S_t^a d\chi_t(\chi_{t-1} / \chi_t)$$

$$\chi_{t+1} = Cy_t^{\gamma} \chi_t^{\beta} p_t^{\phi} / b_t^{\gamma^{\phi}}$$

These equations show that birth rates b depend on income y; and that income depends on past income and political conditions, x. h shows the generational feedback on the creation of human capital, while political instability, S, has a temporal feedback

POFED MODEL

Insights
Economic Fragility is self reinforcing
Growth leads to stability creating further growth
Political Capacity accelerates growth
High & rising Capacity insures stability
Low & declining Capacity generates instability

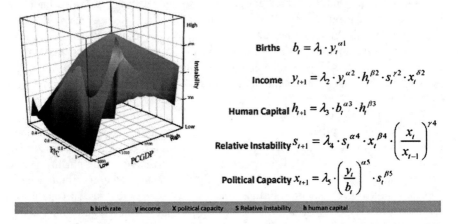

Births $\quad b_t = \lambda_1 \cdot y_t^{\alpha 1}$

Income $\quad y_{t+1} = \lambda_2 \cdot y_t^{\alpha 2} \cdot h_t^{\beta 2} \cdot s_t^{\gamma 2} \cdot x_t^{\delta 2}$

Human Capital $h_{t+1} = \lambda_3 \cdot b_t^{\alpha 3} \cdot h_t^{\beta 3}$

Relative Instability $s_{t+1} = \lambda_4 \cdot s_t^{\alpha 4} \cdot x_t^{\beta 4} \cdot \left(\dfrac{x_t}{x_{t-1}} \right)^{\gamma 4}$

Political Capacity $x_{t+1} = \lambda_5 \cdot \left(\dfrac{y_t}{b_t} \right)^{\alpha 5} \cdot s_t^{\beta 5}$

b birth rate   **y** income   **X** political capacity   **S** Relative instability   **h** human capital

Fig. 3 POFED model components

and depends on external policy p_s. Similarly, political capacity, x, depends on per capita income y, external policy p, and births b.

The intuitive logic of POFED is as follows. In addition to well-established economic determinants, the fundamental political variable of political capacity alters fertility decisions, human capital accumulation, and economic development. In fact, fragile developing societies are defined by a decline in per capita income, by the potential for falling into the poverty trap, and by the low or declining capacity of governments (Guillaumont and Jeanneney 1999, Kugler and Tammen 2010). Robust societies with higher levels of political capacity extract more than anticipated from their economic endowment and allocate such resources efficiently to advance the government's priorities; fragile societies that fall below average political capacity levels of similarly endowed societies fail to do so. Some of the key general policy prescriptions are summarized as follows:

- Sufficient political capacity is a necessary precondition for income growth for poor countries.
- Income growth is self-reinforcing: when birth rates fall, human capital rises and political instability declines.
- The poverty trap is self-reinforcing; when birth rates rise, political instability increases and income falls.
- External aid, policy interventions, and domestic policies can increase stability and promote income growth.

- Increasing political capacity and change in political capacity reduce political instability.
- Income falls and birth rates rise when political capacity is lower and political instability is higher.
- There are thresholds of political capacity and political instability driven by economic performance that can cause a state to fail.

3.3 Micro Individual Approaches

The preceding sections have examined how conflict and cooperation between, among, and within nations can be modeled based upon structural theories using the nation-state as the unit of analysis. The effects of structural variables, such as fertility, unemployment, or public opinion, on political outcomes, however, are often realized gradually over time. If the analytical question being modeled concerns the near future, or if modeling the subnational interactions of individuals or groups, then a micro-level theoretical approach is best applied.

The major difficulty in the political analysis of individuals is that human behavior is inherently difficult to predict, and we are all different from diverse cultures, religions, and political persuasions. Thus, our political preferences are not universal. Although physical security is the fundamental objective of all states, the relative priority of policy considerations ancillary to security differs widely between societies and between individuals. The deeper analytical challenge, however, is in modeling not only the competition between individual preferences, but also the origin of those preferences. For instance, how do individuals order their political preferences in the face of risk? How do individuals determine the utility of different actions amidst uncertainty? Focusing on individuals as primary actors in policymaking, several theories of microeconomics have been employed for micro-level modeling under the rubric of positive political theory. For example, the idea that risk and uncertainty may play a pervasive role in economic analysis was originally suggested by Frank Knight in his 1921 study of insurance markets. In *Risk, Uncertainty, and Profit*, Knight observed that the distinction between risk and uncertainty was based on whether risk can be expressed in a specific mathematical probability. If so, then risk becomes insurable, and if not, it becomes an unmeasurable probability (i.e., uncertain). This notion has also been incorporated into the study of political modeling, leading to explanations of differences between individuals' calculations about the utility expected from taking certain actions.

Jonn von Neumann and Oscar Morgenstern (1944) investigated the motives of an individual making a decision under risk in the *Theory of Games and Economic Behavior*. In their theory, individuals are assumed to be facing a "choice set" of alternative probability distributions (or, in another expression, lotteries). Von Neumann and Morgenstern enable us to model the agent's preference over alternative probability distributions by differentiating between a gamble and a lottery. In a state of nature, an individual's utility is dependent on uncertainty as well as on the monetary payoffs. However, under uncertainty, the decision-maker is forced into a gamble in which it is impossible to ensure that every decision maximizes his utility. Conversely, a lottery enables individuals to calculate the probabilities assigned to choice. A simple lottery is denoted as follows:

$$p \times x + (1 - p) \times y$$

The above equation can be translated as "the individual receives prize x with probability p and prize y with probability $(1-p)$." The prizes may be money, bundles of goods, or even further lotteries. The expected utility property says that the utility of a lottery is the expectation of the utility from its prizes. We can compute the utility of any lottery by taking the utility that would result from each outcome, multiplying that utility times the probability of occurrence of that outcome, and then summing over the outcomes. However, when the probabilities are assumed to be subjective instead of objective, then probabilities are degrees of belief in a proposition rather than a set of events that is inherent in nature. Thus, individuals should update the calculation of probabilities in light of evidence. Bayes's theory of subjective probability nullifies Knight's distinction by reducing all uncertainty to risk through the use of beliefs expressible as probabilities (Earman 1992). This theory argues that even if states of the world are not associated with recognizable, objective probabilities among gambles, decision-makers still behave as if utilities were assigned to outcomes, probabilities were attached to states of nature, and decisions were made by taking expected utility.

After the axiomatization of the expected utility hypothesis by von Neumann and Morgenstern, Milton Friedman and Leonard Savage (1948) advanced the concept of univariate risk propensity by analyzing economic issues within an expected utility framework. The spectrum of risk (with risk aversion at one end and risk acceptance at the other) is demonstrated once again by a gamble. For risk-averse individuals, the utility of the expected value (i.e., the "ante") is higher than the expected utility of the gamble due to the inherent risk of loss. A level of certain wealth provides the same utility as does participating in this gamble. We call this the "certainty equivalent" of the gamble: the amount a person would take for certain rather than play the gamble. The individual will be willing to pay anything up to some value relative to the ante to avoid participating in the gamble. We call this the "risk premium," the amount that a person would pay to avoid playing the gamble. This explains why people buy insurance. Even when these costs are paid, the risk-averse person is as well off as he would be if forced to face the world (or the gamble) uninsured. At the opposite end of the spectrum is the risk-loving individual who prefers a lottery to its expected value.

Kahneman and Tversky (1979), however, showed that individual decisions are made evaluating gains and losses separately rather than in consideration of aggregate totals. This occurs because people perceive improvements or deterioration in their welfare differently; individuals may also misperceive the probabilities underlying their decisions. The two main propositions of this theory are (1) that individuals make decisions based on changes in wealth rather than their total wealth (which is in direct contradiction to expected utility), and (2) that risk aversion does not universally prevail, as some individuals are risk-seeking regarding loss.

The perception of utility in risk assessment demonstrates how expected utility theory can be applied to models of decision-making amid uncertainty. Bueno de Mesquita (1985) asserts that once a crisis develops, calculations of net gain accurately account for the escalation and termination of disputes. He shows that once a crisis starts, the analysis of a nation's net gains by an individual leader distinguishes between asymmetric and symmetric wars, anticipates when wars will be limited and when they will escalate, determines when confrontations will remain bilateral and when they will become multilateral, and indicates how a war will terminate.

One approach to modeling political phenomena amidst the uncertainty produced by competition between individuals with diverse preferences is the "stakeholder" family of models. Stakeholders are individuals who either have the power to influence an outcome of a decision or are deeply interested and thus active in an issue. Stakeholder models assume that individuals are "utility-maximizing rational agents" – in which utility-maximizing means that individuals will seek to enact their preferences, and rational means that those preferences are ordered (i.e., outcome A > outcome B > outcome C). One of the earliest predictive stakeholder approaches was the Prince model, originally constructed by William Coplin and Michael O'Leary (1972). A rational agent model, Prince attempts to predict political events based on the interests of the parties significant to the outcome of the event. The Prince model requires an informed observer to evaluate the orientation toward certainty of position regarding, power over, and salience of an issue to stakeholders capable of influencing event outcomes. This pencil-and-paper model lacks the fidelity of later extensions, but it was the first reasonable way to diminish uncertainty regarding the outcomes of political competition between individuals and their respective policy preferences.

3.3.1 Senturion: A Micro Dynamic Model of Politics

Several of the approaches discussed in previous sections have been formalized into dynamic or computational models (Bueno de Mesquita 1985; Bueno de Mesquita and Stokman 1994; Kugler and Feng 1997). Here, we discuss in detail one example of such an approach, Senturion (Abdollahian et al. 2006), a tool that can help policymakers and analysts predict political events and anticipate domestic or international political stability levels, as well as analyze specific investment decisions in which political matters affect outcomes. Senturion is a simulation system that analyzes the political dynamics within local, domestic, and international contexts and predicts how the policy positions of competing interests will evolve over time. The underlying methodology relies on several micro-level theoretical blocks. The set of rules used by Senturion synthesizes several classes of political science and microeconomic theories drawn from game theory, decision theory, spatial bargaining, and microeconomics. Unlike a statistical or probabilistic approach to predictive modeling, Senturion employs a set of micropolitical algorithms in sequence. Each theory provides a functional component for modeling how agents interact to model the "pulling and hauling" of political processes.

Given a particular issue, such as the attitude of stakeholders toward providing government-subsidized health care or the attitude of stakeholders toward U.S. military operations in Afghanistan, the Senturion approach facilitates subject matter expert (SME) identification of the positions of critical stakeholders on policy issues, weighs their potential influence, and assesses the strength of their commitment or advocacy of a policy position. This SME-generated data input captures a *snapshot* of the current political landscape. Given a particular landscape, several theoretical building blocks are useful to simulate complex human behavior and animate that landscape forward for predicting politics. We can build models of "heavy" agents, those with several initial political attributes, with the following six qualities adapted from Gilbert and Troitzsch (1999):

Knowledge and beliefs: Agents have priors on the political environment in which they are situated. In the Senturion approach, the initial data on the political landscape generated by stakeholder attributes, such as opponents' and supporters' policy positions and potential to influence, is known among all other stakeholders (Coplin and O'Leary 1972).

Inference: Agents can also make inferences from their knowledge about which potential actions to take and which ones are more credible than others as well as anticipating how other agents will react. Here, notions of risk are used to drive potential misperceptions of agent inferences, as social modeling necessitates the inclusion of political perceptions and misperceptions.

Social models: Senturion uses the notions of Black's political median (1958), Arrow-Pratt risk aversion (Pratt, 1964), and game-theoretical models (Bueno de Mesquita 1985; Lalman and Bueno de Mesquita 1989) to model various types of stakeholder interaction games theoretically at different points.

Knowledge representation: Agents update their beliefs about their own political effectiveness based on how successful their efforts are with other agents.

Goals: Each stakeholder has a preferred policy outcome that he or she is trying to achieve. Senturion assumes that agents are rational utility maximizers trying to achieve their desired political outcomes subject to being part of a winning coalition.

Language: Senturion uses the medium of political proposals, moving or shifting from one political position to another, based upon real or perceived political pressure to represent the language of agents' interactions.

The Senturion approach models the intuition behind each stakeholder's political calculus in political discussions by breaking down the process into subelements that can be modeled. Each element models a particular part of the decision process, and by combining the elements sequentially, the approach can anticipate how all stakeholders will interact to arrive at a particular decision or political outcome. The approach is a dynamic and recursive estimation of how stakeholders will interact and the resulting compromises and coalitions that will form in response. Table 4 lists Senturion's component theories and their various attributes.

Figure 4 provides an overview of Senturion. The initial stakeholder environment is defined as a policy issue of political interest; for example, the range of feasible levels

Table 4 Senturion theories and elements

Unit of analysis	Model type	Assumptions	Key variables
Individual decisionmaker	Microeconomics Individual decision making Rational choice Expected utility Spatial bargaining Median voter theorem Risk theory	Individual leader of society is a key stakeholder who can produce outcome Such actors maximize net gains in confrontations Risk is a variable connected to individual decision-makers Divergent preferences for competing goals held with varying degree of commitment are at the root of war Bounded rationality prevents decision-makers from maximizing expected utility	Stakeholder's position Potential power to I influence over the political outcome Salience of particular political issue relative to other concerns Group Importance Issue continuum

Structural environment	Core logic	Implication	Weaknesses
For commercial purposes, Sentia Group released Senturion, the integrated EU computational solution for political, economic, and business analysis	Senturion is a computational solution stakeholder analysis The stakeholder model embedded in Senturion is an agent-based model powered by expected utility equation basis Stakeholders' position, influence, salience data is required Based on those data, agent based stakeholder modeling is performed to predict bargaining outcomes	Senturion can provide a consistent framework for objective analysis of stakeholder politics, rather than relying solely on individual expert opinions about political outcomes	Reliance on experts to extract data regarding stakeholders' position, influence, and salience

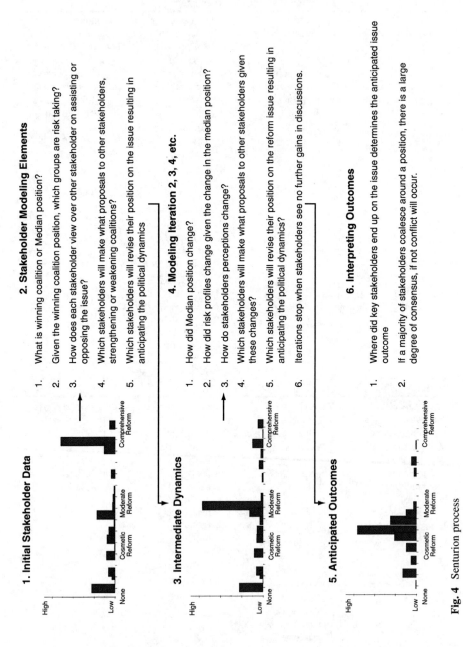

1. Initial Stakeholder Data

2. Stakeholder Modeling Elements

1. What is winning coalition or Median position?
2. Given the winning coalition position, which groups are risk taking?
3. How does each stakeholder view over other stakeholder on assisting or opposing the issue?
4. Which stakeholders will make what proposals to other stakeholders, strengthening or weakening coalitions?
5. Which stakeholders will revise their position on the issue resulting in anticipating the political dynamics

3. Intermediate Dynamics

4. Modeling Iteration 2, 3, 4, etc.

1. How did Median position change?
2. How did risk profiles change given the change in the median position?
3. How do stakeholders perceptions change?
4. Which stakeholders will make what proposals to other stakeholders given these changes?
5. Which stakeholders will revise their position on the reform issue resulting in anticipating the political dynamics?
6. Iterations stop when stakeholders see no further gains in discussions.

5. Anticipated Outcomes

6. Interpreting Outcomes

1. Where did key stakeholders end up on the issue determines the anticipated issue outcome
2. If a majority of stakeholders coalesce around a position, there is a large degree of consensus, if not conflict will occur.

Fig. 4 Senturion process

of budgetary allocation in dollars that stakeholders will compete to influence. This is where individuals vie and compete to influence the ideas and actions of others to support their own claims and political positions. This one-dimensional environment is populated with agents that represent stakeholders that have a potential to influence the particular political issue. These can be individuals, political parties, governments, or members of society. Each of these stakeholders has different attributes, such as a preferred political outcome that locates them in the spatial context, as well as a separate potential to influence that outcome weighted by their salience on the particular issue. This creates a *snapshot* of the political landscape that is quite similar to polling data in American politics or consumer preferences in market surveys.

The second step is to apply micro rules and equations to the agents given their individual attributes, the knowledge and beliefs they form given that particular snapshot of the landscape to influence their social interactions to animate the landscape and ultimately their anticipated behavior. Thus, Senturion first starts with locating the political center of gravity, called the median from the spatial bargaining context. Intuitively, the political center outlines the place where compromise can most likely occur. If one knows what the winning compromise position will be, then we can begin to deduce several other key ideas.

Here, the median position is recognized as the safest position politically, while positions far from the median are more risky. If it is known which stakeholders are willing to take risks, they may be willing to take bigger gambles to get what they want or "hold out," while if they are not willing to take risks, they may be more willing to "sell out." This assumes that more extreme stakeholders are willing to take risks while stakeholders near the political center are willing to make deals in order to achieve an agreement.

Risk-taking propensities subsequently distort how stakeholders will view each other. With these distortions, Senturion estimates the pulling and hauling of the political process by a behavioral game tree. The game structure looks at the anticipated gains or losses of every pair of stakeholders on the particular issue, identifying where offers or compromises will be exchanged between two stakeholders. It then looks at the entire network of proposals among all stakeholders given the pairwise game-theoretical interactions in order to anticipate which stakeholders will revise their positions to produce the third step of iterative dynamics.

Given that stakeholders' positions may change, how has this changed the median? If the median has changed, how have risk profiles changed, with associated impact on perceptions, proposals, and resulting position shifts? Senturion iterates the process to simulate the evolution of political dynamics over time.

One benefit of this approach is that it provides a consistent framework for objective analysis of stakeholder perceptions rather than relying solely on individual expert opinions about political outcomes. Moreover, as with any simulation tool, the specific dynamics of stakeholder proposals surrounding particular political issues can be examined in order to first gauge whether outcomes are politically feasible, second to determine possible strategic options for optimizing political outcomes using knowledge about the stakeholder dynamics, and third to anticipate unintended consequences (second and higher order effects) of actions.

3.3.2 Defining the Political Landscape and Generating Data

The process starts with representing agents in a political state space as opposed to physical environments (Gilbert and Troitzsch 1999). Adopted from economics and positive political theory, Senturion draws from spatial analysis the unidimensional issue(s) that comprise a particular political or strategic problem (Luce and Raiffa 1957; Riker and Ordeshook 1968; Ordeshook 1986). Following Feder's (1994) and Stokman's (2000) processes in collective decision-making, Senturion decomposes any strategic decision problem into its requisite parts in order to define one or multiple issue spaces to populate with agents. Agents are then populated on the landscape with varying attributes given subject matter inputs as described below.

- Desired issue position
- Potential power or influence over the political outcome
- Salience or importance of particular political issue relative to other concerns.

As described above, issues are unidimensional ranges of political outcomes, such as support for a particular reform policy, levels of preferred taxation, or stability. Power is defined as an actor's capability to affect outcomes, position is each actor's desired issue outcome, while salience measures the importance or how much of the actor's agenda the issue occupies (Coplin and O'Leary 1972; Feder 1994; Bueno de Mesquita and Stokman 1994; Kugler and Feng 1997). Thus, stakeholders now have particular influence, importance and positional attributes that could be assigned and scaled to arrive at a relative ranking of political viability but not actual political outcomes. This

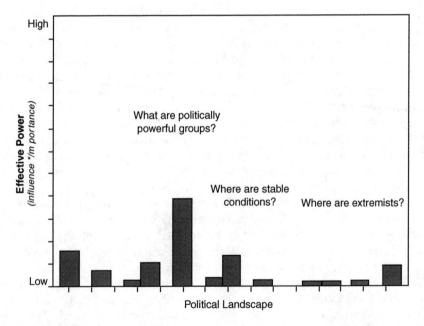

Fig. 5 Stakeholder data

"snapshot of the political landscape" (Fig. 5) shows stakeholders' initial attributes and is subsequently processed and animated by computational processes.

3.3.3   Overview of Senturion Algorithms

Given the generation of the stakeholder political landscape, the Senturion algorithm computes several key components used in various steps. It computes Votes and Forecast, Risk, and Power, expected utility values in a game tree, the resulting Perceptions, subsequent Proposals, Learning from interactions and finally a discount function to determine if agents will continue to interact. Figure 6 outlines the general algorithm and process.

Votes and Forecast capture the support that every stakeholder gets from every other stakeholder. These are used to compute the Median position that is the safest position politically. Stakeholder votes are simply computed by weighing each stakeholder's potential to influence multiplied by their particular salience or importance of the issue to the stakeholder. Thus, the Forecast is the most preferred

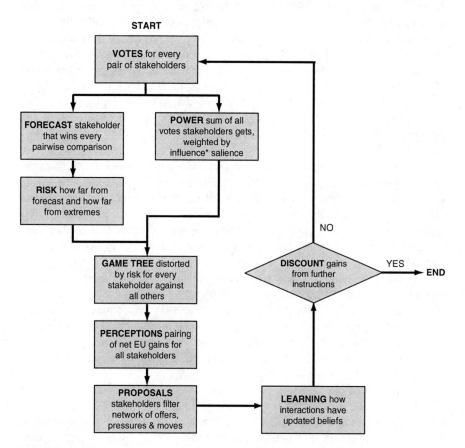

Fig. 6 Senturion algorithm

stakeholder position given all votes. Black (1948) originally proposed the median voter theorem, which identifies the median as the winning position on unidimensional continua among all other alternatives. Enelow and Hinich (1984), Bueno de Mesquita (1985), and Hinich and Munger (1997) suggest how to weight agents' votes in the context of policy applications.

Risk is a key concept that introduces distortions among stakeholder knowledge of the particular political landscape and allows for the incorporation of perceptions to help drive different political dynamics. Risk is computed for each agent to determine the perceptual prism through which the agent views other individuals and introduces distortions to the way individuals will interact. Simon (1955) outlines the evaluation of alternatives in terms of gains and losses relative to a reference point such as the status quo. Thus, risk-taking attitudes can be different above or below this point. Risk-taking propensity is assumed to be individually symmetric around losses and gains. If an agent is risk-acceptant on gains, he or she is also risk-acceptant on losses, maintaining the same risk tendency on either side. Newman (1982) shows how to calculate risk for multiple stakeholders. Thus, every agent balances his or her interests of obtaining policy satisfaction versus the security of being part of a winning coalition (Morrow 1986; Lamborn 1991). It then follows to connect risk propensities back to the status quo, or in this particular case, Black's weighted median for a particular distribution of the political landscape. Thus, stakeholders with positions farther from the median tend to be risk-acceptant while agents close to the median tend to be risk-averse.

Power measures the level of influence of each stakeholder given the likelihood of third-party support. Power is a dyadic value established by Singer et al. (1972) that introduces the notion of relative influence of stakeholders compared to all other stakeholders in a particular political process. Stoll and Ward (1989) explore this concept in detail with alternative measurements that produce effective relative measures of capabilities.

A generalized game tree of political interactions is specified and solved given the expected gains or losses for every pair of stakeholders from each stakeholder. Game theory allows us to specify the social model of political interactions among agents (Harsanyi 1968; Camerer 2003). Kadane and Larkey (1983) and Shubik (1983) show how actors choose to maximize utilities in a rational manner; a potential solution arises in decomposing a large n-person game with n-parallel two-person games. Senturion employs a generalized game that all stakeholders face in their interactions with all other stakeholders. Every agent interacts with every other agent and considers the possibility to challenge or not to challenge his or her opponent depending on the relative expected gains. When two agents decide not to challenge each other, the result is a stalemate. When two agents decide to challenge each other, the result is conflict. When one agent decides to challenge the other while his opponent does not want to challenge, then the result is a potential political offer that may or may not be accepted by the other stakeholder, as shown in Fig. 7.

Perceptions map out the net gains or losses in every dyad of agents, anticipating stakeholder interactions as peaceful, mutually conflictual, or in favor of one or the other. Based on the assumption that agents act according to their perceptions of a given political environment, each agent's perceptions are paired together to produce the anticipated

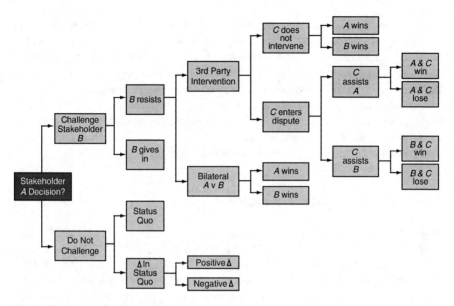

Fig. 7 Game tree

behavioral interaction between two stakeholders. Inspired by Jervis's (1978) work on misperceptions, both Lalman (1988) and Morrow (1986) define a continuous outcome approach to map behavioral interactions. Thus, different game-theoretical outcomes are translated into a perceptual mapping that identifies the behavior relations of every stakeholder versus every other stakeholder on a particular political issue.

Proposals translate the particular stakeholder perceptions back onto the specific policy landscape as offers, pressures, and moves to which stakeholders are subject. As stakeholders may have any combination of positive and negative net EU gains given our perceptual mappings, we must sort through the network of all behavioral relations to identify the push and pull of political dynamics. This is the kind of communication that takes place during agent interactions in an agent-based computational approach. Lalman (1988), Bueno de Mesquita and Stokman (1994), and Kugler and Feng (1997) outline various conditions for stakeholder interactions. An offer is made when a stakeholder believes that there is some positive gain to be made, although this offer may or may not be credible based upon differences in risk perceptions. An offer is made when the driver perceives himself as being able to secure some gains by imposing on or bargaining with the target. In the former case, the driver makes the target move all the way to his position. In the latter case, the driver makes the target move closer but not all the way.

Stakeholders may also learn or update their beliefs about the political landscape given their interaction with all other pairs of stakeholders. Given stakeholder dynamics, information is transmitted through proposals that may create learning among all stakeholders. Stakeholders subsequently can update their beliefs about the political landscape, such as which offers were successful, that can affect future proposals depending on various rules.

Finally, a discount function determines whether the process is iterated again depending on whether stakeholders see gains from further interactions. As stakeholder dynamics may cause some stakeholders to move, their positions on the issue continuum are changed, and thus Senturion animates the evolution of the political landscape through simulated time over several iterations. This process could iterate indefinitely, but that would not accurately mimic the dynamics of political processes, as rules for termination of ABMs vary widely depending on the specific application area. Intuitively, stakeholders will stop the political process when they see no further value from continued interactions (Laibson 1997).

3.3.4   Case Study: Iraq Elections, 2005

This section details the findings from a project focused on support for the January 2005 elections in Iraq, using only open-source data from SMEs from the intelligence community and the Washington Institute for Near East Policy (Abdollahian et al. 2006). Table 5 summarizes Senturion predictions based on data collected by the end of December 2004 and compares them to actual events that unfolded over the following 2 months.

Senturion predicted that a change of approach that made neutral Iraqis feel safer, by either coalition forces or the insurgency, would have allowed either the United States or insurgents to gain the support of neutral Iraqis. A major question facing analysts and decision-makers before the election was the role that other nations in the region might play, and how to assess the reactions to the election of major players in the international community. France, Russia, and Germany were expected to coalesce and increasingly support the election, but their impact would be minimal on Iraqi attitudes.

In assessing the insurgency in Iraq, Senturion provided two conclusions. Zarqawi and other foreign insurgents had very little leverage to undermine support for the election at this point. On the other hand, domestic insurgents, composed mainly of former regime elements, had most of the leverage in this situation in the months before the election. However, they did not recognize the extent of their potential influence. As with any simulation tool, this approach can also be used to test alternative political courses of action. The assumptions, policies, and tactics of U.S. stakeholders can be simulated to identify first- and second-order consequences and then adjusted to find the optimal approach to a particular situation. Moreover, because this approach calculates the perceptions of stakeholders, it can also identify circumstances when perceptions of key stakeholders are inaccurate. At times, such knowledge may form the basis for a plan of action to exploit such limits of perception.

To anticipate political reactions in advance of the 2005 elections, several courses of action to improve the situation were explored. First, a way to persuade Sunni tribal elements to moderate their opposition to the election was identified. Second, a way to obtain support from some former regime elements was sought. Finally, the implications of adjusting the force structures in Iraq were also explored by varying U.S. power levels. A reduced coalition military presence in Iraq would not have appreciably affected the attitudes of Iraqi stakeholders. However, increased coali-

Table 5 Senturion predictions compared to actual events

Predictions (based on 12/30/2004)	Actual events	Date of actual event
Insurgents will continue scope and pace of attacks	Repeated attacks by insurgents continued through the elections	1/31/2005
Strong supporters of the elections, particularly Sistani's followers and secular Shia, will participate in the election	Sistani's supporters and secular Shia voted in large numbers in the election	2/1/2005
Sadrists will be indecisive about supporting the election despite positive signs during January	Sadrists straddle both sides of the election issue, neither boycotting nor actively opposing the process	1/31/2005
Secular Sunnis and Sunni tribal elders will remain neutral toward the election	Sunnis disproportionately stayed home during the election, while not actively opposing the process	2/1/2005
Kurds will strongly support the election	Kurds turned out for the election in large numbers 1/31/2005	2/1/2005
Tension will remain high between Kurds and Shia	Tension between Kurds and Shia on future of Iraq appears to remain high despite the election	1/31/2005
Zaarqawi and foreign insurgents will have little success in undermining support for election in January	Election went forward with high Shia participation, despite attacks by insurgents	2/1/2005
World Bank and IMF will pull back support of the election	Timing and willingness of World Bank and IMF reconstruction efforts in Iraq unclear	1/28/2005
France, Russia, and Germany will increasingly support the election	France and Germany praise the Iraqi election. Russian response ambiguous	2/2/2005

tion military strength in Iraq would have improved the attitudes of Iraqi stakeholders toward the election in the short term, by making them feel more secure.

4 Practical Advice

- Begin modeling by determining the specific political issue, event, or risk in question. For instance: Is it the relations between two countries or a large bloc as a whole? Or, is it the emergence of insurgent groups in general or the effects that a particular group may have on domestic capital formation in a specific region? Recognize that there will always be political variables or phenomena that are outside the project's scope.
- Consider the purpose of the model in order to select the modeling approach. If the client faces an investment decision abroad and wishes to know the probability of an armed conflict in the region, an interstate or structural approach is applicable. If the concern is with the potential enactment of a specific policy that may affect the business, then an agent-based, micro-level approach may be more likely to succeed.

- Determine the desired tradeoff between the level of predictive accuracy and explanatory power. A simple, parsimonious model with a few theoretical building blocks may not be best for predictive accuracy, while an elaborate, complex model that combines multiple theoretical building blocks may increase predictive accuracy at the expense of explanatory power.
- Use Tables 1 and 2 in this chapter to select competing and complementary building blocks for the model. Competing blocks explain similar phenomena with different assumptions and are difficult to combine in the same model. Complementary blocks can be combined in parallel or in series.
- When building a model, construct a flowchart of the political or decision-making process of interest to help visualize and sequence the theoretical building blocks.
- When integrating and testing a model, beware of nonlinear subprocesses whose feedback can drive and overcome the output not only of the next building block but also of the entire system. Here, scaling of variables can be useful to dampen the potential impact across building blocks.

5 Summary

Political modeling generally falls into three categories: micro-level dynamics (the expected actions and interactions of individuals, groups, or governments), intranational structural dynamics (the subnational, structural factors that politically propel a nation, such as economic prosperity, democratization, or other national indicators), and international structural dynamics (the cross-national comparisons of national factors). Each of these categories is best assessed using a particular methodology, e.g., agent-based modeling, dynamic modeling, or game modeling. Our classification system helps assess the applicability of various political theories. It includes unit of analysis, model type, assumptions, key variables, structure of the environment, and the core logic of how the variables are related, in addition to the main implications, empirical support, and shortcomings of these theories. Macro interstate approaches include models that investigate the dynamics of political competition by using coupled differential equations, e.g., deterministic relationship between two states based on action and reaction, in which a small buildup by one side would lead to a larger counter by the opponent. An example of a macro intrastate approach is the POFED model, a dynamic general-equilibrium model based on the intersection of political and economic maximization function: individuals seek to maximize their lifetime utility by choosing how much to consume, while policymakers choose the tax rate and the amount of public investment and military spending to maximize their chances of remaining in power. Micro individual approaches focus on individual decisions of key political actors, leaders, and organizations, where models often use expected utility and must address the issue of risk. Senturion is an example that combines several micro-level theoretical blocks drawn from game theory, decision theory, spatial bargaining, and microeconomics. This simulation-based tool is used to predict political events, anticipate domestic or international political stability levels, and analyze specific investment decisions where political matters affect outcomes. In one case study, Senturion analyzed the 2005 Iraqi elections, and correctly predicted that an increase in Coalition military forces would improve Iraqi attitudes toward the election.

6 Resources

1. Pointers to collections of data that can be used to initialize and validate political modeling of conflict and cooperation

   The Correlates of War Project (COW)
   Cross-National and Cross-Time Conflict Dataset Hosting Program
   http://www.correlatesofwar.org/

   The Groningen Growth and Development Center (GGDC)
   Economic Historical Statistics (Angus Maddison)
   http://www.ggdc.net/

   World Development Indicators (WDI)
   Economic Statistics including more than 800 indicators. (The World Bank)
   www.worldbank.org/data

   International Financial Statistics (IFS)
   International Statistics on All Aspects of International and Domestic Finance.
   (International Monetary Fund)
   http://www.imfstatistics.org/imf/

   Government Finance Statistics (GFS)
   Annual Finance Statistical Data on General Government and Its Subsectors.
   (International Monetary Fund)
   http://www.imfstatistics.org/imf/

   UN data
   Gateway to Statistical Information from Databases of the UN and Member States. (United Nations Statistics Division)
   http://data.un.org/

   EUGene
   Expected Utility Generation and Data Management Program (D. Scott Bennett and Allan C. Stam, III)
   http://eugenesoftware.org/

   Global Terrorism Database (GTD), University of Maryland
   Information on Terrorist Events around the World since 1970 (National Consortium for the Study of Terrorism and Responses to Terrorism, START)
   http://www.start.umd.edu/start/data/

   Polity IV Project
   Dataset on Political Regime Characteristics and Transitions, 1800–2008
   http://www.systemicpeace.org/polity/polity4.htm

   Center for the Study of Civil Wars, PRIO

Data on Armed Conflict:
http://www.prio.no/CSCW/Datasets/Armed-Conflict/

Geographical and Resource Datasets
http://www.prio.no/CSCW/Datasets/Geographical-and-Resource/

Economic and Socio-Demographic Data
http://www.prio.no/CSCW/Datasets/Economic-and-Socio-Demographic/

Data on Governance
http://www.prio.no/CSCW/Datasets/Governance/

Uppsala University
Uppsala Conflict Database Project (UCDP)
http://www.pcr.uu.se/gpdatabase/search.php

James Fearon and David Laitin, Stanford University
Ethnicity, Insurgency, and Civil War (replication data)
http://www.stanford.edu/~jfearon/data/apsr03repdata.zip

James Fearon, Stanford University
Ethnic and Cultural Diversity by Country
http://www.stanford.edu/~jfearon/data/egroupsrepdata.zip

The Minorities at Risk (MAR) Project, University of Maryland
Dataset on Conflicts of Politically-Active Communal Groups
http://www.cidcm.umd.edu/mar/data.asp

The Kansas Event Data System (KEDS) Project, University of Kansas
Political Event Data focusing on the Middle East, Balkans, and West Africa
http://web.ku.edu/~keds/index.html

2. Useful books, guides, handbooks, collections of instructional materials relevant
   to political modeling
   Ronald Tammen et al., Power Transitions: Strategies for the twenty-first
   century
   http://www.cqpress.com/product/Power-Transitions-Strategies.html

   Manus Midlarsky, Handbook of War Studies I, II. The Interstate Dimension *III:*
   The Intrastate Dimension
   http://www.press.umich.edu/titleDetailDesc.do?id=348477

   Yi Feng, Democracy, Governance, and Economic Performance: Theory and
   Evidence
   http://mitpress.mit.edu/catalog/item/default.asp?tid=9932&ttype=2

   Marina Arbetman, J. Kugler, Political Capacity and Economic Behavior
   http://www.amazon.com/Political-Capacity-Economic-Behavior-Interdependence/
   dp/0813333644

   Stathis Kalyvas, Logic of Violence in Civil War
   http://www.cambridge.org/us/catalogue/catalogue.asp?isbn=0521854091

   P. Collier, Nicholas Sambanis, Understanding Civil Wars (v 1+2)
   http://extop-workflow.worldbank.org/extop/ecommerce/catalog/product-detail?
   product_id=3995594&

Barbara Walter, Reputation and Civil War
http://www.cambridge.org/us/catalogue/catalogue.asp?isbn=9780521763523

Douglas Lemke, Regions of War and Peace
http://www.cambridge.org/us/catalogue/catalogue.asp?isbn=0521809851

Andrew Kydd, Trust and Mistrust in International Relations
http://press.princeton.edu/titles/8091.html

3. Professional or academic organizations, NGOs, and foundations that are relevant
   to the political modeling
   American Political Science Association
   http://www.apsanet.org

   International Studies Association
   http://isanet.ccit.arizona.edu/

   Peace Science Society (International)
   http://pss.la.psu.edu/

   Inter-University Consortium for Political and Social Research (ICPSR),
   University of Michigan
   http://www.icpsr.umich.edu/icpsrweb/ICPSR/

   Military Operations Research Society·(MORS)
   http://www.mors.org/

   Center for the Study of Civil Wars (CSCW), PRIO
   http://www.prio.no/CSCW

   The MacMilan Center, Program on Order, Conflict Violence (Yale University)
   http://www.yale.edu/macmillan/ocvprogram/

   Uppsala Conflict Data Program
   http://www.pcr.uu.se/research/UCDP/

4. Pointers to journals, newsletters, and other periodic publications particularly relevant to political modeling
   American Political Science Review
   http://journals.cambridge.org/action/displayJournal?jid=PSR

   American Journal of Political Science
   http://www.ajps.org/

   International Interactions
   http://www.tandf.co.uk/journals/titles/03050629.asp

   International Studies Quarterly
   http://www.wiley.com/bw/journal.asp?ref= 0020-8833

   Journal of Conflict Resolution
   http://jcr.sagepub.com/

   Journal of Peace Research
   http://jpr.sagepub.com/

   Conflict Management and Peace Science
   http://cmp.sagepub.com/

5. Pointers to conferences and workshops that study political modeling

International Studies Association Annual Meeting
http://www.isanet.org/

Peace Science Society (International) Annual Meeting
http://pss.la.psu.edu/

American Political Science Association Annual Meeting
http://www.apsanet.org/

Midwest Political Science Association Annual Meeting
http://www.mpsanet.org/

References

Abdollahian, M. (1996) *In Search of Structure: The Nonlinear Dynamics of International Politics.* Ph.D. dissertation Claremont Graduate School.

Abdollahian, M. and Kang, K. (2008) In Search of Structure: The Nonlinear Dynamics of Power Transitions. *International Interactions*, 34(4), 333–357.

Abdollahian, M., Kugler, J., Baranick, M. (2006) *Senturion: Predictive Political Simulation Model* (2006) Defense and Technology Paper, 32, Center for Technology and National Security Policy, National Defense University.

Abdollahian, M. and Alsharabati, C. (2003) Modeling the Strategic Effects of Risk and Perceptions in Linkage Politics. *Rationality and Society*, Winter.

Abdollahian, M., Nicholson, B., Nickens, M. and Baranick M. (2009) A Formal Model of Stabilization and Reconstruction Operations. *Military Operations Research Journal*, 14(3), 250–281.

Allison, G. T. (1971) *Essence of Decision: Explaining the Cuban Missile Crisis.* Boston: Little Brown.

Aron, R. (1966) *Peace and War.* Princeton University Press.

Arbetman, M. and Kugler, J. (1997) *Political Capacity and Economic Behavior.* Boulder, CO: Westview.

Arbetman, M. and Johnson, K. (2008) Power Distribution and Oil in the Sudan: Will the Comprehensive Peace Agreement Turn the Oil Curse into a Blessing? *International Interactions*, 34(4), 382–401.

Axelrod, R. (1986) An Evolutionary Approach to Norms. *American Political Science Review*, 80(4), 1095–1111.

Axtell, R., Axelrod, R., Epstein, J. M. and Cohen. M. D. (1996) Aligning Simulation Models: A Case Study and Results. *Computational and Mathematical Organization Theory*, 1(2), 123–141.

Barnett, T. (2004) *The Pentagon's New Map.* New York: Putnam.

Battalio, R. C., Kagel, H. and Jiranyakul, K. (1990) Testing between Alternative Models of Choice under Uncertainly: Some Initial Results. *Journal of Risk and Uncertainty*, 3, 25–50.

Bennett, S. and Stam, A. (2000) A Cross-Validation of Bueno de Mesquita and Lalman's International Interaction Game. *British Journal of Political Science* 30, 541–561.

Benson, M. and Kugler, J. (1998) Power Parity, Democracy, and the Severity of Internal Violence. *Journal of Conflict Resolution*, 42(2), 196–209.

Black, D. (1948) On the Rationale of Group Decision-making. *Journal of Political Economy*, 56, 23–34.

Brinton, C. (1952) *The Anatomy of Revolution.* New York: Vintage.

Brodie, B. (1959) *Strategy in the Missile Age.* Princeton, Princeton University Press.

Brown, C. (1995) *Chaos and Catastrophe Theories.* Thousand Oaks, CA: Sage Publications.

Bueno de Mesquita, B. (1980) An Expected-Utility Theory of International Conflict. *American Political Science Review*, 74(4), 917–931.

Bueno de Mesquita, B. (1985) The War Trap Revisited: A Revised Expected Utility Model. *The American Political Science Review*, 79(1), 156–177.

Bueno de Mesquita, B. and Riker, W. (1982) Assessing the Merits of Selective Nuclear Proliferation. *Journal of Conflict Resolution*, 26(2), 283–306.

Bueno de Mesquita, B. and Lalman, D. (1992) *War and Reason: Domestic International Imperative*. New Haven: Yale University Press.

Bueno de Mesquita, B. and Stokman, F. N. (eds). (1994) *European Community Decision Making: Models, Applications, and Comparisons*. New Haven: Yale University Press.

Buhaug, H. and Gates, S. (2002) The Geography of Civil War. *Journal of Peace Research*, 39(4), 417–433.

Buhaug, H. and Lujala. P. (2004) *Terrain, Resources and Civil War — Does the Level of Measurement Matter?* Paper presented at the 45th Annual International Studies Association Convention, Montreal, Canada, March.

Cacey, J. T. (1994) Buyers' Pricing Behavior for Risky Alternatives, Encoding Processes and Preference Reversals. *Management Science*, 40, 730–749.

Camerer, C. F. (2003) *Behavioral Game Theory*. Princeton: Princeton University Press.

Camerer, C. F. and Lowenstein, G. (2003) Behavioral Economics: Past, Present, Future. In C. F. Camerer, G. Lowenstein and M. Rabin (eds). *Advances in Behavioral Economics*. Princeton: Princeton University Press.

Cavusoglu, N. and Tebaldi, E. (2006) Evaluating Growth Theories and their Empirical Support: An Assessment of the Convergence Hypothesis. *Journal of Economic Methodology*, 13(1), 49–75.

Cederman, L. E. (2004) *Articulating the Geo-Cultural Logic of Nationalist Insurgency*. Workshop on Origins and Patterns of Political Violence I: Violence in Civil Wars, Santa Fe Institute, January.

Collier, P. and Hoeffler, A. (1998) On Economic Causes of Civil War. *Oxford Economic Papers*, 50, 563–573.

Collier, P. and Hoeffler, A. (2002) On the Incidence of Civil War in Africa. *Journal of Conflict Resolution*, 46(1), 13–28.

Collier, P. and Hoeffler, A. (2004) Greed and Grievance in Civil War. *Oxford Economic Papers*, 56, 563–595.

Coplin, W. D. and O'Leary, M. K. (1972) *Everyman's Prince: A Guide to Understanding Your Political Problems*. North Scituate, MA: Duxbury Press.

Crenshaw, M. (1995) Democracy and Demographic Inheritance: The Influence of Modernity and Proto-Modernity on Political and Civil Rights, 1965 to 1980. *American Sociological Review*, 60, 702–718.

Davies, J. C. (1962) Toward a Theory of Revolution. *The American Sociological Review*, 27, 5–13.

Diamond, L. (1992) The Globalization of Democrarcy. In R. Slater (ed).*Global Transformation and the Third World*. Boulder, CO: L. Rienner Publishers.

Downs, A. (1957) *An Economic Theory of Democracy*. New York: Harper.

Doran, C. F. (2003) Economics, Philosophy of History, and the Single Dynamic of Power Cycle Theory: Expectations, Competition, and Statecraft. *International Political Science Review*, 24(1), 13–49.

Earman, J. (1992) *Bayes or Bust? A Critical Examination of Bayesian Confirmation Theory*. Cambridge: MIT Press.

Edwards, K. D. (1996) Prospect Theory: A Literature Review. *International Review of Financial Analysis*, 5, 18–38.

Elbadawi, I. and Sambanis, N. (2002) How Much War Will We See? Explaining the Prevalence of Civil War. *Journal of Conflict Resolution*, 46(3), 307–334.

Enelow, J. and Hinich, M. (1984) A *Spatial Theory of Voting*. Cambridge: Cambridge University Press.

Engerman, S. and Sokoloff, K. (2002) Factor Endowments, Inequality, and Paths of Development Among New World Economies. *NBER Working Paper No. 9529*.

Fearon, J. D. (1994) Signaling versus the Balance of Power and Interests: An Empirical Test of a Crisis Bargaining Model. *Journal of Conflict Resolution*, 38, 236–269.

Fearon, J. D. (2004) Why Do Some Civil Wars Last So Much Longer Than Others? *Journal of Peace Research*, 41(3), 275–302.

Fearon, J. D. and Laitin. D. D. (2003) Ethnicity, Insurgency, and Civil War. *American Political Science Review*, 97(1), 75–90.

Fearon, J. D. and Laitin, D. D. (2002) Selection Effects and Deterrence. *International Interactions*, 28(1), 5–29.

Feder, Stanley (1994) Declassified Factions and Policon: New Ways to Analyze Politics. *Studies in Intelligence*, Central Intelligence Agency.

Feng, Y., Kugler, J. and Zak, P. (2000) The Politics of Fertility and Economic Development. *International Studies Quarterly*, 44(2), 667–693.

Feng, Y., Kugler, J., Swaminathan, S. and Zak, P. (2008) Path to Prosperity: The Dynamics of Freedom and Economic Development. *International Interactions*, 34(4), 423–441.

Friedman, M. and Savage, L. (1948) Utility Analysis of Choices Involving Risk. *Journal of Political Economy*, 56(4), 279–304.

Guevara, C. (1968) *Guerrilla Warfare*. New York, NY: Monthly Review Press.

Gilbert, N. and Troitzsch, K. G. (1999) *Simulation for the Social Scientist*. New York: Open University Press.

Gilpin, R. (1981) *War and Change in World Politics*. UK: Cambridge University Press.

George, A. L. and Smoke, R. (1989) Deterrence and Foreign Policy. *World Politics*, 41(2), 170–182.

Goldstein, J. (1988) *Long Cycles: Prosperity and War in the Modern Age*. New Haven: Yale University Press.

Grossman, H. (1991) A General Equilibrium Model of Insurrections. *American Economic Review*, 81, 912–921.

Guillaumont, P. and Jeanneney, S. (1999) How Instability Lowers African Growth. *Journal of African Economies*, 8, 87–107.

Gurr, T. R. (1974) Persistence and Change in Political Systems, 1800–1971. *American Political Science Review*, 68(4), 1482–1504.

Harsanyi, J. C. (1968) Games with Incomplete Information Played by 'Bayesian' Players. Part III: Basic Probability Distribution of the Game. *Management Science*, 14(7), 486–502.

Hegre, H. and Sambanis, N. (2004) *Sensitivity Analysis of the Empirical Literature on Civil War Onset*. Paper presented at the General Meeting of the European Union Polarization and Conflict Project, Oslo, Norway, July.

Hendrix, C. and Glaser, S. (2007) Trends and Triggers: Climate, Climate Change and Civil Conflict in Sub-Saharan Africa. *Political Geography*, 26(6), 695–715.

Hinich, M. and Munger, M. (1997) *Analytical Politics*. Cambridge: Cambridge University Press.

Humphreys, M. (2005) Natural Resources, Conflict and Conflict Resolution. *Journal of Conflict Resolution*, 49(4), 508–537.

Huntington, S. (1968) *Political Order in Changing Societies*. New Haven: Yale University Press.

Huth, P. and Russett, B. (1990) Testing Deterrence Theory: Rigor Makes a Difference. *World Politics*, 42(4), 466–501.

Inglehart, R. (1997) *Modernization and Postmodernization*. Princeton: Princeton University Press.

Intriligator, M. and Brito, D. (1976) Formal Models of Arms Race. *Journal of Peace Science*, 2, 77–88.

Intriligator, M. and Brito, D. (1981) Nuclear Proliferation and the Probability of War. *Public Choice*, 17, 247–260.

Intriligator, M. and Brito, D. (1984) Can Arms Races Lead to the Outbreak of War? *Journal of Conflict Resolution*, 28, 63–84.

Jervis, R. (1978) Cooperation under the Security Dilemma. *World Politics*, 40(1), 167–214.

Jones, C. (1995) Time Series Tests of Endogenous Growth Models. *The Quarterly Journal of Economics*, 110(2), 495–525.

Kadane, J. B. and Larkey, P. D. (1983) The Confusion of Is and Ought in Game Theoretic Contexts. *Management Science*, 29(12), 1365–1379.

Kadera, K. (1995) The Conditions and Consequences of Dyadic Power Transitions: Deductions from a Dynamic Model, In J. Kugler and D. Lemke (eds). *Parity and War: A Critical Evaluation of the War Ledger*. Ann Arbor: The University of Michigan Press.

Kahneman, D. and Tversky, A. (1979) Prospect Theory: An Analysis of Decision under Risk. *Econometrica*, 47(2), 263–292.

Kleindorfer, P. R. and Kunreuther, H. (1982) Misinformation and Equilibrium in Insurance Markets. In Jörg Finsinger (ed). *Issues in Pricing and Regulation*. Lexington: Lexington Books.

Kim, W. and Morrow, J. D. (1992) When Do Power Shifts Lead to War? *American Journal of Political Science*, 36(4), 896–922.

Kim, W. (1996) Power Parity, Alliance, and War from 1648–1975. In Jacek Kugler and Douglas Lemke (eds). *Parity and War*. Ann Arbor: University of Michigan Press.

Knight, F. H. (1921) *Risk, Uncertainty and Profit*. Boston, MA: Houghton Mifflin.

Kugler J., Abdollahian M. and Arbetman M. (2005) *Identifying Fragile States*. DARPA PCAS Sentia Group, Inc. Washington DC.

Kugler J. and Lemke, D. (eds). (1996) *Parity and War*. University of Michigan Press.

Kugler, J. and Feng, Y. (eds). (1997) The Expected Utility Approach to Policy Decision Making. *International Interactions*, 23(3–4), 233–274.

Kugler, J. and Tammen, R. (eds). (2010) *Political Performance of Governments*. Forthcoming.

Kugler, J. and Zagare, F. C. (1987) Risk, Deterrence, and War. In J. Kugler and F. C. Zagare (eds). *Exploring the Stability of Deterrence*. Boulder: Lynne Rienner.

Laibson, D. (1997) Golden Eggs and Hyperbolic Discounting. *Quarterly Journal of Economics*, 112(2), 443–477.

Lalman, D. (1988) Conflict Resolution and Peace. *American Journal of Political Science*, 32(3), 590–615.

Lalman, D. and Bueno de Mesquita, B. (1989) The Road to War is Strewn with Peaceful Intentions. In Peter Ordeshook (ed). *Models of Strategic Choice in Politics*. Ann Arbor: University of Michigan Press

Lamborn, A. C. (1991) *The Price of Power: Risk and Foreign Policy in Britain, France, and Germany*. Boston: Unwin Hyman.

Lemke, D. and Reed, W. (1996) Regime Types and Status Quo Evaluations. *International Interactions*, 22(2), 143–164.

Lemke, D. and Werner, S. (1996) Power Parity, Commitment to Change, and War. *International Studies Quarterly*, 40 (2): 235–260.

Lemke, D. (2002) *Regions of War and Peace*. New York: Cambridge Univ. Press.

Lemke, D. (2010) Power Politics and the Violent Creation of Order. *American Journal of Political Science* (forthcoming)

Levy, H. and Levy, M. (2002) Experimental Test of the Prospect Theory Value Function. *Organizational Behavior and Human Decision Process*, 89, 1085–1081.

Levy, J and Thompson, W. (2010) *Causes of War*. Malden, Mass: Wiley-Blackwell.

Luce, R. D and Raiffa, H. (1957) *Games and Decisions: Introduction and Critical Survey*. New York: Wiley.

Mearsheimer, J. (2001) *The Tragedy of Great Power Politics*. New York: Norton.

Midlarsky, M. (ed). (2000) *Handbook of War Studies II*. Ann Arbor: The University of Michigan Press.

Midlarsky, M. (ed). (2009) *Handbook of War Studies III*. Ann Arbor: The University of Michigan Press.

Morrow, J. D. (1986) A Spatial Model of International Conflict. *American Political Science Review*, 80(4), 1131–1150.

Morrow, J. D. (1987) On the Theoretical Basis of a Measure of National Risk Attitudes. *International Studies Quarterly*, 31(4), 423–438.

Muncaster, R. and Zinnes, D. (1988) The War Propensity of International Systems. *Synthese*, 76(2), 307–331.

Nicholson, M. (1998) *Formal Theories in International Relations*, Cambridge Studies in International Relations v. 3. Cambridge University Press.

O'Neill, T. (1993) *All Politics is Local and Other Rules of the Game*. Hollbrook MA: Random House.

Ordeshook, P. (1986) *Game Theory and Political Theory*. Cabmridge: Cambridge University Press.

Organski, A. F. K. (1958) *World Politics*. New York: Alfred Knopf.

Organski A. F. K. and Kugler J. (1980) *The War Ledger*. Chicago, IL: University of Chicago Press.

Poggi, G. (1990) *The State: Its Nature, Development and Prospects*. Palo Alto, CA: Stanford University Press.

Powell, R. (1987) Crisis Bargaining, Escalation, and MAD. *American Political Science Review*, 81, 717–735.

Pratt, J. (1964) Risk Aversion in the Small and in the Large. *Econometrica*, 32, 122–136.

Raleigh, C. (2004) *Political Geography of Civil War: Patterns of Insurgency in Central Africa, 1960–2005*. Ph.D. diss., University of Colorado.

Rapoport, A. (1957) Lewis F. Richardson's Mathematical Theory of War. *Journal of Conflict Resolution*, 1(3), 249–292.

Rasler, K. and Thompson, W. (1994) *The Great Powers and Global Struggle, 1490–1990*. Lexington: University of Kentucky Press.

Richardson, L. F. (1960) *Arms and Insecurity*. Chicago: Quadrangle Books.

Riker, W. and Ordeshook, P. (1968) A Theory of the Calculus of Voting. *American Political Science Review*, 62(1), 25–42.

Rosenau, J. (1964) *International Aspects of Civil Strife*. Princeton, NJ: Princeton University Press.

Salehyan, I. and Gleditsch, K. (2004) *Refugee Flows and the Spread of Civil War*. Paper presented at the annual meeting of the American Political Science Association, Hilton Chicago and the Palmer House Hilton, Chicago, IL, Sep 02, 2004.

Saperstein, A. (1994) Chaos As A Tool For Exploring Questions of International Security. *Conflict Management and Peace Science*, 13(2), 149–177.

Schelling, T. (1960) *The Strategy of Conflict*. Cambridge: Harvard University Press.

Shubik, M. (1983) Comment on The Confusion of Is and Ought in Game Theoretic Contexts. *Management Science*, 29(12), 1380–1383.

Simon, H. (1955) A Behavioral Model of Rational Choice. *Quarterly Journal of Economics*, 69(1), 99–118.

Singer J. D., Bremer, S., Stuckey, J. (1972) Capability Distribution, Uncertainty, and Major Power War, 1820–1965. In B. Russett (ed). *Peace, War, and Numbers*. London: Sage Publications, pp. 19–48.

Small, M. and Singer, J. D. (1982) *Resort to Arms: International and Civil Wars, 1816–1980*. Beverly Hills, CA: Sage Publications.

Smoker, P. (1964) Fear in the Arms Race: A Mathematical Study. *Journal of Peace Research*, 1(1), 55–64.

Sokoloff, K. L. and Engerman, S. L. (1994) *Factor Endowments: Institutions, and Differential Paths of Growth Among New World Economies: A View from Economic Historians of the United States*. NBER Working Paper h0066.

Stoll, R. J. and Ward, M. D. (eds). (1989) *Power in World Politics*. Boulder: Lynne Rienner.

Tanter, R. (1966) Dimensions of conflict behavior with and between nations, 1958–60, *Journal of Conflict Resolution*, 10, 41–64.

Tammen, R., Kugler, J., Abdollahian, M., Alsharabati, C., Efird, B., Stam, A. and Organski, A. F. K. (2000) *Power Transitions: Strategies for the 21st Century*. New York: Chatham House.

Tilly, C. (1975) Reflections on the History of European State-Making. In Charles Tilly, Gabriel Ardant (eds). *The Formation of National States in Western Europe*, Princeton: Princeton University Press.

Von Neumann, J. and Mongerstern, H. (1944) *Theory of Games and Economic Behavior*. Princeton University Press.

Wagner, H. (1991) Nuclear Deterrence, Counterforce Strategies, and the Incentive to Strike First. *American Political Science Review*, 85(3), 727–749.

Waltz, K. (1979) *Theory of International Politics*. Massachusetts: Addison Wesley Publications.

Waltz, K. and Sagan, S. (1995) *The Spread of Nuclear Weapons: A Debate*. New York: W.W. Norton.

Welzel, R., et al. (2003) The Theory of Human Development: A Cross-Cultural Analysis. *European Journal of Political Research*, 42 (3), 341–380.

Zagare, F. and Kilgour, M. (2000) *Perfect Deterrence*. New York: Cambridge University Press.

Zagare, F. (2004) Reconciling Rationality with Deterrence: A Re-Examination of the Logical Foundations of Deterrence Theory. *Journal of Theoretical Politics*, 16(2), 107–141.

Zinnes, D. and Gillespie, J. (1976) *Mathematical Models in International Relations*. New York: Praeger Publishers, pp. 179–217.

Chapter 4
Economics and Markets

Alan K. Graham

Economic matters are often entangled with interventions. Aid agencies need to understand where they can have the highest leverage, and where aid may cause harmful economic distortions. Humanitarian interventions in crises will be more effective if the economic and social root causes of the crisis are addressed as well. The root causes of insurgencies often include economic issues, particularly economic discrimination. Planners for military operations in a country need to know the economic side effects of military activities, including the effects of withdrawal. Government agencies trying to bring developed-nation investors into a developing country must understand, along with the potential investors, what the economic prospects of the economy are, and how safe an investment is (or is not). Economic modeling and analysis can assist in each of these cases.

It is easy to wish for a general-purpose economic model that not only correctly forecasts future economic behavior but also accurately predicts the impact of any given action. However, there is no economic model built yet that encompasses the entire range of potential issues and impacts for *any* country, let alone the many countries an analyst might need to understand. A model *of* a country will seldom be an adequate model *for* a given very specific purpose and use.

Moreover, uniformly good forecasts are neither possible nor usually required. What is actually required often is an analysis indicating which plan of action is more desirable, irrespective of precise conditions, which implies that choosing a model (or to purpose-build one) very much depends on the exact use to which it will be put, as opposed to a general topic area (e.g., "exchange rates").

This chapter excludes consideration of the related topics around a country's domestic economic policies, which would speak to a different audience and involve a different profile of the economic issues dealt with. The topic of models for monetary and fiscal policy is narrower and deeper than the topics considered here. Monetary and fiscal policy models almost always need to be fully quantitative, well-verified versus time series, and at least medium sized. They also simplify

A.K. Graham (✉)
Massachusetts Institute of Technology, Cambridge, MA, USA
e-mail: grahamak@aol.com

A. Kott and G. Citrenbaum (eds.), *Estimating Impact*,
DOI 10.1007/978-1-4419-6235-5_4, © Springer Science+Business Media, LLC 2010

away many of the givens that are missing or partially functioning in developing economies, such as a nationwide financial system. So monetary and fiscal policies are generally taken as givens in what follows, and the models are not evaluated on the basis of their ability to say what happens when such policies change.

There are many more possible combinations of model purpose (decisions to be influenced), economic issues that impact outcomes, and geographies than there are extant models, and in turn many more extant economic models are available than is practical to catalog. Therefore, this chapter focuses on principles for determining a match among a user's purposes, the economic setting being analyzed, and a given existing model or proposed model creation or modification.

This chapter does not provide step-by-step instructions for building or using economic models but rather is more of a navigation guide, allowing an analyst to slice through the different combinations of purpose, economic behaviors, and geographies to:

Locate data that characterize the relevant economic situation, and

Identify which types of economic analysis or modeling are most appropriate to the purpose, geography, and modeling resource constraints.

For analysts whose background and resources foreclose the active use of a quantitative model, a later section gives a case study of using diagrammatic modeling and scoring to tease apart a complex resource allocation problem with a satisfactory degree of confidence.

For analysts whose background and resources allow them to choose building or modifying a quantitative model, this chapter describes an economic model with the focus on design choices appropriate to a model purpose and validation tests available to a model builder. The same discussion also illustrates by example why "general purpose" models are hardly ever directly suitable for specific uses in international economic situations.

A critical organizing concept in what follows, especially in the context of scoping an analysis, is the idea of a behavior mode. The concept itself originates from feedback systems mathematics. In a complex system, such as the interlinked markets that make up an economy, the system is capable of a variety of somewhat distinct behaviors, a few of which will dominate the behavior at any one time. For example, developed economies tend to show a somewhat irregular cycle (the "business cycle"), with intervals of expansion alternating with intervals of contraction, with peaks typically 3–7 years apart. Sterman (2000, Part V, "Instability and oscillation") provides an extensive analytical framework for such cyclical phenomena.

Behavior modes can be thought of as analogs of medical definitions of diseases or syndromes: patterns of behavior that arise from a distinctive set of causes, which may or may not occur in isolation. Behavior modes are associated with behavior of specific quantities, here called "indicator variables," just as a physician will decide whether a patient has a flu infection by looking for elevated temperature, headache, body ache, and nausea. Similarly, much of the initial scoping involved in economic analysis is arriving at sensible hypotheses for which behavior modes are occurring or are likely to occur.

Before proceeding to the standard behavior modes of macroeconomics, working with developing economies may require some understanding of another set of behaviors that arise and persist due to systemic causes but are not part of the standard economic corpus. For convenience, these can be called "near-economic" behavior modes, and we begin the cataloging of behavior modes and indicator variables with them.

1 "Near-Economic" Behavior Modes

Table 1 lists several behavior modes that represent departures from the developed country norms or ideals: corruption, discrimination, insurgency, and economic controls. These become important to the reader when they materially impact the economy in question.

Each of these behavior modes is maintained by a self-sustaining interaction among multiple stakeholders. For example, a corrupt political leader may take payoffs, whose proceeds are sufficient to bribe prosecutors and judges to suppress law enforcement and permit election-rigging, which maintains political power, which includes the ability to appoint, e.g., judges. Graham (2009a) further discusses such self-sustaining behaviors; Bueno de Mesquita et al. (2002, 2004) provide solid empirical support. One consequence of the self-sustaining nature of such behavior modes is that they

Table 1 Near-economic behavior modes

Behavior mode	Description
Corruption in political processes	Rigged elections, widespread patronage, political control of media
Corruption in law enforcement	Judiciary and enforcement controlled to permit, e.g., drug trade
Corruption in economic processes	Routine confiscation or near-confiscation of private property, extensive bribes needed for construction or business operation
Discrimination in law enforcement	Routine and egregious violation of civil rights of ethnic or racial groups
Economic and social discrimination	Explicit or implicit limitations on education and hiring with respect to race or ethnic group, sometimes by an economic elite, sometimes by the majority ethnic group
Insurgency	Attempts to resist or control government by violent means. "Groups with gripes" are often a consequence of behavior modes above.
Protectionism	Prohibitions or tariffs on imported goods and services. Legal limits on the ability of domestic companies or persons to make investments outside the country. Also limits on foreign investment in domestic corporations, and/or limits on repatriation of domestic profits from foreign-owned assets.

Table 2 Sources for indicators of near-economic behaviors

Organization	Web address
United Nations Universal Human Rights Index	http://www.universalhumanrightsindex.org/
World Bank Governance Indicators	http://info.worldbank.org/governance/wgi/sc_country.asp
Internet Center for Corruption Research	http://www.icgg.org/corruption.html
Transparency International	http://www.transparency.org/
Freedom House	http://www.freedomhouse.org
Human Rights Watch	http://www.hrw.org/
Amnesty International	http://www.hrw.org/
Gallup Political Stability Index	http://www.voice-of-the-people.net/ContentFiles/files/ VoP2005/VOP2005_Democracy%20FINAL.pdf

generally persist. Anticorruption crusades seldom succeed, and if they succeed, they succeed slowly – not over months and years but over years and decades.

The "good news" in the very persistence of near-economic behavior modes is that indicator variables are reasonably trustworthy about future prospects. There are several organizations that produce indexes of corruption, economic freedom, and political freedom, as shown in Table 2:

In addition, as described in the Resources section of this chapter, there are broader research surveys available that emphasize the economic climate as well as significant amounts of international data.

2 Economic Behavior Modes

Tables 3–6 describe standard textbook economic behavior modes.[1] For each, the table gives a brief description, a reference for further reading, and economic variables to look at to determine where a given economy lies on the spectrum from "that behavior is happening right now" through "it isn't happening now but it could" to "it's unlikely to happen." They, along with the near-economic behavior modes, allow a modeler to say what is going on economically in the region of interest.

For clarity, the table information is divided into three types of behavior modes. Table 3 describes "business as usual" behavior modes, which go on all the time, in developing and developed economies alike. Table 4 describes some departures from business as usual that create moderate economic vulnerabilities. Tables 5 and 6 describe severe economic crises and the conditions under which countries are vulnerable to them. The taxonomy is more for convenience than for reflecting any fundamental distinctions.

In brief, then, a modeler or modelers will start from general surveys of a country's Political, Military, Economic, Social, Infrastructure and Information (PMESII) situation. From there, gathering data about each economic behavior mode should

---

[1]Chart of economic behavior modes adapted from Graham et al. (2008a, b), © PA Consulting Group, Inc. Used with permission.

Table 3 Economic behavior modes that are "business as usual" in developing and developed economies

Behavior mode	Description	Indicator variables	Commonly modeled with
Secular industrialization	Long-term (i.e., over decades and centuries) accumulation of physical and financial capital, and technological know-how leads to increasing use of capital plant and equipment broadly across the economy	Long-term trend in real production per capita (adjusted for inflation, and minus income due to natural resource extraction)	Usually, modeled as a function purely of passage of time in productivity equations. See (Solow 1956, 1957) for this classic treatment, and (Romer 1990) for an analytical model of technology investment, technological progress and increased productivity
Demographic transition	Often hand-in-hand with later stages of industrialization, accumulation of physical capital (matched by financial savings), which allows retirement supported by savings, rather than support from many children	Decreasing average family size, increasing household wealth and income per capita (financial and nonfinancial) approaching low end of developed country levels	Implicit in parameters of quantitative models, in population growth and savings targets. See (Caldwell 1976; Caldwell et al. 2006)
Business cycles	"Course correction" by corporations to manage inventories through production and employment, which tends to overshoot. Normal random disturbances turn this into approximately cyclical behavior with 3–7 year period	Timeline of GDP growth Explicit business cycle leading and concurrent indicators, particularly employment and inventory investment relative to long-term trend (see ECRI in references)	Stochastic models (Eckstein 1983; Fair 2004), system dynamics models (Mass 1975; Forrester 1982; Forrester 1989). For business cycle statistics in the US see (NBER 2009) and internationally (ECRI 2009)
Trade balancing through exchange rates	When, e.g., the US imports more than it exports (a negative balance of trade), it gives out more dollars than it takes in other country's currencies, so the value of the dollar should fall. That makes imports more expensive and should reduce them, and likewise exports, which reduce the imbalance, should be increased (may be "stuck open" by currency exchange mercantilism, below)	Stability of exchange rate relative to weighted basket of currencies Gross imports and exports and their difference (net exports) as fractions of GDP	Analytical models (Gandolfo 2002, Ch. 15) Stochastic models (which may leave exchange rates or export demand exogenous (i.e., future values assumed rather than determined by model dynamics), as in (Fair 2004)

Table 4 Economic behavior modes that create moderate vulnerabilities or persistent undesirable conditions

Behavior mode	Description	Indicator variables	Commonly modeled with
Currency exchange mercantilism	A country (e.g., China) keeps exchange value of its currency low by buying, e.g., dollars and accumulating them, making their exports cheap in international markets. So they export more than they import, and continue to accumulate dollars	Growing foreign currency and debt holding of government and central bank relative to GDP, relative to other countries	Analytical models such as (Gandolfo 2002, Ch. 15) could be adapted. See (Curry 2000, Ch. 8) and (Burgess et al. 2009, Sections 1.2 and 3.1) for more detailed descriptive analysis
Natural resource export dependence	Natural resource extraction paying for imports, with relatively few goods and services being produced for domestic consumption or export (natural resource exports may disguise what would otherwise be a failed market economy. Such economies are very vulnerable to world commodity price fluctuations, hence very volatile)	Exports of specific commodities relative to GDP (e.g., oil, iron ore)	Macroeconomic statistical models with endogenous importing/exporting and foreign exchange rates
Import dependence and stagflation	If a country depends on importing a critical commodity (food or oil) whose price increases, consumers buy at the higher price, leaving less money for other goods. The domestic economy starts to decline. The government attempts to maintain consumption by making money easily available, which creates inflation simultaneous with the economic stagnation	Imports of a specific commodity (e.g., oil, coal, food, iron ore, and steel) relative to domestic consumption	Stochastic models (Eckstein 1983; Fair 2004), analytical models
Capital flight	When domestic investment opportunities offer only poor returns, business investment and personal savings leave the country. With little capital investment, the economy does not grow and the incentives for capital flight persist	As fractions of GDP: Domestic capital investment Net capital account balance Investment owned outside the country	If at all, implicit in parameters describing capital investment in macroeconomic statistical models (Eckstein 1983; Fair 2004; University of Maryland 2009)

Table 5 Economic behavior modes of crisis originating from domestic financial issues

Behavior mode	Description	Indicator variables	Commonly modeled with
Debt deflation spiral	Rising asset prices (stocks, housing) secure large debts relative to income. If any event stops the rise, households and businesses use money to reduce debt rather than consuming or investing. So the economy shrinks, and asset prices fall further. Defaults reduce the value of assets held by financial institutions, which further reduces lending and consumer purchasing (usually, follows a period of asset price inflation, and will persist over multiple business cycles) Debt deflation spirals are usually thought to explain the worldwide Great Depression of the 1930s, Japan's Lost Decade of the 1990s, and to some extent the recession of 2007–2009	Consumer Price Index (CPI) decline Housing price index (adjusted for CPI) decline Mortgage foreclosure rate (%/year), whole population, increases sharply Bond default rate (%/year), whole market shows sharp increase Household debt relative to income Medium-term stock market decline (adjusted for CPI)	Analytical models summarized in (von Peter 2005). For extensive discussion as "balance sheet recessions" see (Koo 2008). For system dynamics modeling in terms of 50-year cycles (a.k.a. "Kondratiev cycles"), see, e.g., (Graham and Senge 1980; Graham 1982; Sterman 1986)
Deficit-lead hyper-inflation	The usual trigger for high inflation is government spending far more than its revenues, and borrowing is not possible. So the government creates more money, and prices start to rise. Once inflation is built in, the government has little choice but to support the economy and continue inflating the money supply. Often accompanied by the inability to acquire foreign currencies (through foreign trade) to repay foreign debt, followed by default	Consumer Price Index or GDP deflator Government deficit and debt as fraction of GDP External debt (i.e., debts to foreign governments or financial institutions) relative to national currency reserves (i.e., "do they have enough money to pay off their debts?") Payments of external debts relative to trade surplus (i.e., "are they taking in enough foreign currency to make payments on foreign debts?")	Analytical models

Table 6 Economic behavior modes of crisis created through foreign currency exchange

Behavior mode	Description	Indicator variables	Commonly modeled with
Currency crisis/ investment boom and bust	A period of rising foreign investment brings foreign currency into the economy, which stimulates spending, importing, and borrowing, including borrowing from foreign sources. The currency exchange rate rises, increasing the value of investments. However, if any event (business cycle downturn, oil price increase, whatever) reduces the investment flow, spending suddenly drops, currency exchange rate falls, foreign debts become too large to repay, and the economy declines sharply	Foreign Direct Investment (FDI) sharply declining External debt (debt to foreigners, in foreign currency) relative to FDI plus currency earnings from exports increases sharply Dollar price (or market basket-weighted price) of that foreign currency declines sharply	Analytical models, e.g., (Krugman 1999), also (Gandolfo 2002, Ch. 16). For very accessible description, see (Krugman 2009)
Currency exchange defense	For many reasons, the value of a country's currency may decline. A country's government can decide to attempt to maintain their currency's value at a high level (usually for reasons of national pride, or keeping imports cheap and available). Maintaining a high value requires them to buy their own currency back in foreign exchange markets, and pay for it out of their reserves of foreign currencies, which are limited. Financial markets have a way of betting that a currency value will fall (short selling) which actually tends to make the currency value fall further. The country runs out of foreign currency, and the currency falls anyway	Exchange rates Announced government policy of defending the value of its currency	Analytical models (Gandolfo 2002, Ch. 16). For very accessible description, see (Krugman 2009)

give a picture of the economic situation that is both broadly scoped and concise. That is the background for modeling decisions: whether creating or borrowing a model at all is justifiable, and if so, what economic behaviors a model will need to deal with.

3 Overview of Modeling Approaches

Tables 7–9 summarize the major varieties of analysis and modeling approaches used or advocated for economic issues. The names of the methods in some cases are not quite what their practitioners would call them; the names were chosen to be meaningful to nonpractitioners.

Strengths and weaknesses of different modeling approaches are sometimes a topic of disputes among practitioners of different methods. In part, the disputes originate from pure misinformation about "foreign" methodologies; for example, inferring characteristics of a whole class of methods from academic teaching examples. An expert will know the power and flexibility of his or her own method but will not easily see the power or flexibility in other methodologies, especially if judging only from published academic work.

In part, the disputes are due to the (often correct) perception that many weaknesses of an approach can be overcome by experts. Expert statisticians have techniques to deal with missing data. System dynamics experts can often build and use models quickly, and so on. That said, the experts' bags of tricks are unhelpful for nonexperts and very unhelpful for analysts trying to determine the extent to which a model built by someone else for a different purpose can in fact be useful for the analyst's purpose.

Perhaps the most contentious disputes arise when an expert in one field has tightly-held assumptions about what modeling is for, what form it is to be delivered in, and what types of validation tests are acceptable to the final audience. Modeling for a different purpose, delivered in unfamiliar ways and using unfamiliar validation tests (and missing the familiar ones) is palpably substandard, at least in perception. Academics develop compact theories that explain things in a teachable way. They rightly prefer results that are objective, reproducible, and publishable in a refereed journal. So there is a strong incentive toward small models, and hence working on problems amenable to analysis via small regression models and analytical models. Modeling practices and preferences and indeed curricula have built up around these methodologies.

Academic model-building often has several disconnects with the needs and purposes of corporate and government decision-making. Many areas of concern are simply not well-studied academically and often lack even elementary data. Academic studies are often limited to cross-sectional comparisons of different countries, with the predictable result that the results are often inconclusive. First-time explorations of problems that include data-poor areas should be conducted differently than well-studied problems. Logically, modeling that pioneers such issues lean toward use of

Table 7 Summary of analytic methodologies for assessing economic effects

Methodology	Description	Typical application	Strengths	Weaknesses
Cross-sectional and historical indicators	Collect concise statistics measuring a behavior from many countries to identify the most troubled (or likely to be troubled) country	Assessing political risks in developing countries; assessing financial risk of, e.g., currency crisis (Gallup Corporation 2009)	Straightforward and easily communicable. Typically, data are widely available from commercial and government organizations	Indicative rather than definitive; usually, does not explicitly account for special circumstances nor quantify the uncertainties involved
Quantitative Systems thinking	With Subject Matter Experts (SMEs), diagram the (usually interlocking) drivers of the behavior of interest, and then use SMEs to score causes and effects to rate alternative actions. Use sensitivity testing to assess robustness of recommendations	Resource allocation in data-challenged situations, e.g., law enforcement, counterinsurgency, complex or very new markets, regulatory issues (Mayo et al. 2001; Lyneis 1999; J8/ WAD 2007)	Makes full use of SME first-hand operational knowledge, supported by statistical evidence. Process is transparent	Judgments on strength and timing of cause and effect are subjective (even though often repeatable and surprisingly accurate as measured by later quantitative modeling) Quantitative validation testing against time series is not possible
Analytical	Few enough symbolic equations (usually without numerical values) that their behavior and implications can be derived with algebra, calculus, or graphs	Publish a journal article testing an hypothesis about a given economic behavior (Krugman 1999; Solow 1956; Romer 1990)	Conclusions derived rigorously and explicitly from equations or graphs. Explanation of behavior is compact	Multiple simplifying assumptions can prevent analysis of more than one behavior of interest, which may make the scope of analysis smaller than that would be useful for a decision-making situation Derivations often require fluency in calculus to follow

Single-equation regression	Explains movements of one variable in terms of a few drivers, with parameters derived by statistical methods	Present an hypothesis about an economic cause and effect in an academic publication (Modigliani and Brumberg 1954)	Confirms or disconfirms hypotheses with objective, repeatable statistical tests. Widely taught in college courses as "econometrics" or "statistical regression" (as opposed to the more theoretical probability and statistics)	For analysis in support of a decision, variables in a regression equation may be too aggregate to be relevant for assessing the impact of a specific action, and too restricted in scope to be trusted in influencing a decision. Without significant expertise and diligence, flaws in data can create apparently good but incorrect results (Sterman 1988)

Table 8 Summary of analytic methodologies for assessing economic effects

Methodology	Description	Typical application	Strengths	Weaknesses
General equilibrium	One of the two major variants of macroeconomic statistical models (i.e., uses many equations to describe an economy, and uses statistical methods to derive parameters for the equations). Uses data on prices and quantities to estimate supply and demand functions, and then compute equilibrium. Analysis is comparing different equilibria resulting from different government policies or external events	Understand the effect of regulatory change, or (in detailed models) understand the economy-wide impact of large sectoral changes (e.g., collapse of the car industry) (Kubler 2008; University of Maryland 2009)	When the primary analytical deliverable is a changed end state (e.g., competition in an industry under a different regulatory scheme), comparing two equilibria shows the consequences of policy changes very clearly (assuming that transitioning from one state to another is benign and secondary for the analysis) Excluding dynamics allows greater focus on detail (e.g., by industry) and scope (with multiple countries)	Limited to markets where high-quality data on price and quantity are available, usually "real" (non-financial) economy Says little about the path from one equilibrium to another, which excludes many dynamic behaviors of interest, e.g., business cycles, foreign exchange currency crises, the 2007–2009 debt-deflation (see Koo 2008). In such economic behavior modes, transitions between equilibria are far from smooth, and usually involve major overshoot and other transient conditions important to stakeholders

| Stochastic | The other major variant of macroeconomic statistical models, likewise using many equations and statistical estimation of their parameters. Known more technically as "dynamic stochastic general equilibrium (DSGE) models". Translation: some markets are assumed to balance instantly ("general equilibrium"), the system is part driven in part by randomness ("stochastic"), and some changes happen over time ("dynamic") | Short-term economic forecasting (Eckstein 1983; Fair 2004) | During "business as usual" periods, best forecasts and known error bounds | Limited to markets where data on price, quantity, debt, interest, inventory, foreign exchange rate, etc. are available, usually "real" (nonfinancial) economy and balance of trade. Generally not capital account dynamics Quickly becomes impossible when data are unavailable Forecasts based on "business as usual" fail when they are most needed – during unusual economic behavior Model testing usually does not emphasize boundary adequacy testing, subsystem testing, extreme condition testing – all of which support the ability to forecast (see Lyneis 2000) |

Table 9 Summary of analytic methodologies for assessing economic effects

Methodology	Description	Typical application	Strengths	Weaknesses
System dynamics	Use SME knowledge to formulate dynamic equations (i.e., present conditions and the equations that govern the change in those conditions over time, equivalent to differential equations), then test against observed behavior (see Section 6.5)	Economic cycles (Mass 1975; Forrester 1982), commodity cycles (Sterman 2000 Ch. 20), public policy and regulation (Graham 1976; Graham and Godfrey 2003), counterinsurgency (DARPA 2006; OUSD Policy 2007), as well as corporate strategy, large project management, and legal disputes (Stevens et al. 2005)	Can integrate economic and non-economic issues, can use SME data to compensate for scarce time series data, allows many types validation, models can analyze extreme conditions and not-yet seen behaviors Counterintuitive behavior and unintended consequences often well treated (Forrester 1971) Well suited to medium-term forecasting (Lyneis 2000)	Serious model-building and policy analysis requires considerable experience, not readily available Models are typically valid and useful for a sharply defined set of questions and issues – clients and academics can be disappointed by the inability to appropriately address the issues outside the defined scope
Game theory	For each of (usually two to five) actors, quantify their decisions, goals and constraints on action. Use mathematical theory to solve the game for viable solutions	Negotiations; auctions (e.g., wireless spectrum). Concepts are probably more widely used than the formal analysis (Dixit and Skeathmore 2004; Williams 2007)	Captures thinking and behaviors very difficult to capture with other methods, e.g., coalitions, "cheating," multiple-move strategies (analogous to chess strategies)	Relative to methods above, practical application lags far behind mathematical sophistication and academic exploration. So software and the skills to use them are rare (game theory is mostly about the form of analysis; it can coexist with different ways of modeling the underlying descriptions of reality, e.g., there is a rich theory of dynamic games)

| Agent-based (AB) simulation | Give a multitude of agents (hundreds or thousands) goals, available actions, and a decision-making algorithm, all randomly distributed. Do Monte Carlo simulations to understand behavior | Where the distribution of behaviors is the central problem: AB is the dominant method to analyze the spread of infectious disease. For social sciences, see (Billari et al. 2006) | Truly unexpected behaviors can emerge. Problems where the "tail" of a distribution is key (e.g., insurgents), Situations involving changing of a taxonomy (e.g., new definition of market segments) | Like game theory, practical application lags behind academic exploration. So software is immature by comparison to other methods, and long application experience is rare Need to detail all decisions of all actors can be considerably more work than other methodologies (Rahmandad and Sterman 2004) |

diagrammatic and system dynamics modeling as more robust approaches in such circumstances, with validation techniques and standards different from those for problem domains in which considerable knowledge and modeling already exist. Similarly, many areas of government and business concern spread across many disciplines and interactions of many types of actors, and analysis of such decisions must weigh the impact in all areas. Analysis in such situations calls for models that are far from compact.

In an unachievable ideal world, everyone responsible for modeling to support government or business decisions would understand the strengths and weaknesses of a variety of methods and be comfortable working with models both small and large. However, realistically, professors have only so much time in MBA or MPA programs to teach about modeling. Learning on the job usually gives in-depth experience with a very limited spectrum of models – in part because success with one type of modeling usually causes repetition of that same type of modeling.

Limited exposure to modeling techniques has left an unfortunate gap in the use of models in corporations and governments, in which there is not only a real lack of practical knowledge about other approaches but also a lack of knowledge about how to go about choosing one approach versus another. Indirectly, the widespread paucity of knowledge or skill in matching models to purpose is why the author is so adamant about using explicit deliverables that define and validate what a model's purpose is to be.

So in approaching the potentially contentious subject of choosing a modeling method, think of Tables 7–9 as describing the "comfort zones" of applicability, where the characteristics of the method are well suited to the problem domain and a body of application experience, a user community, and perhaps good software support have developed.

Tables 7–9 can also be used as a first cut at understanding where a given model might be weak, when a given problem setting seems well outside the comfort zone of a given technique. Tables 7–9 cannot *prove* that a given approach *cannot* be used to provide useful answers to a given question, let alone whether a given approach is "better" than another in any absolute sense. But it is a guide.

4 Case Study: Resource Allocation in Law Enforcement

The following case study is offered both because it shows a methodology for tackling complex issues that is a great deal easier to use than, say, macroeconomic modeling or system dynamics simulation, and because it is also an example of the preparatory steps for more thorough quantitative modeling. This method is an extension of causal diagramming (Sterman 2000, Ch. 5; Senge 1990) that arose when the London Underground needed a very rapid assessment of different ways that it could be privatized (Mayo et al. 2001). The resulting method relied on a diagram and scoring by subject matter experts (SME). There are many similar situations in which formal quantitative equation-writing is not possible, whether due to time pressure, budget pressure, or lack of time-series data. Such situations often still allow adequate modeling, and robust, usable recommendations can be obtained.

Specifically, the setting is the situation of Transport for London (TfL), the government agency responsible for all public transport in greater London. Like most major cities, London suffered acutely from traffic congestion. Increasing the public's use of public transport, especially buses, was highly desirable, but research indicated that the public feared crime and violence around bus stops and aboard buses and disliked the length and unreliability of the ride. Controlling such crime and violence is the responsibility of the Metropolitan Police Service (MPS). However, MPS had limited resources, and more urgent issues, such as antiterrorism and street and gun crimes, inevitably took priority over transport policing.

The traveling public was about to receive a powerful incentive to ride buses rather than take private cars when the congestion charging scheme, starting February 2003, would impose a sizable cost on taking a private car into downtown London.

There was a plan to deal with the fear and unreliability deterrents to bus travel, but the plan involved many uncertainties. A pilot, then a fully-funded branch of London's MPS, would be dedicated to transport issues, the Transport Operational Command Unit (TOCU). This unit needed to be in place and doing a good job by the time the congestion charging scheme went into effect. The human, economic, and political costs of failing to improve the safety and reliability of bus transportation would be huge.

Here, the bang for the buck discussions emerged. There are dozens of major categories of activities to which the TOCU efforts might be applied, including educating the public about bus lanes, ticketing bus lane or "yellow box" (stopping in an intersection) offenses, placing cameras for remote ticketing, centralizing traffic control (which also coordinates with enforcement against vandalism), riding buses both to reduce fare evasion and to protect against on-board crime (as opposed to crime at the bus stop), and so on. These activities have both direct and indirect impacts on fear and reliability. Activities affect other facets of the transport ecology with diverse delay times, ranging from minutes to months or years. Activities interact with one another (e.g., better monitoring has no impact if there are no resources with which to act on what the monitoring shows) and so on.

How was TOCU to quickly find an effective mix of activities in time for a pilot and full-scale deployment prior to the onset of the congestion charging scheme?

TfL asked a team of modelers[2] to conduct a semiquantitative analysis of the resource-allocation problem, that is, to diagram the key interactions and interventions and use SMEs to score each potential use of resources. The modeling activity mostly comprised meetings among consultants, TfL representatives, and SMEs (TfL, MPS, and others). The meetings always focused on one of a sequence of graphical representations of the problem. The first representation, a block diagram shown in Fig. 1, is useful for discussing and defining the scope of the issues involved:

---

[2]The modelers were from PA Consulting Group, which also employed the author. Text and figures adapted from various materials copyright © PA Consulting Group, Inc., and are used with permission.

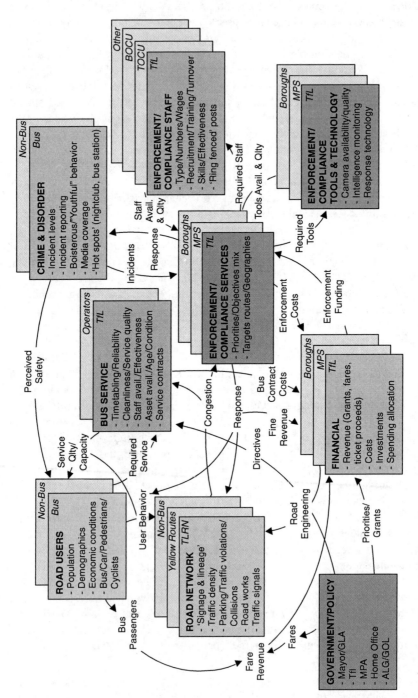

Fig. 1 Block diagram for TfL law enforcement resourcing analysis

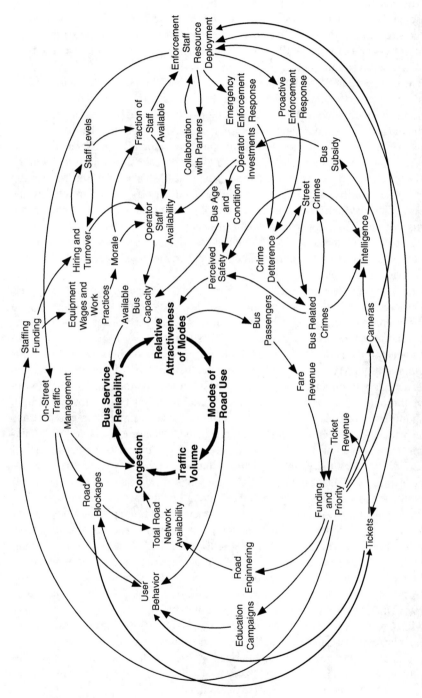

Fig. 2 Causal diagram for TfL enforcement resourcing analysis

The second representation transformed the block diagram into specific cause-and-effect relationships on a *causal diagram*, as shown in Fig. 2. In meetings (in contrast to publications like this book), such diagrams are shown to people as step-by-step buildups so that each piece of the diagram can be discussed and digested before adding more complexity.

The diagram (and its buildup) emphasizes feedback loops, which can have major impacts on the operation of the system and the effectiveness of activities. For example, balancing loops (a.k.a. negative loops) like that highlighted in the middle of the diagram tend to compensate for interventions. A larger number of bus riders leads to more buses and more traffic, which reduces reliability and deters more riders. The operation of such a loop will tend to control ridership relative to what the roadways can carry. By contrast, the other type of feedback loop, self-reinforcing loops (a.k.a. positive loops, or vicious or virtuous circles), tends to amplify the effects of interventions. So a larger number of riders would somewhat deter crimes at bus stops and on buses, which would increase safety and thus ridership.

The causal diagram must also show drivers for outcome measures of interest – what the client is trying to improve, in this case, two measures: perceived safety and bus service reliability.

The modelers sat with SMEs to score each link on the diagram: high, medium, or low strength, and short or long time to impact. (Sometimes, this step is skipped in the interest of time, especially SMEs' time. The omission leads to more uncertainty and to the need for discussions in later steps.)

In general, the next step is for SMEs to trace impacts of the actions under consideration through the various links to the outcome measures of interest. In a simple case, actions will directly change one variable on the diagram, and the SMEs can trace through the impacts on outcome measures of interest and score (High, Medium, Low, or 0–5, or some other scoring scheme). In more complex settings (which includes this case), actions can affect more than one variable on the diagram. In this case, the scoring task can be divided in two: scoring the impact of TOCU activities on the variables, then scoring the impact of variables on the outcome measures. Thus, potential TOCU activities were mapped onto the causal diagram, as shown in Fig. 3. Then, SMEs were asked to trace the path from impact points to outcome measures, looking at the strength and timing ratings of each link along the way, to score each combination of impact point and outcome measure. This scoring was done in collaboration with modelers, who knew from working with all the SMEs what the variables of the diagram were supposed to represent.

In a separate task, the cost of doing each activity at the present levels was estimated so that the cost of each activity at different levels of intensity could be known. The raw scores were first aggregated into a cost-benefit measure of single activities, which were then reviewed and analyzed for sensitivity of weightings, etc. Then, portfolios of enforcement activities at different mixes of intensity were scored and again reviewed, as summarized by Fig. 4, for an assumed increase in total budget. Notice, in that figure, that it is possible to spend more money and get worse results, a reassuringly plausible outcome.

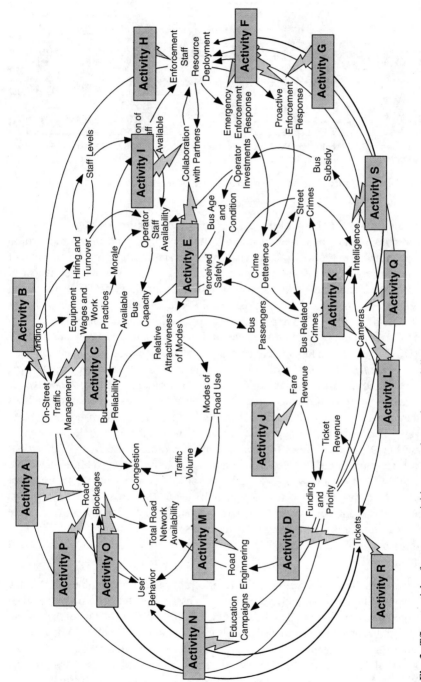

Fig. 3 TfL potential enforcement activities mapped onto the causal diagram

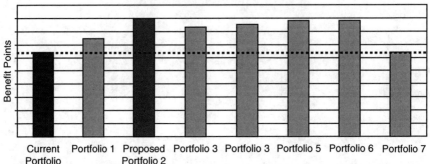

Overall Portfolio Benefit Score

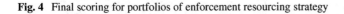

| Current | Portfolio 1 | Proposed | Portfolio 3 | Portfolio 3 | Portfolio 5 | Portfolio 6 | Portfolio 7 |
| Portfolio | | Portfolio 2 | | | | | |

Fig. 4 Final scoring for portfolios of enforcement resourcing strategy

In addition to simplicity of discussion, another reason for looking at portfolios of activities is that some require organizational change, both of who reports to whom, and who works with other divisions. For example, changing the numbers of people working at various activities means changing the number of supervisors and possibly the number of layers of organization. In addition, TOCU staff doing enforcement on buses and on the street would need procedures and points of contact to interact with the regular police. As in corporations, such changes are impractical to execute piecemeal; they are accomplished in a single reorganization.

TOCU decided on a mix of activities and implemented them as it launched itself as a separate unit. The impact of TOCU's activities was immediate and impressive, relative to the undesirable and static situation that had persisted for many years. It increased bus reliability and reduced levels of lost mileage due to traffic problems on the routes it covered. Customer satisfaction improved significantly according to surveys. There was a 14% aggregate growth in the number of people traveling by bus in London in the 2 years following implementation. There were 1,700 arrests by the new unit in the first year for crimes, including robbery, theft, assault, and possession of drugs and weapons. And these arrests have been made in areas that were formerly very lightly patrolled if at all by the MPS. For comparison, there are somewhat over 700 bus routes (Transport for London 2009). Managers at TfL were very pleased with the analysis and the actual results.

If an analyst decides to use this approach, these are the areas to which special attention must be paid:

1. Control the level of detail and the level of abstraction of variables so that the diagram fits on one page and yet captures the essence of what goes on well enough to get the agreement of up to dozens of SMEs. SMEs are able to unleash a torrent of detailed operational knowledge. It is up to the modeler to abstract details into concepts. For example, there are doubtless dozens of different types of crime. A causal diagram, however, should have at most a handful, and perhaps just one.
2. Practice making and discussing several causal diagrams before the final one, or work with someone who has that experience. When one has a facile skill of knowing how positive and negative feedback loops operate, one can identify

important loops during conceptualization (i.e., drawing the causal diagram) and understand their impact during the scoring process.
3. Do a dry run. This prepares the modelers to deal with many of the concepts and questions that will come up during sessions with SMEs and gives the modelers time to think through a scoring system appropriate to the situation at hand: Score links or not? Separate scoring for actions and impact points or not? Experience suggests that the scoring process usually changes slightly from case to case and from problem domain to problem domain.

This case will come back later, in the discussion of model validation. It turns out that the block diagrams and causal diagrams are important and useful steps in quantitative modeling to obtain expert feedback about model purpose, scope, and detail so that such questions can get worked out before the modelers begin writing detailed equations.

That said, we turn now to a different kind of economic modeling, fully quantitative, with the discussion emphasizing the design choices that make models well matched or poorly matched to a specific purpose and use.

5   Case Study: COMPOEX Economic Model

5.1   Model Setting, Purpose, and Methodology

The Conflict Modeling, Planning and Outcomes Experimentation (COMPOEX) project has produced an eponymous planning tool (Kott and Corpac 2007). COMPOEX explicitly accounts for PMSEII effects (Political, Military, Social, Economic, Infrastructure, and Informational effects) through a number of models run together automatically, with a general and uniform facility for testing actions and analyzing results in the aggregate system. Most critically, the methodology for each model is chosen to be appropriate for its own problem domain. In effect, COMPOEX combined a "politician in a box," an "economist in a box," and so on by linking a political model with an economic model, and so on. The COMPOEX Economic Model (hereafter, Economic Model), the topic of this case, attempts to mimic an "economist in a box" by predicting the reactions of markets and actors in the region of interest to U.S. government (USG) operations.[3] Infrastructure is also represented in the Economic Model; investment in infrastructure is an important economic process.

The Economic Model must deal with issues specific to developing countries for which far fewer data are available. Many elements of infrastructure such as electricity, water, and health are sporadically available, and the availability is an important component of both the response to U.S. actions and of limitations and modulators of economic responses. Domestic government responses, particularly biases in distributing and spending among regions and activities, had to be explicitly modeled;

---

[3]This example is adapted from "Economic Model Capability Description Document" by PA Consulting, March 31, 2008, copyright © PA Consulting Group, Inc. 2008, used with permission.

economists look for redistribution effects, in which availability of aid may reduce the need for the domestic government to spend money. In some regions, financial markets were nonexistent. In some regions, the illicit economy (e.g., growing poppies, white slavery, and smuggling) was a key element in the economy and security planning, over and above the shadow economy of transactions for normal goods and services hidden from government taxation.

The team managing COMPOEX development approached methodology choice essentially with two lists: One list specified a set of PMESII variables of interest derived from various planning documents and prior experience. The other list was modeling teams and people and their methodologies thought relevant to the exercise. Several rounds of discussion assigned responsibility for defining each variable to a modeling team or person.

Data issues, the need for ubiquitous modification of standard economic assumptions, and the availability of system dynamics modelers made that methodology a relatively uncontroversial choice for the Economic Model.

5.2 Scope and Level of Detail

Outputs of other COMPOEX models affect the Economic Model. For example, the Corruption Model influences incentives for capital investment and the government's ability to collect taxes. The Rule of Law Model (representing police and the criminal justice system) and the Military Model take a portion of government revenues to support.

Similarly, the Economic Model affects other COMPOEX models. When the Rule of Law Model or the Military Model invest funds to purchase plant and equipment (representing buildings, vehicles, and hardware), the Economic Model tracks the accumulation and depreciation of that plant and equipment, which goes back to influence the effectiveness of law enforcement and military operations. The Economic Model tracks employment in the Rule of Law Model, the Military Model, and in the illicit activities sector inside the Economic Model, accounting for all labor in the country.

An important question is which economic sectors should be modeled separately. The needs of the other specialized models suggest separate representation of Rule of Law, Military, and illicit activities. Beyond these, planners have particular interest in actions that impact different types of infrastructure, particularly those supported by the government, or mixed public and private infrastructure (such as education and health care). Therefore, the Economic Model has separate representations for the industries that produce education, health care, transportation infrastructure, power and telecommunications, shelter, water and sanitation, and other government services. The "private sector" includes only two industries: agriculture (a major factor in most developing nations) and all other goods and services. This mix of detailed public sectors and highly aggregated private sectors is nearly the reverse of typical macroeconomic models of developed countries.

The importance of education and technological expertise in developing economies has given rise to the "human capital" school of development economics (Eatwell et al. 1987 v1, pp. 818–825; Sen 1985; UN Development Programme 2009). The Economic Model represents human development relatively simply, by separately tracking blue-collar and white-collar labor, with the mix determined over time by a combination of education, work experience, and work opportunities. The mix and experience base of labor impacts productivity and production of each type of good or service (including education).

Economic sectors interact in several ways. They all participate (along with the household sector) in a labor market, with variable wages. The government and consumers allocate spending among economic sectors. The government collects taxes from households and private sectors and makes spending decisions for the public goods. Households decide whether to spend or save, and to varying extents, businesses are able to turn to external financing, ultimately from the householders.

The financial sector and treatment of financing in this model is simpler than would be the case for an economic model whose purpose was monetary and fiscal policy. A single financial instrument aggregates debt and equity, and there are no explicit financial intermediaries so that the household holds net assets and the public and private sectors hold net liabilities. Tax revenues and an acceptable level of deficit spending govern public borrowing. This simplistic treatment is sufficient to allow the model to be the "economist in a box" and provide warning if, e.g., an action produces crowding out in the financial markets (if there *are* financial markets). While imports and exports for each industry are explicit, exchange rates are assumed constant, since for various reasons they were not at issue for the regions to which the Economic Model has been applied.

Figure 5 shows a top-level view of the Economic Model. Ovals are groups of equations (often themselves organized hierarchically into groupings that are more detailed). The heavier lines show the simple linkages of textbook economics: The Population Needs combine with available Finances to create Demand (and Importing and Exporting) of products. That demand both creates Employment and justifies investment in Physical Infrastructure. These combine to create Production. Employment creates wages that feed back into Finances. Production creates tax revenues, which also feed back into (government) finances.

The Macro Finance group is where interest rates strike the balance between household saving and industry and government need to borrow. The Macroeconomic Performance group, its name somewhat to the contrary, is where measures of macroeconomic performance such as GDP and its components are *computed* from activities performed elsewhere.

The Economic Model has around 1,300 equations, many of them repeated (with different parameters) for different geographic subareas and different economic goods and services. The simulation starts at the year 1995 and simulates forward in time steps of ¼ week. The simulation was compared to time-series data generally between 2001 and 2006 (due to poor data availability both before and after that time period). The analysis used the simulation from the beginning of 2008 to the beginning of 2011. Standalone, the 1995–2011 simulation took about 9 s on a laptop.

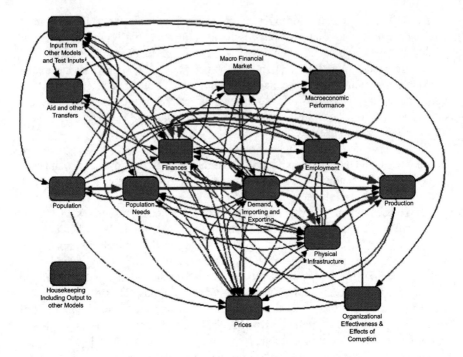

Fig. 5 Top-level view of the COMPOEX Economics Model

5.3   Numerical Example

Both as an entrée to the system dynamics methodology and as a simplified example
of one portion of the Economic Model, the Employment Group, this section gives
an extremely simplified example of how a system dynamics simulation model
works numerically. The commentary will describe some of the issues that differentiate the simple example from what is actually in the model. To give a measure of
perspective, the example here is perhaps 20 times simpler than the corresponding
piece of structure in the economic model, and the labor sector is perhaps a tenth of
the whole model.

5.3.1   Flow Diagram

Figure 6 illustrates several of the key concepts used to structure system dynamics
models. The rectangles (Agricultural Employment and Unemployed Workers) represent stock variables. These variables can, in principle, be counted or measured at
any moment in time. The "bow tie" variables (Separation Rate, Hiring Rate) represent flow variables, which have no instantaneous meaning and are defined only as

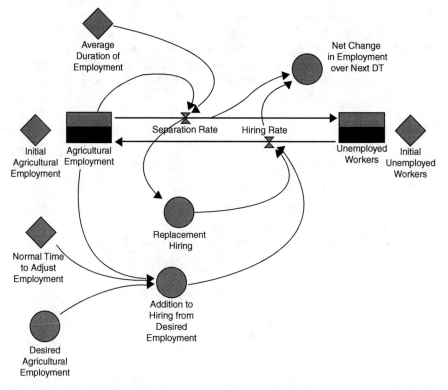

Fig. 6 Flow diagram for numerical example

occurring over an increment of time. Over time, flows accumulate to change stocks, and in a feedback system, the values of the stocks influence the flows.

In Fig. 6, the diamond symbols are constants, which have the same value throughout a simulation. The round symbols represent arithmetic computation performed at a single instant in time. Although they could be folded into the flow equations, they are separated both to simplify the presentation of the algebra and to allow a priori estimation of the constants involved.

The constants next to the stock variables (Initial Agricultural Employment and Initial Unemployed Workers) set the initial values of the respective stock variables. For simplicity, the information flow from initializing constants to stocks are not shown, since they have no further effect once the simulation is started.

Even at the level of flow diagram, several simplifications relative to a full macroeconomic model are evident: there is no feedback from Unemployed Workers. A more complete model would have Unemployed Workers (or their absence) affecting the speed of hiring (constant here, characterized by the Normal Time to Adjust Employment). And Unemployed Workers (or their absence) would also change wages, which eventually would have an impact on the number of workers desired (Desired Agricultural Employment). Of course, that would depend on whether other *factors of production* (such as better irrigation systems, better fertilization and pest control, mechanized planting and harvesting, allowing land to lay fallow, etc.) are available and cost-effective to substitute for labor. Finally, each distinct economic sector (education, power and communications, military, etc.) would need to keep track of its own stocks of employees, both white-collar and blue-collar.

5.3.2 Equations for Stocks

$$Agricultural\ Employment_{(t+dt)} = Agricultural\ Employment_{(t)} +$$
$$(Hiring\ Rate_{(t)} - Separation\ Rate_{(t)})*dt$$

$$Agricultural\ Employment_{(t0)} = Initial\ Agricultural\ Employment$$

Unit : people

Initial Agricultural Employment = 1000

Unit : people

$$Unemployed\ workers_{(t+dt)} = Unemployed\ workers_{(t+dt)} +$$
$$(-Hiring\ Rate_{(t)} + Separation\ Rate_{(t)})*dt$$

$$Unemployed\ workers_{(t0)} = Initial\ Unemployed\ Workers$$

Unit : people

Initial Unemployed Workers = 100

Unit : people

Knowing the initial condition for each stock and the rates of flow, the equations above compute the value of the stock at the point in time, dt ("delta time") later.

The rates of flow are measured in terms of people per week, so multiplication by dt gives the change that should occur over the time interval of dt. This convention allows convenient units of measure (weeks, months, years) separate from the computation interval. The assumption here is that the dt is short enough that it is a reasonably accurate assumption that the flow rate does not change in the time from t to t+dt. This is a testable assumption.

Those with mathematical background may recognize this arithmetic as simulating the integral form of an ordinary differential equation by Euler integration. But the mathematical terminology does not change the fact that, at heart, this simulation is a straightforward process.

Estimating the initial stocks of people in a real economic sector is usually straightforward; virtually, all national governments estimate the size of their population, keep track of unemployment, and estimate rough distribution of employment by occupation. (For other economic quantities, the Economic Model started from physical units of measure such as square feet of housing, kilowatts of electrical generating capacity, calories per day of food, and so on to facilitate a priori estimation of quantities present in the economy. The monetary transactions regarding these goods and services then had separate equations and were measured in currency units per week.)

5.3.3   Hiring and Separation Rates

$$Hiring\ Rate_{(t)} = Replacement\ Hiring_{(t)} +$$
$$Addition\ to\ Hiring\ from\ Desired\ Employment_{(t)}$$

Unit : people / week

$$Replacement\ Hiring_{(t)} = Separation\ Rate_{(t)}$$

Unit : people / week

$$Separarion\ Rate_{(t)} = Agricultural\ Employment_{(t)} /$$
$$Average\ Duration\ of\ Employment$$

Unit : people / week

Average Duration of Employment = 50

Unit : weeks

The hiring rate represents a simple heuristic: Hire to replace people who leave plus a bit more if more employees are desired.

The Average Duration of Employment is set to have a round number for computations rather than to be realistic. The equation implies that each week, 1/50th of the workers separate (quit or are fired), for an average duration of slightly less than a year. In industrial economies, the figure is closer to a 2-year average. In the setting of a developing economy, 50 weeks implies separation on average after every season of agricultural employment, which is probably too short. But 50 weeks is a nice round number for the numerical example coming shortly.

5.3.4   Hiring to Change Employment

The component of hiring that deals with a need to change the number of agricultural employees is the simplest possible formulation: If there is a difference between desired and actual employees, hire proportional to that difference:

$$Addition\ to\ Hiring\ from\ Desired\ Employment_{(t)} =$$
$$(Desired\ Agricultural\ Employment_{(t)} - Agricultural\ Employment_{(t)}) /$$
$$Normal\ Time\ to\ Adjust\ Employment$$

Unit : people / week

Normal Time to Adjust Employment = 2

Unit : weeks

The division by a time constant is usually preferable to multiplication by some number. Here, it is easier to understand "on average it takes 2 weeks to get the number of employees to where you want" than it is to attach any intuitive meaning to "multiply by 0.5." Two weeks is probably short for changing employment in an aggregate economy, but it is conveniently short for the numerical example.

5.3.5   Exogenous Input

In this simple system, Desired Agricultural Employment is considered to be determined outside the dynamic system (exogenous = "born outside"), and so is modeled as a simple function of time:

Desired Agricultural Employment$_{(t)}$ = If t < 2 then 1,000, else 1,050

Unit : people

In the Economic Model, the Desired Agricultural Employment is determined by demand for agricultural products, wage rates, productivity,and other factors.

5.3.6   A Supplementary Equation

Sometimes, one defines a variable merely to monitor it, even though it is not needed to conduct the simulation itself. Here, the numerical example will break out the last part of the equation for Agricultural Employment as a separate quantity:

Net Change in Employment over Next DT$_{(t)}$ =

(Hiring Rate$_{(t)}$ − Separation Rate$_{(t)}$)* dt

Unit : people

5.3.7   Specifying the Simulation Routine

Finally, to conduct a simulation, one defines the parameters of the simulation process: the time when the simulation starts

$$t0=0$$

The time interval ("delta time") that occurs between simulation steps

$$dt=1$$

The time at which to stop the simulation

$$LENGTH=20$$

and (for simulation software), how often (i.e., at what interval) to save the results for later examination.

$$SAVPER=1$$

The equations above completely specify a simulation, which starts at the initial conditions and uses the flow equations to compute the stocks at the next time interval, which are the basis for computing the flows, and so on through simulated time until the end.

Figure 7 shows the computation specified by the equations above carried out in a spreadsheet format.

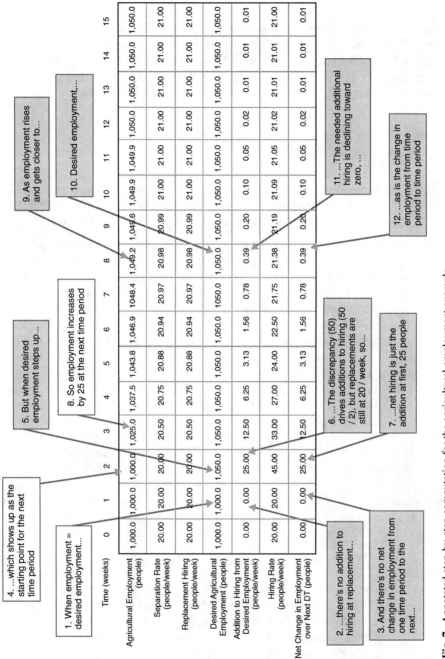

Fig. 7 Annotated step-by-step computation for the agricultural employment example

The numerical simulation will be familiar to some spreadsheet users, especially those who compute evolution of a balance sheet over time. Spreadsheets can and have simulated models formulated according to system dynamics standards. But for complex models, specialized system dynamics simulation software is vastly easier to use, and creates models that are far more transparent. In addition, the formulas that create future behavior must be the same formulas that create past behavior, which helps make a powerful validation statement: The rules used to project forward are exactly those that created (roughly) the past behavior.

5.4 Validating the Model

In textbooks and academic papers, validation is often pass-fail: The hypothesis fits the data well or not. Modeling to support real-world decisions is more complex. The modeler must (implicitly or explicitly) test three generic hypotheses:

1. That the modeler understands what the user (client) wants to achieve, and what means are available to achieve them – this is the model purpose. Unless the intended model use is for short-term budgeting, forecasts as such are rarely the actual use. More typically, the use is taking action to change one of the behavior modes we discussed earlier or to assess the medium- and long-term investment climate. Both these call for models that deal with the economic (and near economic) behavior modes relevant to a given region. Principles of specifying model purpose are discussed in Sterman (2000, Section 3.5.1). Graham (2009b, Section 2.1) presents diagrammatic tools for validating model purpose.
2. That the modeler has captured in the modeling how the real-world system works, in terms and at a level of detail and scope appropriate to the purpose. Sterman (2000, Ch. 21) describes validation tests of model structure and baseline behavior. Classic statistical tests are a subset of these.
3. That the modeler understands the impacts of the actions being analyzed by the simulation experiments and why they happen, particularly those with favorable outcome that the analysis finally recommends. Such understanding can become a "model of the model" that explains in the simplest possible terms why some actions turn out to be desirable and others do not. Sterman (2000, Ch. 21) describes validation tests of model analysis-based recommendations.

There is one additional wrinkle to system dynamics modeling, which is, the three validations above are often done twice, once with a diagrammatic systems thinking "model" in which SMEs go through scoring exercises to start to quantify the model and policy impacts and once again with a fully quantified simulation model. Lyneis (1999) gives real examples of the whole process in corporate strategy. Mayo et al. (2001) provide a particularly well-worked out public policy example.

Figure 8 gives a stylized view of simulation model construction and validation. Each column represents one round of hypothesis testing and evolution of the analysis. The unshaded headings (steps 1–3) describe the three hypothesis tests for the

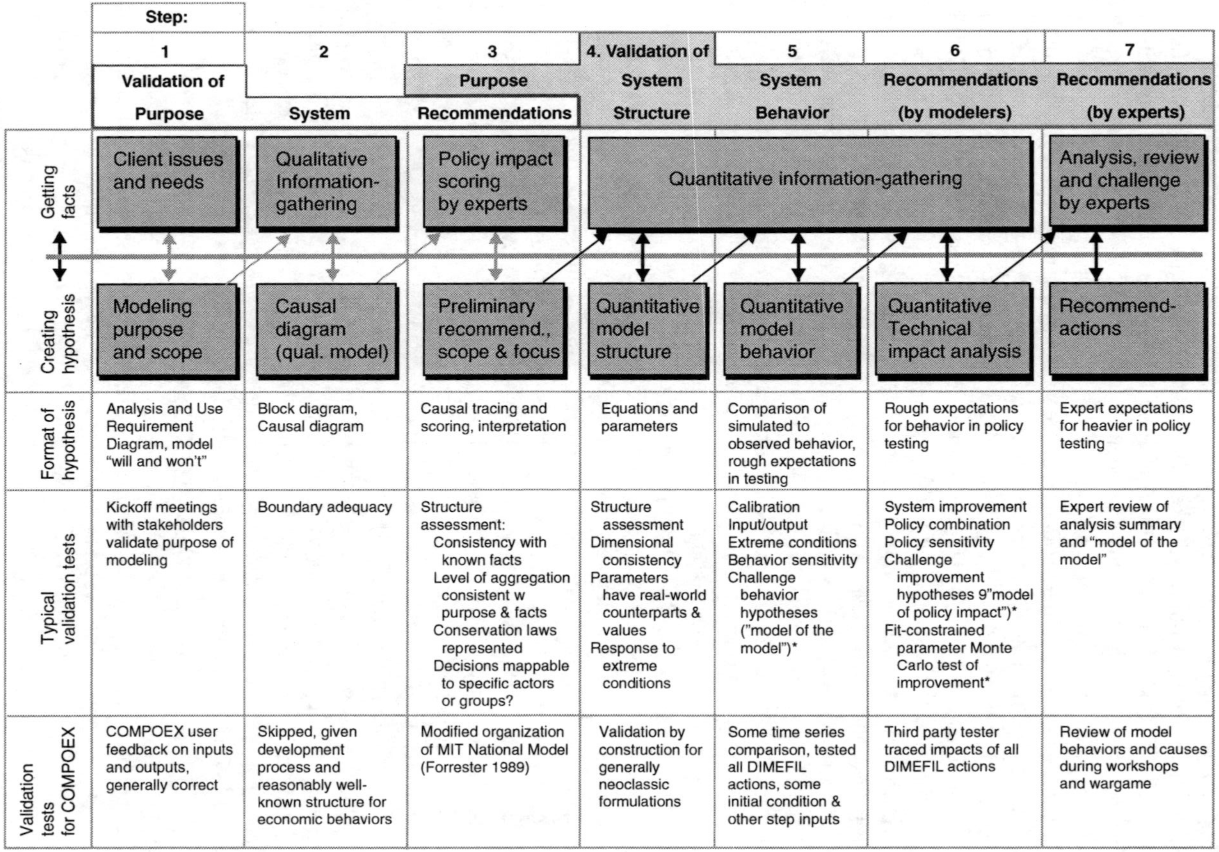

Step:	1	2	3	4. Validation of	5	6	7
	Validation of Purpose	System	Purpose Recommendations	System Structure	System Behavior	Recommendations (by modelers)	Recommendations (by experts)
				Quantitative information-gathering			
(boxes)	Client issues and needs; Modeling purpose and scope	Qualitative Information-gathering; Causal diagram (qual. model)	Policy impact scoring by experts; Preliminary recommend., scope & focus	Quantitative model structure	Quantitative model behavior	Quantitative Technical impact analysis	Analysis, review and challenge by experts; Recommend-actions
Getting facts ⟷ Creating hypothesis — Format of hypothesis	Analysis and Use Requirement Diagram, model "will and won't"	Block diagram, Causal diagram	Causal tracing and scoring, interpretation	Equations and parameters	Comparison of simulated to observed behavior, rough expectations in testing	Rough expectations for behavior in policy testing	Expert expectations for heavier in policy testing
Typical validation tests	Kickoff meetings with stakeholders validate purpose of modeling	Boundary adequacy	Structure assessment: Consistency with known facts; Level of aggregation consistent w purpose & facts; Conservation laws represented; Decisions mappable to specific actors or groups?	Structure assessment; Dimensional consistency; Parameters have real-world counterparts & values; Response to extreme conditions	Calibration; Input/output; Extreme conditions; Behavior sensitivity; Challenge behavior hypotheses ("model of the model")*	System improvement; Policy combination; Policy sensitivity; Challenge improvement hypotheses 9"model of policy impact")*; Fit-constrained parameter Monte Carlo test of improvement*	Expert review of analysis summary and "model of the model"
			Modified organization of MIT National Model (Forrester 1989)	Validation by construction for generally neoclassic formulations			
Validation tests for COMPOEX	COMPOEX user feedback on inputs and outputs, generally correct	Skipped, given development process and reasonably well-known structure for economic behaviors			Some time series comparison, tested all DIMEFIL actions, some initial condition & other step inputs	Third party tester traced impacts of all DIMEFIL actions	Review of model behaviors and causes during workshops and wargame

Fig. 8 Steps of simulation model evolution, with validation activities

systems thinking model and analysis by scoring. Each shaded column (steps 3–7) represents the hypothesis tests of purpose, system, and analysis for a simulation model. Step 3 shows an overlap between the analysis validation step for the diagrammatic model with the purpose validation step for the simulation model. Step 3 is a decision point, whether the additional effort of creating a simulation model is useful and appropriate. And if a simulator is useful, the diagrammatic exercise will likely have changed the stakeholder's ideas about the model scope and what policies should be analyzed.

Steps 4 and 5 break the validation of the model system into two parts for convenience, since the tests of pieces of the model (step 4) can be quite different from tests of the larger model (step 5). Similarly, steps 6 and 7 break the validation of the analysis into those conducted by the modelers from those conducted by SMEs, again because the tests are quite different.

Each step of hypothesis testing includes a form of fact gathering and summarizing those facts into a hypothesis, which can be tested. These are the first three rows in Fig. 8. The fourth row describes typical tests, generally using the vocabulary of Sterman (2000, Ch. 21). The previous discussion in this chapter has given examples of three of the formats for hypotheses (block diagram, causal diagram, and equations).

Few analysis efforts do all possible validation tests. Indeed, only simulation analysis for legal disputes comes close to doing everything (Stephens et al. 2005). The last row describes the validation testing used for the Economic Model operating within the COMPOEX system. For the COMPOEX purpose of demonstrating sensible and generally plausible and useful results for an economic model working in harmony with other, equally specialized, models, these validation activities seemed sufficient. "Validation by construction" here means using pieces that have been validated individually elsewhere. The COMPOEX economic model was able to borrow heavily from earlier research on economic dynamics, both in overall architecture and organization (Mass 1975; Sterman 1982; Forrester 1989) and more specifically for labor markets (Runge 1976), financial markets (Low 1977), and capital investment (Senge 1978).

One type of validation test is often misunderstood, as it differs fundamentally from a common and seemingly related statistical process. In standard statistical regression, as commonly taught in MBA programs and initial economics courses, one specifies an equation form and uses data (and mathematics embedded in software) to find parameter values that best fit the data.

In system dynamics modeling, the equations and parameter values are both set by a priori information: firsthand knowledge of cause and effect in the real system, derived from background reading and interviews with SMEs. "Calibration" or "behavior reproduction" validation tests are passed when the simulator, driven by a handful of exogenous variables, steps through simulated time to create a simulated history that *independently* reproduces the behavior of time-series data that correspond to model variables. This test is extremely useful in detecting formulation and parameter errors (and data errors).

However, there is one circumstance in which this test can be misleading: Cyclical behavior (such as the business cycle, described briefly in Table 3) that is generated by systems responding primarily to many unknown random events need to be tested differently. To simplify: Simulations of such systems will not have the same inputs as the real system, so simple comparison of simulation to time series data is not meaningful. The economic business cycle and some commodity markets have this characteristic.

For such cases, one must either test for invariant characteristics (e.g., phase and amplitude relationships among variables) (Forrester 1961, Section 31.5) or use some variant of weighted least squares (which includes full-information maximum likelihood) using Kalman filtering (Schweppe 1973), a mathematical technique generally restricted to models in the form of state-space dynamic systems, which system dynamics simulators do have. Graham (2009b) further discusses this issue in model testing.

Figure 9 shows (disguised) behavior reproduction of GDP for three countries in one version of the COMPOEX Economic Model (clearly not dominated by cyclical behavior).

Of course, the simulation will not match any of the time series at first. But the only changes allowed to the modeler are cause-and-effect relationships and parameter values, both to be constrained by the a priori knowledge of what is plausible and what is not. Achieving such consistency is usually a nontrivial challenge. In a feedback system, if one variable differs from what happened in real life, the variables that it drives will be off, which throws off the variables they drive, and so on. Moreover, because any one change usually affects multiple variables, simple curve-fitting is not possible.

Passing a single validation test, even a difficult one like the behavior comparison test, does not prove that the model is correct. The only thing that any validation test ever does is fail to disprove a hypothesis. Therefore, a successful behavior

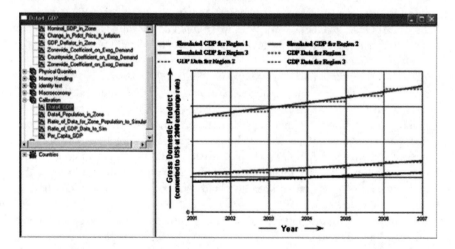

Fig. 9 Example of one (typically of many) validation plots comparing the historical time series for a variable to the independently-produced simulation of that same variable, here gross domestic product (GDP)

comparison test fails to find inconsistencies between the a priori plausible structure and parameter values and the multiple time series of observed behavior.

5.5   Findings: Use of COMPOEX Model

The COMPOEX system has been implemented several times, representing a few different regions of interest. These systems have been tested in workshops involving planners and SME's in hypothetical planning exercises, in addition to formal system testing and review by SMEs. It was also used in a war-game exercise to support white team (referee) adjudications that determined consequences of actions by the various teams after each round.

The experiment might be called reasonably successful, in that the Economic Model functioned much as experts do, pointing out unintended consequences and unexpected (or not entirely expected) results in the exercises. A few of these were:

- *Scale*. Many developing countries have populations in the hundreds of millions. Many have economies that are supported by oil revenues or other natural resource exports. An aid project measured in mere millions of dollars will generally be too small to have a substantial impact on the economies. Yet too often, the planning and intention of aid programs still proceed under the assumption that the U.S. aid in the millions will have an effect that will be noticed and appreciated.
- *Coupling actions to media coverage*. Although too small to have measurable effect on the overall economy, actions can be turned into media events, which are opportunities for a message to reach national audiences. It is all too often the case that the U.S. good works and constructive engagement do not achieve public visibility nor on message communication with the public of the host nations. In COMPOEX, planners were able to couple media actions with economic actions to achieve larger and much broader effects than could be achieved without media effects.
- *Reducing corruption and increasing government effectiveness*. In general, corruption increases investment risk and the effective cost of capital and siphons off cash. Corruption consumes especially white-collar time and productivity in making new capital investments. Moreover, corruption reduces the ability of the host government to collect taxes and thus to finance development. Government lack of effectiveness functions in some aspects similarly to corruption; indeed they can often be closely linked. Actions that reduce corruption or increase the effectiveness of government spending, and government regulation can have a major economic impact.
- *Price or wage feedback*. Giving aid money to one sector will increase wages and prices in the targeted sector. This reduces both the effectiveness of the aid and the free-market demand due to higher prices. For example, a country experiencing high food prices may attempt to get more food for its population by prohibiting food exports. This works partially, but it lowers the incentive to domestic farmers

to produce food, so the additional food obtained for the domestic population will be less than what was exported formerly. Therefore, despite government actions, food would remain a critical issue. This is a textbook example of feedback loops defeating the intent of policy interventions (Forrester 1971).

- *Substitution.* To the extent that giving aid to provide a good or service actually alleviates the shortage of that good or service, the incentives for the host nation government, local corporations, or private individuals to provide that good or service is reduced. Less money will be spent on the targeted goods and services, and more money will be spent in other places. To some extent then, even targeted aid will have a diffuse effect across many goods and services through substitution of aid for other sources of support.
- *Arithmetic for complex situations.* Aggregate economic performance measures can contain hidden surprises. For example, in some circumstances, increased hiring in a sector may cause average wages to go down. This happens when white-collar employees are scarce and their wages are substantially higher than those of blue-collar workers. If the additional hiring is mainly blue-collar workers, even though wages for both white- and blue-collar employees go up, average wages will go down because the mix changes.

6   Practical Tips

- Even if quantitative modeling seems unlikely, collect the available indicator data early in the project; one needs to know where problems are likely to arise. Deep is nice, but broad is necessary. Indicator data helps to control (restrict) the scope of subsequent quantitative modeling. In terms of project management, data collection is usually the longest single task and should be started early, even if all specifics are not nailed down.
- Resist a common request from the modeling client for a broad, unfocused forecast. Such an open-ended mission is bound to fail. The resulting model will forecast the overall economy poorly and miss details and behaviors that are critical to the specific decisions that the model should support. Eventually, good analysis will require searching for specific conditions and events that could undermine the conclusions and recommendations of the analysis. If the modeler does not know what recommendations the analysis will be used to support or reject, it will be impossible to do a good job of sensitivity analysis.
- Involve stakeholders throughout the modeling process and in their terms. Modeling effort is wasted if it does not influence action, which means the modeling must result in believable conclusions for the end user. Certain techniques, such as benchmarking, expert seminars, or war gaming, are compelling for analysts but not for decision-makers. Choose an analysis method with an understandable audit trail from the evidence to conclusions. One benefit of a quantitative systems thinking exercise (described in this chapter) is that all of the process is visible and comprehensible to stakeholders.

- Integrate available knowledge with "quantitative systems thinking" (as in Table 7 and the first case) first, even if quantitative modeling follows. Resist the temptation to dive into equation writing. Identifying an instance of too much detail is far easier to fix on a diagram than in equations.
- Check the capability of candidate models against the behaviors of interest to a client. For instance, in making an investment, currency crises may be important, especially if indicator variables show vulnerability to currency crises. Thus, if a model is to serve as support for investments, it must be able to represent the dynamics of runaway currency crises and indicate whether vulnerability is likely to increase or decrease.
- Use "what would have to be true" analysis to mitigate uncertainty of economic models. An important way to create robust conclusions that help real decision-makers is to use the model to find the assumptions (usually extreme) that would invalidate the recommendation. For example, suppose a billion-dollar investment in a developing country was hedged with certain financial arrangements. And, suppose the model of the investment and its hedges shows that the currency would have to decline to 1% of its current value before the investment lost money. If this decline were judged to be well outside a plausible range, then the investment decision would be robust, even though currency exchange rates are very unpredictable.

7 Summary

To select, and effectively employ the model best suited for a specific economic problem, one starts with understanding the purpose (particularly, the intended decisions) of modeling and analysis, then gathers data on the economies being studied to identify actual and potential "behavior modes," i.e., known patterns of economic behavior and their cause-and-effect origins. They define the elements that the analysis needs to deal with. We classify such behavior modes and analytical and modeling approaches along with zones of applicability for each.

In one case study, the Transport for London (TfL) agency wished to optimize allocation of the available funds, in order to increase public transportation safety and service reliability and to cope with congestion charges on automobile use in central London. Using quantitative systems thinking approach, the analysts and SMEs developed a block diagram and then a causal diagram. Then, SMEs scored the strength and time delays of each link in the causal diagram, and the direct impacts of each potential action on causal diagram variables. Finally, the SMEs traced all major paths from each action to the two outcome measures, safety and reliability. The resulting portfolio of recommendations was implemented, and measures of safety and reliability were substantially improved.

The second case is economic modeling in COMPOEX, a system that helps government planners understand consequences of interventions in a region of interest. The economic model represents supply and private and government demand in government-funded or mixed-funded sectors such as health, power and communications, water and sanitation, and education; also the illicit goods and services sector, which draws revenue from corruption and direct sales. Economic performance in the economic model also feeds the power struggle model and the population satisfaction model within the COMPOEX system.

Validation of decision-oriented economic models occurs during the entire modeling process, e.g., in seven phases for system dynamics models. Validation testing should address three fundamental hypotheses: (1) the purpose is validated; (2) the model is validated; and (3) the results are validated.

8 Resources

8.1 Economic Behaviors and Data

1. "Near-economic" behavior modes

   Theories of political power: (Bueno de Mesquita et al. 2002, 2004)

   Statistics:

   United Nations Universal Human Rights Index
   http://www.universalhumanrightsindex.org/

   World Bank Governance Indicators
   http://info.worldbank.org/governance/wgi/sc_country.asp

   Internet Center for Corruption Research
   http://www.icgg.org/corruption.html

   Transparency International
   http://www.transparency.org/

   Economist Intelligence Unit Country Reports. London: The Economist Group

   Freedom House
   http://www.freedomhouse.org

   Human Rights Watch
   http://www.hrw.org/

   Amnesty International
   http://www.hrw.org/

   Gallup Country Stability Index (Gallup Corporation 2009)

2. Secular industrialization, demographic transition

   (Caldwell 1976; Caldwell et al. 2006)

   Statistics:

   United Nations Economic and Social Development theme
   http://www.un.org/esa/

   UN Development Programme, Human Development Report and Human Development Indices
   http://hdr.undp.org/en/humandev/hdi/
   (IMD various years)

3. General macroeconomics

   General economic theory: (Samuelson and Nordhaus 2004)

   International economic theory: (Curry 2000; Krugman and Obstfeld 2008; Gandolfo 2002)

   Statistics:
   Fedstats (aggregation of US government data) http://www.fedstats.gov/

   International Monetary Fund, Washington, DC
   http://www.imf.org/external/pubs/ft/weo/2008/02/weodata/index.aspx

   United Nations

   World Bank, Washington, DC
   http://www.worldbank.org/

   Federal Reserve System, Flow of Funds Accounts of the United States
   http://www.federalreserve.gov/releases/z1/

4. Business cycles

   Theory: (Mass 1975; Forrester 1982; Forrester 1989; Sterman 2000)

   Statistics:

   US: National Bureau of Economic Research, Cambridge, Massachusetts.
   http://www.nber.org/
   World: Economic Cycle Research Institute, New York and London.
   http://www.businesscycle.com/

5. Trade balancing through the exchange rates

   (Curry 2000; Gandolfo 2002)

6. Currency exchange mercantilism

   (Burgess et al. 2009; Das 2009)

7. Import dependence and stagflation

   Models: (Bernanke and Blinder 1988; Godley and Lavoie 2006)

8. Capital flight

   (Shibuya 2001)

9. Debt-deflation spiral

   Models: (Von Peter 2005; Sterman 1986)
   Descriptive analysis: (Graham and Senge 1980; Graham 1982; Koo 2008)

10. Deficit-lead hyperinflation

    Models: (Taylor 1991)
    Description: (Krugman and Obstfeld 2008)

11. Currency crisis/investment boom and bust, currency exchange defense

    Models: (Krugman 1999)
    Descriptive analysis: (Krugman 2009)

8.2   *Analytical Methods*

1. Economic models generally

   Journal of Economic Literature, which publishes only review articles on economic topics, will intermittently include macroeconomic models.

2. Quantitative systems thinking

   (Lyneis 1999; Mayo et al. 2001; Sterman 2000, Ch. 5)

3. Analytical models

   (Samuelson and Nordhaus 2004; von Peter 2005; Krugman 1999; Solow 1956, 1957)

   http://en.wikipedia.org/wiki/Exogenous_growth_model

4. General equilibrium

   Large numerical GE: (Inforum 2009; Kubler 2008)

   Small analytical GE:

   ISLM analysis in any macroeconomics textbook, e.g., (Samuelson and Nordhaus 2004)

   More modern small GE models are actually market equilibrium and steady-state growth (Solow 1956, 1957; Foley and Sidrauski 1971; Romer 1990)

5. Single-equation regression and stochastic models

   (Theil 1971; Eckstein 1983; Fair 2004.
   http://fairmodel.econ.yale.edu/)

   Discussion of strengths and weaknesses: (Sterman 1988)

6. System dynamics

   General:

   (Forrester 1961; Alfeld and Graham 1976; Sterman 2000)

   System Dynamics Society
   http://www.systemdynamics.org/

   Well-known cases: (Mayo et al. 2001; Lyneis 1999)

   Methodological cases: (Graham et al. 2002; Graham and Ariza 2003; Lyneis 2000)

   Mathematical foundations:

   (Schweppe 1973)

7. Game theory

   (Dixit and Skeathmore 2004; Williams 2007)

8. Agent-based

   Social science applications: (Billari et al. 2006)

   Methodological discussion:

   (Rahmandad and Sterman 2004)

References

Alfeld, L., & Graham, A. (1976). *Introduction to Urban Dynamics*. Waltham, Mass.: Pegasus Communications.

Bernanke, B., & Blinder, A. (1988). Credit, money and aggregate demand. *American Economic Review*, 78(2), 435–439.

Billari, F., et al. (2006). Agent-Based Computational Modeling: Applications in Demography, Social, Economic and Environmental Sciences (Contributions to Economics). Heidelberg, Germany: Physica-Verlag.

Bueno de Mesquita, B., Morrow, J., Siverson, R., & Smith, A. (2002). Political institutions, policy choice and the survival of leaders. British Journal of Political Science, 32, 559–590.

Bueno de Mesquita, B., Smith, A., Siverson, R., & Morrow, J. (2004). The Logic of Political Survival. Cambridge, MA: MIT Press.

Burgess, G., Chong, S., Summers, G., Strobl, M., Graham, A., Horne, S., Torre, C., & Kreider, B. (2009). PA&E Global Economics Study Final Report: A reconnaissance of economic issues impacting DoD. Washington, DC: Department of Defense Directorate for Program Analysis and Evaluation.

Caldwell, J. (1976). Toward a restatement of demographic transition theory. Population and Development Review, 2, 321–366.

Caldwell, J., Caldwell, B., Caldwell, P., McDonald, P., & Schindlmayr, T. (2006). Demographic Transition Theory. Dordrecht, the Netherlands: Springer Science & Business Media.

Curry, J. (2000). A Short Course in International Economics. Novato, CA: World Trade Press.

DARPA. (2006). Towards a system dynamics model of insurgency in Fallujah: Integrating knowledge to understand underlying drivers and identify high leverage intervention points, Final Deliverable for Contract HR0011-06-C-0128. For Official Use Only.

Das, D. (2009). The evolution of renminbi yuan and the protracted debate on its undervaluation: An integrated review. *Journal of Asian Economics*, 20(50), 570–579, September 2009. Order article at http://www.sciencedirect.com/

Dixit, A., & Skeathmore, S. (2004). Games of Strategy, 2nd ed. New York, NY: WW Norton.

Eatwell, J., Milgate, M., & Newman, P., (Eds.) (1987). The New Palgrave: A Dictionary of Economics. London, UK and New York, NY: Macmillan and Stockton.

Eckstein, O. (1983). The DRI Model of the US Economy. New York, NY: McGraw-Hill.

ECRI (Economic Cycle Research Institute). (2009). New York and London. Retrieved from http://www.businesscycle.com/.

Fair, R. (2004). Estimating How the Macroeconomy Works. Cambridge, MA: Harvard University Press. http://fairmodel.econ.yale.edu/.

Foley, D., & Sidrauski, M. (1971). *Monetary and Fiscal Policy in a Growing Economy*. New York: Macmillan.

Forrester, J. (1961) Industrial Dynamics. Waltham, Mass.: Pegasus Communications.

Forrester, J. (1971). Counterintuitive behavior of social systems. Technology Review, 73(3), 52–68. Retrieved from http://sdg.scripts.mit.edu/docs/D-4468-2.Counterintuitive.pdf.

Forrester, N. (1982). A dynamic synthesis of basic macroeconomic theory: Implications for stabilization policy analysis (PhD Thesis). Cambridge, MA: Alfred P. Sloan School of Management, Massachusetts Institute of Technology.

Forrester, J. (1989). The System Dynamics National Model: Macrobehavior from Microstructure. In Computer-Based Management of Complex Systems: International System Dynamics Conference. Springer, Berlin.

Gallup Corporation. (2009). Retrieved from http://www.voice-of-the-people.net/ContentFiles/files/VoP2005/VOP2005_Democracy%20FINAL.pdf.

Gandolfo, G. (2002). International Finance and Open-Economy Macro-economics. New York, NY: Springer.

Godley, W., & Lavoie, M. (2006). *Monetary Economics: An Integrated Approach to Credit, Money, Income, Production and Wealth*. London: Palgrave-Macmillan.

Graham, A. (1982). The long wave. Journal of Business Forecasting, 1(5), 69–74.

Graham, A. (2009a). Four grand challenges for system dynamics. In Proceedings of the 2009 International System Dynamics Conference. Albuquerque, NM.

Graham, A. (2009b). Methodological changes needed to meet the world's Grand Challenges. In Proceedings of the 2009 International System Dynamics Conference. Albuquerque, NM.

Graham, A., & Ariza, C. (2003). Dynamic, strategic and hard questions: using optimization to answer a marketing resource allocation question. System Dynamics Review, 19(1), 27–46.

Graham, A., & Godfrey, J. (2003). Achieving Win-Win in a Regulatory Dispute: Managing 3G Competition. In Proceedings of the 2003 International System Dynamics Conference. Albany, NY: System Dynamics Society. Retrieved from http://www.systemdynamics.org/.

Graham, A., Moore, J., & Choi, C. (2002). How robust are conclusions from a complex calibrated model, really? A project management model benchmark using fit-constrained Monte Carlo analysis. Proceedings of the 2002 International System Dynamics Conference, Palermo, Italy.

Graham, A., & Senge, P. (1980). A long-wave hypothesis of innovation. Technological Forecasting and Social Change, 17, 125–142.

Graham, A., Mayo, D., & Pickels, W. (2008a). OSD/PA&E Economic Analysis Study: Conceptual Framework. Cambridge, MA: PA Consulting Group.

Graham, A., Mayo, D., & Pickels, W. (2008b). OSD/PA&E economic analysis study: Conceptual framework and discussion items. Presentation for "Global Economics and Finance Game Design Planning Seminar" 2008. Cambridge, MA: PA Consulting Group.

J8/WAD. (2007). Analysis of Network Enabled Stabilization & Reconstruction Operations. Final deliverable for contract: W74V8H-04-D-0051, December 7, 2007.

Koo, R. (2008). The Holy Grail of Macroeconomics: Lessons from Japan's Great Recession. New York, NY: Wiley.

Kott, A., & Corpac, P. (2007). Technology to assist leaders in planning and executing campaigns in complex operational environments: conflict modeling, planning and outcomes experimentation program (COMPOEX). In Proceedings of the 12th International Command and Control Research and Technology Symposium. Newport, RI. Retrieved from http://www.dodccrp.org/events/12th_ICCRTS/CD/html/presentations/232.pdf.

Krugman, P. (1999). Balance sheets, the transfer problem, and financial crises" in Isard, P., Razin A. & Rose, A. (eds). International Finance and Financial Crises: Essays in honor of Robert P. Flood, Jr. Norwell, Mass.: Kluwer.

Krugman, P., & Obstfeld, M. (2008). International Economics: Theory and Policy (8th ed.). Reading, Mass.: Addison-Wesley.

Krugman, P. (2009). The Return of Depression Economics and the Crisis of 2008. New York, NY: W. W. Norton & Co.

Kubler, F. (2008). Computation of general equilibria (new developments). In Durlauf, S. N., & Blume, L. E., (Eds.), The New Palgrave Dictionary of Economics. Second Edition. New York, NY: Palgrave Macmillan.

Low, G. (1977). Financial market dynamics: an analysis of credit extension and savings allocation (PhD Thesis). Cambridge, MA: Massachusetts Institute of Technology, Sloan School of Management.

Lyneis, J. (1999). System dynamics for strategy: a phased approach. System Dynamics Review, 15(1), 37–70.

Lyneis, J. (2000). System dynamics for market forecasting and structural analysis. System Dynamics Review, 16(1), 3–25.

Mass, N. (1975). Economic Cycles: An Analysis of Underlying Causes. Waltham, MA: Pegasus Communications.

Mayo, D., Callaghan, M., & Dalton, W. (2001). Aiming for restructuring success at London Underground. System Dynamics Review, 17(3), 261–289.

Modigliani, F., & Brumberg, R. (1954). Utility analysis and the consumption function: an interpretation of cross-section data. In Kurihara, K. K., (Ed.), Post-Keynesian Economics. New Brunswick, NJ: Rutgers University Press.

NBER (National Bureau of Economic Research). (2009). Cambridge, MA. Retrived from http://www.nber.org/.

OUSD Policy Forces Transformation and Resources. (2007). System Dynamics Modeling of Stability Operations. Final Deliverables for contract: W74V8H-04-D-0051.

Rahmandad, H., & Sterman, J. (2004). Heterogeneity and Network Structure in the Dynamics of Diffusion: Comparing Agent-Based and Differential Equation Models. Sloan School of Management Working Paper Series ESD-WP-2004-5. Cambridge, MA: Massachusetts Institute of Technology.

Romer, P. (1990). Endogenous technological change. Journal of Political Economy, 98(5), S71–S102. Retrived from http://www.jstor.org/stable/2937632.

Runge, D. (1976). Labor-market dynamics: an analysis of mobility and wages (PhD Thesis). Cambridge, MA: Massachusetts Institute of Technology, Sloan School of Management.

Samuelson, P., & Nordhaus, W. (2004). Economics (18th ed.). New York, NY: McGraw-Hill.

Schweppe, F. (1973). Uncertain Dynamic Systems. Engelwood Cliffs, NJ: Prentice-Hall.

Sen, A. (1985). Commodities and Capabilities. Oxford, UK: Oxford University Press.

Senge, P. (1978). The system dynamics national model investment function: a comparison to the neoclassical investment function (PhD Thesis). Cambridge, MA: Massachusetts Institute of Technology, Sloan School of Management.

Senge, P. (1990). The Fifth Discipline. New York, NY: Doubleday.

Shibuya, H. (2001). Economic takeoff and capital flight". *ESRI Discussion Paper Series* no. 8. Tokyo: Economic and Social Research Institute Cabinet Office, Government of Japan. Available online at http://www.esri.go.jp/en/archive/e_dis/abstract/e_dis008-e.html

Solow, R. (1956). A Contribution to the theory of economic growth. Quarterly Journal of Economics, 70(1), 65–94.

Solow, R. (1957). Technical change and the aggregate production function. Review of Economics and Statistics, 39(3), 312–320.

Sterman, J. D. (1982). The energy transition and the economy: A system dynamics approach (PhD Thesis). Cambridge, MA: Massachusetts Institute of Technology, Sloan School of Management.

Sterman, J. D. (1986). The economic long wave: theory and evidence. System Dynamics Review 2, 87–125.

Sterman, J. D. (1988). A Skeptic's guide to computer models. In Grant, L., (Ed.), Foresight and National Decisions. Pages 133–169. Lanham, MD: University Press of America.

Sterman, J. D. (2000). Business Dynamics: Systems Thinking and Modeling for a Complex World. New York, NY: McGraw-Hill/Irwin.

Stephens, C., Graham, A., & Lyneis, J. (2005). System dynamics modeling in the legal arena: meeting the challenges of expert witness admissibility. System Dynamics Review. 21(2): 95–122.

Taylor, M. (1991). The hyperinflation model of money demand revisited. *Journal of money, Credit and Banking*, 23(3), 327–51. Available through JSTOR at http://ideas.repec.org/a/mcb/jmoncb/v23y1991i3p327-51.html

Theil, H. (1971). *Principles of Econometrics*. New York: John Wiley & Sons.

Transport for London. (2009). Retrived from http://www.tfl.gov.uk/corporate/modesoftransport/1548.aspx.

UN Development Programme. (2009). Human Development Report and Human Development Indices. Retrived from http://hdr.undp.org/en/humandev/hdi/.

University of Maryland. Inforum (2009). Retrieved from http://inforumweb.umd.edu/.

Von Peter, G. (2005). Debt-deflation: concepts and a stylized model. BIS Working Papers, 176. Basel, Switzerland: Bank for International Settlements.

Williams, J. (2007). The Complete Strategist: Being a Primer on the Theory of Games of Strategy. Santa Monica, CA: RAND Corporation.

Chapter 5
Media and Influence

William H. Bennett

The public information media provides information on current events (news), entertainment (programming), and opinions offered by trusted public sources (e.g., business, academic or religious spokespersons, journalists, and government officials). Consequently, it is a major force in shaping a populace's attitudes toward significant social issues and of great interest to intervention planners. The chapter attempts to provide modelers and intervention analysts alike with sufficient understanding of media mechanisms and current research that they can begin contributing to, and benefiting from this important area of study.

The chapter begins by exploring the effects that accrue from information disseminated through public media, and the conceptual mechanisms that may contribute to these effects. The discussion also introduces the terminology needed to characterize media influence.

The chapter then explores the evolution of models for analyzing the influence of media-based communication on public attitudes. It discusses key theories that have been developed to understand how media influences public attitudes and illustrates how media influence theories have evolved in an attempt to keep pace with the expansion of media and its public reach.

Next, the chapter surveys computational models and methods that have been developed to explore media influence. As is stressed, these models – and any current media models – should be viewed as exploratory in purpose. Each was developed to enable controlled, computational experiments to help understand and characterize mechanisms that are thought to contribute to media influence.

The subsequent section provides a detailed look at one model that the author worked on, the Media Influence Model (MIM) (Bennett 2009; Waltz 2008), and chapter concludes with a brief look at analytic cases which illustrate the use of MIM.

W.H. Bennett (✉)
BAE Systems, Inc., 6 New England Executive Park, Burlington, MA 01803, UK
e-mail: william.bennett@baesystems.com

A. Kott and G. Citrenbaum (eds.), *Estimating Impact*,
DOI 10.1007/978-1-4419-6235-5_5, © Springer Science+Business Media, LLC 2010

1   Media Influence Mechanisms and Effects

1.1   Defining Key Concepts

An informed public forms opinions toward political, social, and economic themes using the information it accesses through a *public information environment* (PIE) (Goidel et al. 1997). A modern, sophisticated PIE provides public access to many types of media through various distribution channels. The public's perception of topical issues can be shaped by its preferred access patterns to media and information provided by influential sources.

The media is said to *influence* public opinion whenever its coverage on topical opinions is followed by a noticeable change in public attitude. Although other factors may also contribute to public attitude, it is the *innovation* or change in attitude due to media that is its effect.

Individual public segments evolve access patterns to information sources through preferred media channels. Media outlets produce content to satisfy information needs of public segments with special interest in political, social, or economic themes. This self-sustaining relationship is the foundation of public trust in media.

Media refers to recorded information produced for distribution to an audience. It may be formatted using print, video, text, audio, or other formats. *Media outlets* produce media content for distribution on one or more channels. The information contained in media is said to be in the *public information domain*.

Media is distributed to the public through channels; e.g., newspaper circulation, television or radio broadcast stations, magazines, and websites. A communication channel is a means of transmitting a message from a source (or sender) to an audience (or receiver). Media channels distribute content designed for consumption by a target audience. Many other forms of communication that do not involve distribution of media exist. For example, a conversation between two or more individuals conducted over a telephone produces no legitimate record of the information that is available to the public. It produces no public record. In contrast, a conversation that takes place over a public radio transmission may produce a transcript that falls within the legitimate public information domain. We will consider media to be any information format that is designed to be accessed by the public.

Broadcast media is produced by professional media outlets for transmission to a broad audience. The production format is often carefully tailored to the distribution channel. Media produced for broadcast on television, radio, or newsprint channels is formatted appropriately. Broadcast media is produced to reach a broad audience by containing themes of broad interest and balance that will maintain audience interest and loyalty.

Generally, both media channels and media outlets are managed by a professional enterprise that must survive in a competitive PIE. Whether they are publicly endowed or commercially funded, they survive by maintaining reach to their targeted audiences. A media channel may employ editorial policies that select content and produce media formats tuned to its target audience.

Other forms of media may also contribute to a PIE. The term *gray media* is often used to describe media that is produced for limited distribution to a narrow audience segment. Gray media may not be viewed as credible or interesting to a broad audience. Examples might include bulletins produced for religious, social, or civic organizations for distribution to its membership. Typically, the content is less professionally produced, more culturally biased, and narrowly targeted. Hence, although it is in the public information domain, its potential influence is typically narrower than broadcast media.

Communication theory provides the general principles for studying media influence. In this chapter, we consider several conceptual models derived from communication theories that describe causal mechanisms of media influence. We then illustrate how to build a computational (simulation) model to perform analysis of influence effects within a PIE.

Let us consider a conceptual model (Fig. 1) that represents several mechanisms believed to contribute to media influence. First, an information source (sender) communicates to a public audience (receiver) by issuing a statement (message) to one or more media outlets. Second, a media outlet produces media content that places the source statements in the context of other public information. During production, the media outlet may incorporate an original source statement together with related statements or news reports to provide balanced coverage of the issue. Media production encodes message meaning by referencing a subject and providing rhetorical emphasis (sentiment) through its editorial process. Third, a media channel shapes the message-related information for distribution to its target audience through its selection, placement, and further editorial emphasis. The editorial,

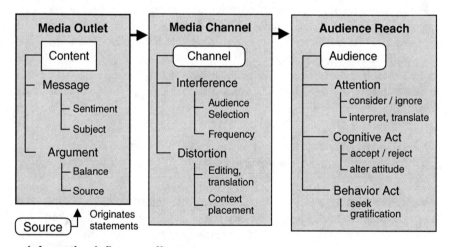

Information Influence effects:
Audience pays attention to message cues
Audience considers message content form strong opinion
Audience segment experiencing dissonance seeks gratification from other information sources

Fig. 1 Media influence communication model

production, and distribution process can either aid in the interpretation of the message or distort the intended meaning. It can also distort original message meaning. A media channel may introduce message interference by editing content presentation or positioning to satisfy its target audience interests. It will adapt distribution frequency within its programming and may place the message adjacent to interfering media content.

An audience that is exposed to media distributed through one or more channels and elects to access the information is said to be *reached*. An audience may then expend some cognitive effort to consider the media content, determine the message, and adopt its position toward the subject. The cognitive process leading to influence is driven by the extent of attention and active consideration given by the audience. The audience cognitive process extracts meaning from the media content. The decoded message (interpreted meaning) may differ from source statement position due to distortion during production and distribution. Audience attitude influence may depend on its acceptance or rejection of the decoded message and its trust in the message source.

An influence effect is evident if the audience changes its attitude toward the message subject in response to the information. Normally, influence must be determined through an independent measurement such as a poll. An influence effect may result in opinion movement toward either agreement or disagreement with the source statement depending on whether the message resonates or conflicts with the audience sentiment toward the subject and the source. Agreement influence can result if alignment of source trust and position are acceptable to the receiver. Disagreement influence may result if either are in conflict with receiver sentiment.

Public segments that rely on information access from multiple channels may experience a range of content emanating from an initial news report. When an issue resonates in a PIE, the expanded coverage can lead to both enhanced coverage and opportunities for message distortion. On one hand, resonance in the PIE provides the public with enhanced access to issue-related information. On the other hand, this mechanism can distort message content and confound the intended influence effect of the message source.

1.2 Media Influence Terminology

Media influence within a PIE can be characterized in terms of variables that represent dynamic changes in media content contributing to public opinion. Public opinion is a variable that represents an aggregated state of opinions contained within a public segment toward a subject. Polls, surveys, and focus group assessments are often conducted to estimate audience segment opinions towards various subjects. An opinion is often measured as a categorical choice (e.g., multiple-choice selection) and is typically characterized as *hard* (strongly held), *soft* (partially formed), or *neutral* (undecided) and as *positive* (supportive) or *negative* (opposing).

A public or audience segment holds a distribution of opinions. *Attitude* is a variable that represents the distribution of opinions held within a public segment toward a subject.

The media influences public opinion by providing access to content. Media content includes messages that intend to inform or persuade. Persuasive messages may intend to shape public behavior by encouraging compliance with guidelines (e.g., "no smoking," "speed limits strictly enforced," or "recycling benefits all"). Persuasive messages may also intend to shape public opinions toward subjects, e.g., "support government officials advocating policies that strengthen family values."

A persuasive message may express intent by encoding meaning directly or indirectly; e.g., using a metaphor. A message carries influence if the receiver can decode the communication and interpret the opinion expressed by its *source* toward the *subject*. A carefully formed message will encode the desired intensity of opinion in its *tone*. Often, message tone is used to gain audience attention by emphasizing sensational aspects of the information. The variability of message *influence* on the formation of public opinion depends on several aspects of the message and its communication that can affect the ability of the public to access, decode, and interpret the message intent. Audience receptivity to message influence often depends on its initial sentiment toward the subject. Audience receptivity can alter the outcome of influence, resulting in attitude trends that may accept, reject, or ignore the message intent.

A message that contains an unambiguous identification of the source – i.e., *source attribution* – is more likely to be viewed as legitimate, and hence improve audience receptivity. A message expressing an opinion by unnamed sources typically will have less influence than a message containing a named source.

Sentiment is the state of existing belief underlying opinions expressed by a public segment in a poll or survey. An opinion can change as a result of exposure to influence and can be measured in polls.

A media channel distributes content associated with themes with varying placement and frequency. *Theme coverage* of a media channel represents the allocation of access and transmission frequency of content expressing opinions about theme subjects. Coverage can be measured per unit of production (e.g., a journal issue) or by unit of time exposure (e.g., per week). A media channel can exhibit a *theme tone* in its coverage depending on the net tone of messages carried about the theme subject. A media channel's strong tone in its coverage of certain themes may indicate bias toward themes' subjects.

Public information access and diversity in a PIE can be characterized in terms of factors and parameters that define the range of subjects, breadth of opinions, and accessibility to the public. A *sophisticated PIE* provides coverage of a broad range of themes and public access to information through a wide range of media channels.

A *public segment* (also called *audience segment* or *target*) is a population group delineated by its demographic profile (e.g., age, education, wealth, political or religious affiliation, etc.), cultural affiliation, or geospatial attributes. One goal of media influence analysis is to identify differences in how opinion formation on common subjects occurs among different public segments. Information campaigns conducted by corporations, political parties, and government agencies often begin by *audience segmentation*. Segmentation seeks to divide public audiences into groups or segments according to demographics or other attributes that help characterize each group's receptivity to information sources, channels, and content.

A *PIE* is often characterized by the persistent themes carried by media channels. A *theme* is a pattern of consistent, recurring emphasis (i.e., coverage) and tone of content expressing opinions toward a set of subjects. A PIE is characterized by the set of all subjects addressed by its persistent media themes. Subjects within a PIE are represented as named entities. Thus, a theme expressing opinion on political legitimacy of an incumbent politician gains influence by focusing an opinion toward the named incumbent.

A *public information source* is a named entity that provides information content within a PIE directed toward one or more public segments. Sources can be actors that contribute content such as public affairs officers, press secretaries, journalists, scholars, authors, charismatic political or social leaders, etc. A PIE is further characterized by the set of its information sources.

A public segment will often develop a dependent relationship with a set of information sources that it views as credible and trustworthy. Information sources tend to tailor content production toward its constituent audience segments. We will refer to such a relationship as a *line of communication* (LOC).

In communication theory, a channel carries a message from a sender to a receiver. We view a PIE as a network of *media channels* that host content from multiple sources and provide access to public segments. A media channel is an information service that is managed to distribute content to certain public segments, viz. subscribers. A media channel can be managed by a corporate enterprise such as a media outlet, news service, or other commercial enterprise, or it can be managed by a cooperative enterprise (e.g., wiki) or by an individual (e.g., weblog). In each case, the entity responsible for the management of the channel exerts some control over content and audience reach. Media channels are managed so as to provide target audiences with access to content from sources and to maintain audience reach. A media channel is said to have *reach* to a public segment if that segment accesses content from the channel. It is a mutually reinforcing relationship that depends on the media channel's ability to provide content that satisfies the audience. A well-managed media channel maintains its reach to its target audience. Hence, we can treat reach as a parameter in the PIE.

Once content is published to the PIE, it is said to be in the public domain, i.e., it is available for redistribution (as long as appropriate attribution policies are followed). The extent to which a media channel tailors its coverage of content in the public domain through editorial selection and rhetorical emphasis is seen as evidence of *media bias*. Often bias (or *spin*) is merely an attribute of how the channel adapts content to its target audience. In a modern PIE, public exposure to content can be so broad that content tailoring reinforces perceptions of bias. The message distortion that occurs when content originating on one channel is repackaged and tailored by adjacent channels can lead to complex reactions in media influence.

The media may distort statements provided by a source in its coverage of themes. Any distortion that alters the message subject or sentiment associated with a media channel or outlet is taken as evidence of bias. One view (Allen 2008) characterizes media bias as purposeful filtering of message transmissions on a public media channel depending on an intrinsic sentiment held by the media channel management.

In the next section, we will review key conceptual models that derive from communication theories of media influence. Among the most important are the concepts of agenda setting, priming, and framing. *Agenda setting* describes the effect of broadcast media on what issues the public addresses in forming opinions (McCombs and Shaw 1972). Media emphasis on themes provides a forcing function for the public to prioritize its consideration of important social, political, or economic issues. This effect is often evident, for example, during political campaigns (Scheufele and Tewksbury 2007). Coverage dominance, placement, and use of peripheral cues are often sufficient to stimulate agenda setting. *Priming* is a related effect in which media content encourages the public to recall aspects of an emerging issue that help the public to reach an informed decision or opinion as the issue is further developed in the media (Goidel et al. 1997). The causal model of priming is a time-sensitive response to media coverage. *Framing* describes the effect of media content on influence. The framing effect is seen when small changes in message presentation lead to significant changes in influence (Chong and Druckman 2007). Another possible goal of public communication is to establish a broad and sustained public understanding and supportive opinion toward a subject entity. In particular, *branding* is a communication strategy to reinforce, sustain, and perhaps extend a desired, positive public opinion held toward a subject. Business marketing practice refers to a brand as the strong association of identity that the public sometimes forms toward a subject. An effective media model should be capable of representing how agenda-setting, priming, message framing, and branding effects contribute to achieving desired objectives and avoiding unintended consequences.

2 Theoretical Underpinnings for Modeling Media Influence

2.1 *Communication Penetration Theory*

Media communication theory has evolved through phases in response to changes in media and public access (McQuail 2005; Perse 2001). As a first step in understanding media influence, the *communication penetration theory (CPT)* (Berlo 1960; Stone et al. 1999), illustrated in Fig. 2, was developed to explain why messages in broadcast media sometimes fail to reach an audience. An audience may fail to pay any attention to media presentation. It may pay attention but reject the message. Even messages containing valuable and important information may be subject to audience negligence

Fig. 2 Communication penetration model

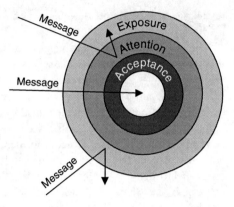

and inattention. The theory evolved to address several questions related to how media influence might fail.

What fraction of media is actually considered by an audience? What factors alter the effectiveness of media to reach certain audiences? Why do some media campaigns fail to get any public attention or consideration? Is the audience unable to understand the message or just unwilling to allocate any effort to consider it? An audience might be overexposed to media and unable to make collective rational decisions on how to allocate attention. CPT offered little understanding of how public attention might be allocated or how to assess media presentation effectiveness in gaining attention other than by elevating rhetorical tone and saturating media coverage.

CPT offers an early, conceptual model of public communication in which the public is a passive and somewhat inattentive receiver of information (Berlo 1960). In this model, the public is viewed as overexposed to media and unprepared to receive and process the information. The capacity of the public to attend to and consider any specific message carried in the media is limited. CPT recognizes that not all media messages reach the desired public audience. Strategies for message reach emphasize placement, frequency of coverage and rhetorical emphasis. Placement may elevate the likelihood that the public will pay attention to the message; e.g., front-page news. Rhetorical tone and emphasis can often affect public priority for selection by appealing to audience strongly held beliefs to gain acceptance.

2.2 Source–Message–Channel–Receiver Model

Another early attempt to understand media influence described the process by which an audience interprets and understands a message contained in media that it has decided to consider. What characteristics of media messages contribute to influence? What are the factors that may impede the effectiveness of media as a form of communication?

Fig. 3 SMCR model of
persuasive communication

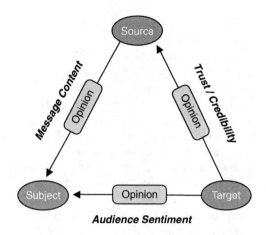

Media influence can be understood as a form of persuasive communication
(Stone et al. 1999). The *source–message–channel–receiver (SMCR)* model of Berlo
et al. (1969) provides a conceptual representation of communication from a sender
to a receiver over a channel. In this conceptual model, a source attempts to send a
message to an audience by producing media that contains the message. Production
encodes meaning in the media using language, presentation, style, and rhetorical
tone. The media may use a blend of stylized text, prose, images, or audiovisual
content to encode the message. The media is transmitted through a channel that
may alter the content that is received by the audience. In this model, the audience
receives and considers the media. The receiver attempts to decode the message
contained in the media by interpreting its meaning. Any difference between the
message as understood by the receiver and the sender constitutes a *communication
channel distortion effect*. SMCR theory posits that the risk of message distortion is
mediated by the common understanding or *coorientation* between the sender and
receiver toward the message subject. Coorientation can be established through the
media content by appealing to common references or by placement of the media
message within a context that aids in establishing a common understanding.
Figure 3 illustrates the SMCR model (Stone et al. 1999; Berlo et al. 1969).

2.3   Opinion Leadership Theory

In public communication, a target audience is a heterogeneous group that is likely
to have a range of sentiment, interest in media communications, and ability to
interpret content. One early attempt to understand the receptivity of a public audience
to a media message examined the difference between social reactions to a message
that can lead to influence. What audience characteristics contribute to the development
and propagation of influence? What audience characteristics contribute to the

differences in receptivity between two audiences? When might the same media create different influences in each of two or more separate audiences?

Opinion leadership theory (OPT) (Katz and Lazarsfeld 1955) describes how influence trends develop and propagate within an audience. Opinion leaders or *gatekeepers* represent the fraction of the audience that adopt the media message influence and elect to promote the innovation within the society. Gatekeepers are often journalists, political activists, or community leaders who have both strong social ties and access to communication channels that reach the public. Gatekeepers contribute to the development of trends within an audience by promoting the message through social relationships that we refer to as an LOC.

Chaiken (1987) studied how public audiences consider and form opinions in response to media influence. The study explored how information consideration is complicated by the pace of modern life and its cognitive load. The resulting model asserted that little time is reserved for consideration of public issues, and a mix of mindful and mindless responses to debate on public issues characterizes public receptivity to media. Different public segments can be more or less receptive to media influence depending on their cognitive capacity to absorb information. Such demographic factors as literacy, education, age, health, wealth, etc., can provide indicators for the psychocultural capacity and desire to process information.

Audience receptivity to media influence can depend on access to gatekeepers. This theory explains the propagation time lag from audience exposure to media content and noticeable change in public attitude. Demographic factors such as education, age, health, etc., have been associated with the prevalence of gatekeepers within an audience and hence the audience's receptivity to attitude innovation. More recently, the diffusion of innovation business model (Rogers 2003) has been developed to describe the process by which early adopters contribute to mass acceptance of innovative commercial products. In OPT, the influence of media is an innovation or change in public attitude.

2.4   Social Judgment Theory

Social influence theory (SIT) (Latané 1981) has evolved to explain the mechanisms by which social interaction within an audience contributes to the change in attitudes. Social judgment theory (SJT) (Sherif et al. 1965) describes the social process of assimilation or rejection between social communities that hold differing attitudes. SJT has evolved to explain the mechanisms leading to either consensus or polarization of opinion within a social community. Initially, both SIT and SJT evolved separately from media influence theory. This line of theoretical work addresses how social structures affect the formation of public attitudes. What attributes of social structure may lead to either consensus or polarization in attitude?

A public community within a nation or a region may be composed of separate cultural segments of the public that access information through a common PIE. Formation and propagation of public opinions can change in response to cross-cultural interaction between public segments that occur through the PIE and the exogenous influence of media. SIT describes social influence as the pressure to alter the attitude of social entities (i.e., individuals or groups) exerted by other social entities. Influence is communicated through social relationships that are either direct or indirect. A direct relationship is an immediate social contact or communication between two entities. An indirect relationship occurs when influence is propagated between two entities that have no direct social relationship through intermediaries.

Weisbuch et al. (2002) extended SIT by adding an agent behavior model of information sharing and decision-making. The theory describes how influence propagates in a social network of communicating agents and explains the phenomenon of opinion polarization in response to influence within a heterogeneous population. In SIT, an agent attitude represents an absolute (binary) decision.

SJT (Jager and Amblard 2004) further develops SIT by describing attitude influence as a decision with several levels of confidence in response to message sentiment. In SJT, an agent that receives a message can elect to assimilate, contest, or defer a commitment to the influence. The innovation decision response of the receiving agent depends on how confidently it agrees or disagrees with the message sentiment. If its agreement is within the agent latitude of indifference, then the message influence is assimilated. If the agreement exceeds the latitude of rejection, then the influence is to contest the influence and reinforce an opposing view. If the agreement exceeds the latitude of indifference but does not exceed the latitude of rejection, then SJT predicts that the receiver will not commit to any influence.

2.5   Media Agenda-Setting Theory

Media sets the agenda for public consideration of issues through its emphasis, placement, and rhetorical tone applied to its coverage of important issues. *Agenda-setting theory* (AST) (McCombs and Shaw 1972) posits that mass media influences public opinion formation by driving the public consideration of issues. They argued that the effect of media was less in telling people what to think and more in telling people what to think *about*. During an election campaign, broadcast media will provide coverage that attempts to elevate and focus public discourse, debate, and consideration of related issues. The agenda-setting effect is then to draw the public's attention toward developing well-informed opinions.

How does the public decide how to allocate cognitive capabilities in utilizing the information provided in media? What issues or themes receive priority in public consumption of media information, and how is this agenda for consideration affected by the emphasis placed in media presentation? AST suggests that the public agenda for consideration of issues is driven directly by media presentation. This theory suggests that the selections made in media presentation drive penetration. It reaffirms the passive nature of audience participation in public communication through media.

Agenda-setting effects have been seen as a significant influence on the outcome of political campaigns in the past (Scheufele and Tewksbury 2007). Recent expansion in media access and growth in diversity and sophistication of the public audience has called into question the influence of broadcast media in agenda setting. Nevertheless, in a less sophisticated PIE the agenda-setting effect can still be a significant driver in influencing public opinion. To what extent agenda setting may extend to a modern PIE involving active public access to information through other, nontraditional media channels such as internet remains an open question.

2.6 Media Priming and Framing Effects

Priming and framing theories were developed to help understand how the audience predisposition toward subject and presentation of a message can affect media influence. Why are certain presentations of a media message more effective in one audience than in another? Why does a small change in the presentation of a media message often lead to significant change in influence? Both priming and framing theories were developed to understand how an audience participates in acceptance or rejection of media influence.

Priming theory (Goidel et al. 1997) posits that the public forms attitudes by drawing on those elements of information that are most accessible at the time. It is a psychosocial theory that describes the cognitive process of decision-making as derived from an aggregate of multiple lines of consideration aligned with an audience belief structure. This model is an application of expectancy value theory (Ajzen and Fishbein 1980). Here, a decision is said to be derived by combining independent lines of consideration according to the value assigned to each.

Media can influence public belief structure by incrementally building arguments along separate lines of consideration in presenting information on an issue. *Priming* is a media effect that activates public awareness of selected information elements that can be used to form an opinion on some issue. It functions by influencing public sentiment along individual lines of consideration that relate to social, economic, or political beliefs toward an issue. Priming effects are more likely to be evident in a sophisticated audience that actively participates in a PIE (Goidel et al. 1997). Both priming and agenda-setting effects in a PIE are driven by the change in theme coverage, message focus, and distribution to audience segments over time.

Framing is a media effect (Chong and Druckman 2007) that is evident when the same message, when produced in a media in two slightly different ways, results in significantly different influence effects. For example, an opinion in support of a new economic policy may be offered by espousing the benefits of the policy. In a second rendition, the same message of support states that the new policy will reduce negative factors such as current unemployment. *Framing theory* describes how frames that appeal to negative aspects are generally more effective than frames with positive appeal (Chong and Druckman 2007). Framing arguments used in media often draw relationships between issues and attributes of current audience appeal. Framing uses language references to build these references. Framing is often associated with the "spin" employed in media that associates new references to the argument or presentation of an issue. Framing effect can shift emphasis between audience lines of consideration and thereby affect influence.

One example of the impact of media framing is the use of "night letters" by the Taliban to influence the political debate in Afghanistan from 2003 to 2006 (Johnson 2007). These strategically posted written communications often argue the illegitimacy of the Karzai government by glorifying the long history of struggle against invaders and occupying forces in Afghanistan. This argument frames public political consideration toward Afghanistan government legitimacy by drawing attention to the international support for the Afghan national government as a negative reference.

Priming and framing effects often interact within a PIE. One way to understand this interaction is to appeal to the concept of expectancy value (Ajzen and Fishbein 1980). In this concept model, the attitude an entity forms toward an issue is a summary of a set of component beliefs held toward the subject. An attitude is derived as a summary of individual lines of consideration aligned with beliefs. Each consideration (or belief) has a sentiment with a strength and valence. The summary attitude is weighted by a salience factor applied to each consideration. A priming effect alters the belief structure by influencing component sentiments. Hence, priming builds memory over time that the public can access to form opinions. Framing (Chong and Druckman 2007) affects the process but through a slightly different mechanism. A framing argument emphasizes relationships between lines of consideration and an issue. A framing effect influences the salience factors that prioritize the contribution along lines of consideration to the attitude influence.

In summary, priming effects influence belief structure (and hence knowledge contained) along the existing lines of consideration. Framing effects influence salience factors used to derive an attitude by building the importance of consideration dimensions into the argument. Both priming and framing effects are seen when a sophisticated audience considers arguments expressed in media content rather than merely reacting to peripheral cues that are used in the media to grab the audience's attention. While agenda-setting theory describes how media effects what issues people think about, priming and framing theories describe how media shapes what people think about issues (Perse 2001, Scheufele and Tewksbury 2007).[1]

2.7 Elaboration Likelihood Theory of Persuasion

Producing media content involves making choices in how to present a message to achieve an effect within a target audience. We have discussed how priming and framing effects may be understood as contributing to how media influences what the public can think about issues. How should we characterize the content of media to achieve these effects? What factors in media message presentation contribute to audience appeal? What factors in media presentation affect how long an attitude change might persist? How might media content mitigate the risk of public inattention to an important message?

---

[1] Scheufele and Tewksbury (2007) provide an insightful comparison of three aspects of media impact: *agenda-setting*, *priming*, and *framing*.

Elaboration likelihood theory (ELT) (Petty and Cacioppo 1986) describes a conceptual model for how an audience processes information obtained through media and its impact on attitude change. ELT posits that an audience can follow one of two cognitive paths in responding to a media communication. In the first path, a receiver chooses to consider carefully the arguments contained in the media presentation. Thoughtful elaboration of the merits presented in support of an opinion leads to a strong, informed basis for attitude change. This path is deemed the *central path*. In contrast, a receiver may elect to limit cognitive processing to consider only the peripheral items in the media presentation of a particular message. Any resulting change in attitude is obtained not with thoughtful consideration of the merits of an argument but rather in response to one or more peripheral cues (e.g., references to an attractive source, simple anecdote, testimonial, or other reference).

The second or *peripheral path* offers an often expedient mechanism for processing information that demands less cognitive effort and can often appeal to an audience that is poorly prepared to consider an elaborate argument. Instead, the attitude change may be induced in response to indirect references that help the audience reach a satisfactory attitude. ELT posits that the effect of media influence results from a combination of both paths that coexist in an audience according to its ability to comprehend the argument contained in the message and its resonance with peripheral cues. The theory describes factors in the message presentation and the audience cognitive state that contribute to the *elaboration likelihood*. A media presentation that is designed to stimulate high elaboration likelihood anticipates that the audience will expend cognitive effort in considering its content. Educational material is often presented using high elaboration style of presentation. In contrast, a message that is designed to achieve a quick impact through short exposure may rely on the use of peripheral cues: testimonials, iconic references, or framing devices that target audience appeal. The theory posits that an influence that is achieved through the central path is likely to persist longer than an influence achieved through the peripheral path.

ELT describes a set of three qualitative factors that characterize media presentation: argument quality, peripheral cues, and attitude. These factors can be used to assess media content potential influence on a target audience. The target audience elaboration likelihood depends on its cognitive ability, subject knowledge, and distraction. Rucker and Petty (2006) used ELT to define a process to create effective (i.e., influential) media presentations.

3 Methods for Modeling Information Influence

3.1 Overview of Models and Methods

We now review several computational models described in the literature and highlight the extent to which these models represent the theories and media effects described in the previous sections. We start by summarizing the underlying theories in Table 1. Table 2 summarizes how exploratory models described in this section represent the theories.

Table 1 Communication theories underpinning media influence

Theory/effect	Hypothesized mechanism or effect
Communication penetration theory (CPT)	Messages contained in media will reach only a limited fraction of public with access to media channels
Source–message–channel–receiver theory (SMCR) (Berlo et al. 1969)	A media message that is received by an audience may be accepted or rejected depending on its agreement with message sentiment and trust in message source
Opinion leadership theory (OLT) (Katz and Lazarsfeld 1955)	Trends in opinion formation are led by a small fraction of gatekeepers within a public segment. Gatekeepers act to select and reinforce media influence within a public segment. Gatekeepers establish receptivity to influence within a public segment
Social judgment theory (SJT) (Jager and Amblard 2004)	Opinion formation within the public is driven by the social interactions between public segments
	An influence on public segment attitude represents collective confidence and strength of opinion held within a social group
Agenda-setting theory (AST) (McCombs and Shaw 1972)	Media coverage of themes encourages public consideration leading to attitude influence within the public
Priming theory (PT) (Goidel et al. 1997)	Media coverage of topics can inform the public and prepare it to reach informed decisions
Framing theory (FT) (Chong and Druckman 2007)	Frames of reference used in form arguments contained in media can alter public influence
Elaboration likelihood theory (ELT) (Petty and Cacioppo 1986)	Both argument quality and use of peripheral cues contained in media message content can influence the acceptance and retention of influence in a public audience depending on its ability to consider argument details. Public attitudes formed through elaborate consideration of media content will persist longer than attitudes formed in reaction to peripheral cues

Table 2 Relations between theories and selected computational models of media influence

Computational model	Influence theories							
	CPT	SMCR	OLT	SJT	AST	PT	FT	ELT
Rational choice model (Weisbuch et al. 2002)			X	X				
Social judgment model (Jager and Amblard 2004)			X	X				
Public education and broadcasting model (Gonzalez-Avella et al. 2005)	X	X	X	X				
Media influence model (Bennett 2009)	X	X	X	X	X	X	X	
Elaboration likelihood model (Mosler et al. 2001)				X	X			X

3.2 Rational Choice Model

If SIT suggests that the attitude of individuals tends to become more similar due to social interactions, then why do not we observe consensus more commonly? The *rational choice model* (RCM) (Weisbuch et al. 2002) is a computational model of social opinion dynamics that was developed to explore the consequences of SIT using a simulation. It represents the influence of social relationships on attitude distribution within a population. The purpose of this model is to explore the conditions leading to either diversity or uniformity of opinion in a social system comprised of agents that interact by exchanging opinions with adjacent social agents within a social network. An agent opinion is represented as a binary, rational choice of agreement or disagreement with adjacent entity influence. The model represents opinion leaders as highly connected entities of the social network.

RCM is implemented as a *multiagent simulation* (MAS) model that represents the behavior of a larger number of social entities in response to interaction events. RCM agent behavior represents how people dynamically adapt their attitude by exchanging information with others. An agent with limited personal knowledge of a subject may rely on opinions already formed by other agents to adopt its attitude; i.e., it exhibits bounded rationality. An agent may also be encouraged to adopt an opinion it perceives as common to the majority of a social group; i.e., it responds to external influences. RCM agent behavior models both bounded rationality and external influences (Weisbuch et al. 2002) using an SIT framework. It explores the impact of factors involving the strength and distribution of opinion and social identity on attitude movement toward either consensus or polarization in a social organization.

An RCM instance is constructed by representing a population as a multiagent system. Agent behaviors are defined by a decision threshold parameter. The model is initialized by setting the attitudes of all agents and the social network links. At each time update agent attitudes are updated and shared with other agents through their social network links. RCM is executed recursively to simulate the dynamic evolution toward opinion consensus or polarization. Typical output identifies the communities of common opinion within the population after a sufficient number of simulation iterations needed to reach a steady-state opinion distribution.

3.3 Social Judgment Model

Social judgment model (SJM) (Jager and Amblard 2004) is an agent-based model that extends the rational choice model by describing an explicit agent message assimilation behavior in response to both pejorative and ameliorative information exchanges about opinions. The purpose of this model is to explore the mechanisms of attitude formation consistent with SJT.

The model uses a multiagent representation in which agents interact within a fixed, regular lattice structure to develop social affiliations based on the impact of common judgments represented as commonality of opinions. Agent interactions represent information sharing between adjacent agents. There is no explicit representation of communication channels.

SJM is implemented as a MAS and instantiated similar to RCM. SJM introduces a continuous variable for each agent representing the strength or confidence of opinion. It requires additional data to establish the attitude decision thresholds of each agent consistent with SJT. SJM produces outputs that are similar to RCM but can extend the analysis to determine the strength of opinion within population segments once steady state is reached. SJM has been shown to produce different consensus and polarization community results than RCM under similar conditions due to the representation of opinion confidence (Jager and Amblard 2004).

3.4 Public Education and Broadcasting Model

The *public education and broadcasting model* (PEBM) (Gonzalez-Avella et al. 2007) is an extension of a model of cultural diversity developed by Axelrod (1997) that represents media influence. PEBM was developed to explore the emergence of communities of common culture under the influence of media communication feedback into the social process. The model uses an MAS that extends the Axelrod model by incorporating media communication feedback.

Axelrod (1997) represents culture as a set of attributes (e.g., cultural values or traits) that are influenced by social interactions. An agent culture state describes the trait assigned to each of its cultural features. Each trait represents an agent attitude toward the cultural feature and can take on one of a finite set of values. Agents with common traits are said to have cultural overlap. A pair of agents can interact according to a probability that increases in proportion to the degree of cultural overlap. Cultural overlap between a pair of agents is computed from the number of common agent cultural feature traits.

PEBM (Gonzalez-Avella et al. 2005) extends the culture model by incorporating a separate agent that represents media. The media agent shares information on culture state among other agents using one of two interaction mechanisms: direct and indirect media influence. Direct media influence represents the impact of information originated by the media that disseminates global information on culture to all agents. Indirect media influence represents feedback from the social entities through media that provide local feedback within communities with a common culture.

The purpose of this model is to explore and compare the impact of local (intercultural) and global (cross-cultural) information sharing on cultural diversity. This model represents an early attempt to explore the implications of media information coverage on cultural diversity. This model treats media as an unbiased agent and neglects communication penetration effects that may limit media reach.

PEBM is instantiated as an MAS with considerations for input data that are similar to both RCM and SJM. PEBM introduces media communication as a new type of agent that can be configured to represent either direct or indirect communication influence. Output data obtained includes the distribution of attitudes of all agents representing the population.

3.5 Elaboration Likelihood Model

In Mosler et al. (2001), a computer-simulation model of ELT is developed. This model formulates a computational model that estimates the elaboration likelihood of a receiving agent that is exposed to a message. The model characterizes a message using variables to represent source attitude, source argument quality, and use of peripheral cues. The model computes the elaboration likelihood and determines the contributions to change in attitude from the central and peripheral paths in accordance with ELT.

In Mosler et al. (2001), the model is instantiated as a two-agent communication interaction through a noisy channel that introduces random distortion to the messages transmitted by each agent. The simulated communication interaction between agents operates as follows. One agent initiates a message offering an opinion with a mix of argument quality and use of peripheral cues. The second agent receives the message, determines its elaboration likelihood, and updates its attitude and argument knowledge. The second agent then issues a message to the first agent using its newly formed argument knowledge and attitude. Messages are transmitted through a noisy channel that perturbs the variables that describe attitude and argument quality. The cycle of bidirectional communication is updated recursively.

This form of simulation model introduces a more complex, psychosocial behavior for each agent and an uncertain result due to a noisy channel. Analysis (Mosler et al. 2001) explores the process leading to consensus or polarization of attitudes as influenced by attributes of the message and audience elaboration behavior. This type of model offers analytical advantages that may aid in developing communication strategies. However, it is typically more difficult to obtain data from opinion polling surveys that can address the attributes that contribute to elaboration ability as described in ELT.

3.6 Media Influence Model

In Bennett (2009), we described a simulation model of media influence that was developed to analyze the effect of persuasive media messages on public attitude change. The model represents the impact of media outlets on message distortion and the dissemination of media through channels that reach certain public audience segments. Media message content is represented as expressing opinion toward a subject. The message contains sentiment and source attribution. Media effects are computed as media channel sentiment on each of several issues or themes, source statement sentiment, and public segment attitude. We discuss this model in more detail in the next section.

4   A Computational Model of Media Influence

Having reviewed briefly several computational models of media, let us take a detailed look at one particular computational model, a *media influence model* (MIM) (Bennett 2009) that represents major media effects in a PIE.

MIM is a part of a planning and analysis tool (Waltz 2008), a suite of computational models designed to represent political, social, military, economic, and information influence effects (Waltz 2009). The MIM employs a hybrid, computational modeling approach that blends MAS of communications with a system-dynamic model of media influence. This approach represents the subjects, sources, audience targets, and media channels that comprise a PIE. It represents the causal flow of influence from source statements that are issued in a PIE to the change in attitude toward subjects and confidence in sources.

Figure 4 illustrates the causal flow represented by MIM. One or more media outlets pick up statements issued by sources that offer opinions toward subject actors. An outlet produces a media message that frames the statement toward the subject by expressing sentiment along social, political, or economic lines of consideration. The message may include source attribution that is either explicit or indirect (e.g., unnamed sources). Media channels select distribution to audiences and placement given competing media content and audience interest. When a message reaches an audience, it may elect to accept or reject the message it extracts from media depending on its existing attitude and source confidence. Accepted influence can result in an attitude innovation that causes a trend to emerge in the audience. A sustained trend will result in a change in audience attitude. A similar causal loop represents how message influence can alter confidence in a source. The four causal loops identified in the figure are described in this section.

4.1   Media Themes

AST (McCombs and Shaw 1972) suggests that mass media plays an important role in setting the agenda for public debate by elevating attention to certain salient themes. For example, media outlets may strive to inform the public about the function and performance of government, industry and civic leadership by providing content that draws attention to the statements and actions of selected leaders. Media

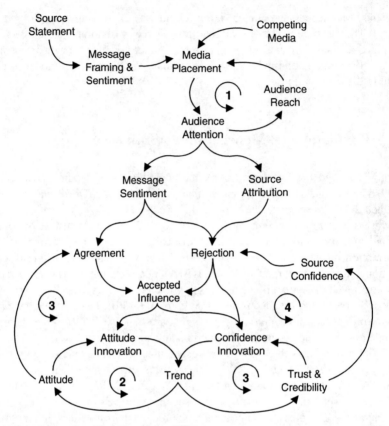

Fig. 4 Causal flow of media influence in a PIE

content emphasizes themes by elevating attention to positions expressed toward subject actors.

A *media theme* (or issue) is a recurring, pervasive, and general category of statements and messages expressing a common position toward one or more subjects that are identifiable to the public. For example, a theme expressing concern for personal security in a region draws attention to the performance of a police or military organization and may raise doubts about its capabilities or performance. A media message is an expression of opinion toward a subject actor having an established identity within the PIE. Media coverage of a theme subject is said to be *on-message* if the net sentiment and framing is consistent with the theme. Professional public communicators often strive to maintain media *voice share*, defined as the fraction of on-message coverage of their desired theme. MIM models a theme as a statement of opinion toward a subject entity. Next, we examine how to represent message content using a computational model.

4.2 Message Sentiment and Framing

MIM represents the sentiment contained in a statement using a five-point scale of intensity. A statement expressing an opinion toward a subject can take one of five possible positions: *Hard Opposition (HO), Soft Opposition (SO), Undecided (UD), Soft Support (SO),* or *Hard Support (HS)*. Media content can contain many statements, references, and cues that contribute to its net sentiment. MIM represents media sentiment as variable taking values on a continuous *attitude scale* as shown in Table 3.

The sentiment an audience holds toward a subject can also be positioned on the attitude scale. An opinion poll could determine the distribution of opinion within an audience. Attitude scale represents consensus of opinion (i.e., strength) that a public segment holds toward a subject. The attitude value represents the opinion intensity and degree of audience consensus.

Integer attitude values in MIM represent opinion consensus as shown in Table 3. Nonintegral attitude values represent balance between audience opinions as shown in Fig. 5. For example, an attitude value of 1.5 implies that 50% of the population group holds an opinion of *SO* while the remaining 50% hold an opinion of *HS* toward the subject. In a similar manner, the attitude expressed in media content represents the balance in statements contained in an article or publication.

MIM represents message framing by expressing sentiment along three independent *lines of consideration* called *legitimacy, affinity,* and *competency*. A statement that argues support for the legitimacy of a subject entity is appealing to a political line of consideration of an audience. Even though an audience may consider an actor a legitimate political candidate, that does not mean they will vote for his election.

Other lines of consideration (Chong and Druckman 2007) often contribute to the net sentiment an audience holds or is willing to express in a poll. An argument may appeal to social affinity of a subject entity by expressing sentiment toward the alignment or social affiliation of the subject with friendly or opposing social groups. Sentiment expressed along the affinity line of consideration attempts to label the

Table 3 MIM opinion and attitude scale

	Hard opposition (HO)	Soft opposition (SO)	Undecided (UD)	Soft support (SS)	Hard support (HS)
Opinion					
Attitude	−2	−1	0	+1	+2

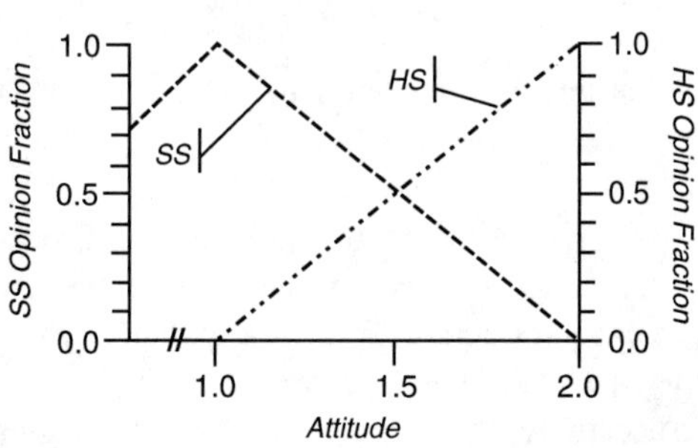

Fig. 5 Attitude representation of opinion consensus

subject as either a trusted ally or an adversary. Even if a subject entity is deemed a friend, he may not be viewed as competent and hence worthy of strong sentiment. Likewise, an antagonist with limited capability to threaten a person is not likely to be the object of strong opposing sentiment. An argument may then also appeal to considerations of the *competency* of a subject entity in expressing sentiment.

A message in MIM is also represented as containing its source attribution. Source attribution can be explicit (e.g., a named entity is quoted in the message statements) or indirect (e.g., an unnamed source is quoted as offering the opinion).

4.3 Communication Penetration

CPT describes how reach is affected by the intensity, framing, and placement of a media message. In any given PIE, the competition between themes, media channel coverage, and information sources sets the threshold for voice share needed to penetrate and gain audience attention. MIM represents communication penetration by comparing media sentiment and audience sentiment for all active subjects associated to issues in PIE. Now let us consider the computational model in more detail in the following development.

MIM computes the attitude $a(P,S) = \{a_c\}_c \in_C$ of a public group P toward a subject S along three independent *lines of consideration* (see discussion on expectancy value theory in Ajzen and Fishbein (1980) and discussion in Chong and Druckman (2007) on its interpretation in framing theory) or *framing contexts*, $C = \{legitimacy, affinity, competency\}$. Each attitude element can take a value in range of numbers $[-2, +2]$. The sentiment of media content is represented as an attitude $a(T,S)$ expressed by a source (transmitter) T toward subject S. A public segment P also holds an attitude $a(P,T)$ that represents its confidence toward the source T. Figure 6 illustrates how message and audience sentiment contribute to penetration. The source's intended message, $a(T,S)$, is the sentiment expressed by the source T toward the subject S.

A message can gain audience attention when the tone of media content is sufficiently strong relative to competing messages in the PIE. The tone of media content is the dominant sentiment expressed toward a message subject along all lines of consideration,

$$\|a(T,S)\| = \max \left\{ c \in C : |a_c(T,S)| \right\} \tag{1}$$

Following SJT, a message whose tone exceeds a *latitude of indifference*,

$$\|a(T,S)\| \geq d_1 \tag{2}$$

can overcome the competition for attention and gain audience attention.

CPT describes how a message's intended influence can differ from audience-accepted influence. When an audience pays attention to and properly interprets media content, it extracts an interpretation of the message to consider. It can choose to accept, reject, or ignore the interpreted message. In SJT, the decision is based on

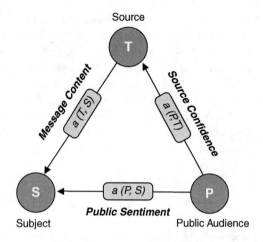

Fig. 6 MIM representation of source-subject-public attitude

the size of the *message innovation* (difference between interpreted message and audience sentiment),

$$e_m(P,S) = a(T,S) - a(P,S) \tag{3}$$

An audience exposed to media messages can completely accept the intended message influence whenever the innovation satisfies a *latitude of acceptance*,

$$\|e_m(P,S)\| \leq d_A, \tag{4}$$

and completely ignore the intended influence whenever the innovation exceeds a *latitude of rejection*,

$$\|e_m(P,S)\| > d_R. \tag{5}$$

Whenever message innovation exceeds the latitude of acceptance but does not violate the latitude of rejection, the accepted influence will be proportionally diluted. MIM computes a proportional message accepted influence as

$$I_A(P,S) = c_A(e_m(P,S))e_m(P,S) + a(P,S) \tag{6}$$

where the *coefficient of partial acceptance* is computed as

$$c_A(e) = \begin{cases} 1, & \|e\| \leq d_A \\ (\|e\| - d_R)(d_A - d_R), & d_A < \|e\| \leq d_R \\ 0, & \|e\| > d_R \end{cases}$$

4.4 Message Rejection and Source Confidence

Source attribution contained in a message has an effect on audience-accepted influence. If an audience has confidence in a source, then it is more likely to accept its influence. On the other hand, if an audience distrusts a source, it is more likely to reject the influence and may adopt an opinion contrary to the message content. Let us explore a computational model that represents how message acceptance and rejection is related to source attribution. The following model implements SMCR theory to estimate the impact of source attribution. It represents confidence as dependent on the attitude held by the audience toward the source.

An audience segment P derives its confidence in source T from the trust and credibility it holds toward the source. Let *trust* $\alpha_T(P,T)$ and *credibility* $\alpha_C(P,T)$ be defined as

$$\alpha_T(P,T) = (1 - C_{TC})a_{\text{Affinity}}(P,T) + C_{TC}\alpha_{\text{Competency}}(P,T)$$
$$\alpha_C(P,T) = (1 - C_{CC})a_{\text{Legitimacy}}(P,T) + C_{CC}\alpha_{\text{Competency}}(P,T)$$
(7)

where C_{TC} is a coefficient of trust dependence on competency and C_{CC} is a coefficient of credibility dependence on competency.

The confidence P holds toward T is a combination of its trust and competency

$$\alpha_S(P,T) = (1 - C_{SC})\alpha_T(P,T) + C_{SC}\alpha_C(P,T)$$
(8)

where C_{SC} is a coefficient of credibility importance to source confidence.

MIM computes accepted message influence due to source confidence by refining (6) as

$$I_A(P,S) = c_R(\alpha_S)c_A(e_m(P,S))e_m(P,S) + a(P,S)$$
(9)

where the *coefficient of message rejection*, c_R, depends on the audience confidence in source α_S:

$$c_R(\alpha_S) = \begin{cases} 1, & \text{if } \alpha_S \geq d_\alpha \\ \dfrac{\alpha_S - d_\rho}{d_\alpha - d_\rho}, & \text{if } d_\rho < \alpha_S < d_\alpha \\ \dfrac{\alpha_S - d_\rho}{d_\rho + d_\rho}, & \text{if } -d_\alpha < \alpha_S < d_\rho \\ -1, & \text{if } \alpha_S \leq -d_\alpha \end{cases}$$
(10)

Source confidence is reinforced when messages are accepted and diminished when message influence is rejected. Thus, message acceptance influences source confidence dynamically. The amount of confidence innovation depends on the magnitude of accepted innovation for the message and impacts confidence toward source. MIM computes influence toward source confidence as:

$$I_S(T,S) = c_R(\alpha_S)|I_A(P,S)|$$
(11)

4.5 Attitude Influence

Public opinion formation evolves over time under the influence of media and social interactions. MIM is designed to explore the causal mechanisms that explain time lag between media coverage and emergence of trends and shifts in public attitude. It represents the dynamic interactions that explain attitude intransigence (unwillingness to change), attitude retention (ability to retain attitude change after media coverage ceases), and influence receptivity (ability to transfer influence acceptance to strongly held attitude that can be verified in public polls and surveys).

MIM uses an MAS to represent a public response to media influence that influences public perception of themes. It extends agent behavior of SJT (Jager and Amblard 2004) to represent the time evolution of attitude change due to audience resistance. MIM computes attitude and trend change at each time update for each audience agent toward each subject agent.

To illustrate the agent attitude model behavior, let us consider a single audience segment and subject. Let $a(t)$ represent the attitude of the audience segment toward the subject at time t. Let $I_a(t)$ represent the accepted influence due to exposure to media content. OPT (Berlo 1960) suggests that the dynamics of opinion change are driven by trends that originate from opinion leaders. Let $\tau(t)$ be the trend at time t, representing the amount of attitude change within the population group over a time interval T_s (say, 1 week). For example, a trend value of 0.1 represents an increasing attitude movement within 10% of the audience since the last update.

Trends develop in proportion to the innovation contained in accepted influence. As accepted influence agrees with existing sentiment, a media fatigue effect sets in and the trend will diminish. The magnitude of a trend is modulated by audience *receptivity R* (also called *coefficient of resistance* (Schumann et al. 1990)). Audience receptivity represents its learning and retention ability. Demographic factors such as age, level of education, literacy, affluence, and culture can affect audience receptivity.

Here, we simplify notation and drop the explicit representation of attitude source, subject, or target. The accepted innovation at time t obtained from a message is

$$e_m(t) = I_A(t) - a(t) \tag{12}$$

and the accepted innovation toward the source confidence is

$$e_s(t) = I_s(t) - \alpha_S(t) \tag{13}$$

MIM attitude propagation model updates the trend and attitude at each simulation time step according to the following pair of equations

$$\tau(t+1) = R_s e_s(t) + R_m e_m(t)$$
$$a(t+1) = a(t) + T_s \tau(t) \tag{14}$$

which are updated for all audience attitudes toward all subjects in response to each influence innovation.

Table 4 summarizes MIM parameters and variables that are updated using (8) and (11)–(14) at each simulation time step.

The model extends the behavior of SJM agents (Jager and Amblard 2004) while retaining two important properties. It is both causal and well posed, i.e., the mathematical properties are qualitatively consistent with the evolution of trends and attitudes in public sentiment analysis. It is scalable to a desired granularity of population segmentation; i.e., any number of source, subject, and target entities can be represented.

Affecting opinion change in a PIE can often be limited by audience exposure to dissonant information content representing conflicting, indecisive, or weakly expressed attitudes. Whenever accepted influence is weak, the trend will be negligible.

Audiences tend toward intransigence in opinion change when confronted with dissonant or indecisive information. MIM represents audience attitude intransigence by augmenting the evolution (13) with an opinion state transition model illustrated in Fig. 7. The model restricts transition between opinion states

Table 4 Attitude propagation model nomenclature

Type	Notation	Description
Variable	I_A	Accepted message influence
	I_s	Influence toward source confidence
	e_m	Innovation accepted toward message subject
	e_s	Innovation accepted toward source confidence
	τ	Trend toward message subject
	a	Audience attitude toward message subject
	α_s	Audience confidence toward source
Parameter	R_m	Receptivity to message innovation
	R_s	Receptivity to source confidence innovation
	T_s	Time step to next model update

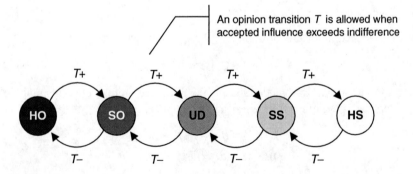

Fig. 7 The intransigence of public opinion resists impact of dissonant influence

whenever the magnitude of accepted influence is less than a *latitude of opinion indifference*, $\|I_A\| < d_I$.

For example, consider an audience that holds a strong positive opinion of HS on some subject. Assume it is exposed to media coverage that carries dissonant sentiment toward the subject with net influence, $I_A = 0$. The audience attitude will drift toward value $a = 1.5$ representing a split consensus of *HS* and *SS*.

4.6   Media Channel Reach

A media channel distributes content to the public. MIM represents media distribution as a communication channel from the source (sender) to the audience (receiver). Each channel is characterized by its reach.

The *reach* of a media channel to a target audience is the fraction of the audience that is exposed to content carried by the channel. Reach is a mutual relationship of reliance between a public segment and a media channel. The public segment relies on a media channel for information, and the media channel relies on the public segment subscribership for its media status. Reach represents the fraction of time or attention that the target audience allocates to the media channel. Media channels function to maintain reach to their target audience. Reach is a key competitive factor in media. Media outlets and channels manage production and distribution to maintain desired reach. MIM assumes that media channels attempt to track their target audience attitudes and will modify coverage to satisfy audience interests and maintain trust.

Consider a hypothetical media poll that measures the interest of a public segment to each of available channels 1, 2, or 3. The poll finds that the audience allocates 30% attention to channel 1, 60% to channel 2, and 10% to channel 3. Then, channel 2 has 0.6 reach to the target audience segment.

Figure 8 illustrates the MIM media channel reach model. The model defines reach for each channel to the target audience segments. MIM uses the audience preference interpretation of media channel reach. For each target audience, the sum of channel reach across all channels to that audience should total less than unity. Each source has a defined access to media channels. The figure illustrates typical channel types. Each channel can provide a maximum exposure frequency to all its subscribers.

4.7   Media Channel Distortion

A media outlet production and distribution can be modeled as a communication channel between a source and an audience. Both production and distribution can alter the intended source message by placing source statements in the context of other statements that may alter the sentiment, framing, or subject. The audience's

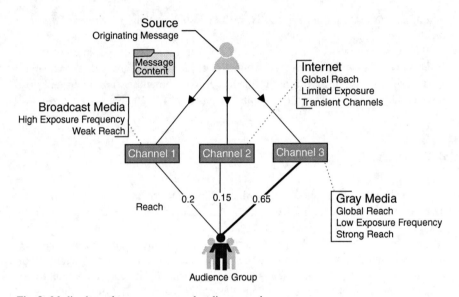

Fig. 8 Media channel source access and audience reach

ability to interpret the message correctly depends on its cognitive ability and exposure to the message. A message can be obscured by its placement in media distribution that is adjacent to similar messages that create interference. For example, a front-page story is often isolated for emphasis, whereas a story buried on page ten may be difficult to find and discriminate from adjacent articles. Often the intensity of the sentiment provides the discrimination that enables the audience to find and interpret the message.

MIM incorporates a hypothetical model for a media channel based on communication theory (Cover and Thomas 1991) to estimate the on-message distortion of a media channel. The following factors are computed to define a media channel message distortion.

A target audience influence depends on the frequency of its exposure to a message carried on a media channel. *Exposure Frequency* (F_e) by a target audience depends on the product of two factors, $F_e = R_C F_M$. *Channel Reach* (R_C) is the fraction of target audience exposed to messages carried on the channel. *Message Frequency* (F_M) is the frequency of distribution of *on-message* media content carried on the distribution channel.

Media channels carry content covering a wide range of messages. A receiver that is concerned about a particular subject must find the information of interest in the media content to correctly interpret the message influence. A receiver who wishes to determine the subject, sentiment, framing elements, and source attribution contained in the message is faced with a problem of reducing his initial uncertainty. He may need to receive and attend to several media transmissions over time to extract meaning and reduce his uncertainty. Communication theory (Cover and Thomas 1991) defines information as the ability of a message to reduce uncertainty measured as the entropy in the receiver's decision. If each media channel carried only one message at a time, there would be no distortion. However, an audience must extract and interpret message meaning from a channel that is broadcasting many simultaneous messages that cause distortion. Communication theory describes information distortion on a communication channel using a very general computational model that quantifies the reduction of given the complexity of interfering messages. A detailed discussion of communication theory and entropy is beyond the scope of this chapter. We refer the reader to Cover and Thomas (1991) for more details. The following summarizes the definitions and computational model used to represent distortion in a media channel.

Message Information (I_M) is the amount of information that must be encoded in media content for the audience to interpret the full meaning of the message. *Channel Noise* (N_c) is the amount of competing information, unrelated to the message, that is carried on a channel at any time that causes interference in interpreting the meaning of the message.

Signal-to-noise ratio (SNR) is a characteristic of a channel and the information content covering a message

$$SNR = \frac{I_M^2}{N_C^2} \tag{15}$$

A large SNR (greater than unity) implies that the message has strong sentiment compared to other messages on the channel.

Channel Capacity (C_C) is the amount of information that can be distributed successfully to audiences at each time. It is computed Cover and Thomas (1991) as

$$C = F_e \log_2 \left(1 + \frac{SNR}{F_e} \right) \tag{16}$$

Information Environment Entropy (E) represents the complexity of the potential content carried on any media channel operating within a PIE. It represents the complexity of all possible combinations of subjects, themes, sources, and message frames that must be discriminated for a receiver to decode the correct interpretation of a message.

Received Message Information (I_{RM}) is the amount of distributed on-message information that can be recovered by the target audience from exposure to media distributed over a channel. The target audience must discriminate the message within the media content it receives from all the competing messages.

According to communication theory, received information may be limited by the channel capacity if its value does not exceed the entropy in the information environment. It is computed as

$$I_{RM} = \min \{ C_C, -E \} \tag{17}$$

Channel Distortion (D_C) is the distortion in interpreted message obtained from media content due to the capacity of a channel and complexity of the message. It is computed as

$$D_C = \frac{2^{-E} - 2^{I_{RM}}}{2^{-E}} \qquad (18)$$

Interpretation Channel Factor (c_c) is the fraction of correctly interpreted message influences that can be decoded by the target audience over a channel. It is computed as

$$c_c = 1 - D_C \qquad (19)$$

MIM models channel distortion of accepted message influence as

$$I_A = c_c c_R (\alpha_S) c_A (e) e + a \qquad (20)$$

Media channel distortion is a conceptual model that is incorporated in MIM to explore the impact of mechanisms that create distortion and confusion in message interpretation whenever media channels are carrying a high volume of distinct media themes and messages.

4.8   Media Outlets

Media outlets can alter the sentiment contained in source statements by its coverage of themes. Media campaigns are often frustrated in their ability to anticipate and account for media distortion and bias. No single theory that describes a simple mechanism to account for media bias has emerged. Allen (2009) describes bias as purposeful filtering of messages carried on media channels (P. D. Allen (2008), "Accounting for Bias in Broadcast Media Message Acceptance," *IO Sphere, the Joint Information Operations Warfare Command*). Other studies have recognized that media bias often exists but have provided no concrete models (Scheufele and Tewksbury 2007; Schumann et al. 1990; McQuail 2005; Perse 2001). Understanding mechanisms for media bias is potentially an important contribution of exploratory analysis. In this section, we describe one hypothetical model that characterizes bias as a behavior of an agent representing a media outlet.

A media outlet manages production to maintain reach to its target audiences. It selects coverage and placement of messages consistent with its priority themes. It also uses editorial processes and policies to adjust tone of its content to satisfy the interests of its target audience. A media outlet will select and emphasize coverage of messages that are aligned with its priority themes and deemphasize other messages.

Media outlet behaviors: MIM models media outlet bias effects by representing several behaviors that are typical of media outlet production and channel distribution policies. MIM represents a media outlet actor as having five distinct behaviors.

Behavior 1: A media channel will balance the sentiment of content it carries in an attempt to satisfy the message-sentiment latitude of acceptance of its target audience.

A media outlet will develop and retain a profile of the attitudes of its target audiences for theme subjects. A media outlet then adapts the tone of its coverage of source statements to comply with its *standard subject tone* $a(M_c, S)$ toward the subject S.

A *media outlet agent* determines its *standard subject tone* as a reach weighted average of its target-audience $P(M_c)$ sentiment toward S:

$$a(M_c, S) = \sum_{k \in P(M_c)} \mu_c(P_k) \cdot a(P_k, S)$$

$$\mu_c(P_k) = \frac{R_c(P_k)}{\sum_{k \in P(M_c)} R_c(P_k)} \tag{21}$$

A media outlet agent will adapt the tone of the content it distributes on a channel MC to balance between the source statement and the media channel content norm as needed to capture the attention of its target audience. MIM computes the sentiment carried on media channel MC as

$$a^{MC}(T, S) = (1 - C_M)a(T, S) + C_M a(M_c, S) \tag{22}$$

where $0 \le C_M \le 1$ represents the fractional shift in message tone or balance factor needed to satisfy the target audience latitude of acceptance

$$\left\| a^{MC}(T, S) - a(M_c, S) \right\| \le d_A \tag{23}$$

Behavior 2: Each media channel will prioritize its coverage and distribution of content according to its intrinsic priority themes. A media channel emphasizes coverage of messages that express opinions toward subjects contained in its intrinsic priority themes. Media emphasizes coverage by elevating placement and exposure frequency.

Behavior 3: A media outlet will carry statements exclusively from a list of its legitimate (authorized) sources.

Behavior 4: Each media channel has a limited capacity of messages it can carry during a time interval. A media outlet will fill the capacity of its media channels with content according to two selection criteria as follows.

A media outlet will prioritize selection of content to:

1. Prefer messages having strong tone to capture audience attention,

$$V_1 = \left\| a^M(T, S) \right\|$$

2. Prefer messages that can impact its target audience

$$V_2 = \left\| a^M(T, S) - a(P_M, S) \right\|$$

A media outlet selects content in order of increasing priority given by:

$$V = (1 - w_2)V_1 + w_2 V_2$$

where $0 \le w_2 \le 1$ is a weighting factor for priority message selection.

Behavior 5: Media channels interact in a PIE creating potential conditions for resonance. Once distributed by a media channel, media is in the public domain and hence accessible by other adjacent media outlets. Whenever a media outlet obtains publicized content offering an opinion toward one of its priority themes, it may adapt the content for its own message distribution. The media outlet may also adapt the tone of the new distribution to satisfy its own target audience interest. Resonance can occur when adjacent media outlet coverage extends the distribution beyond the target audience and distribution time of the originating channel. Resonance can involve message distortion as each media channel adapts its own content and selects its own coverage. This resonance behavior is represented in MIM when a channel picks up content from an adjacent public channel and issues new content carrying coverage of the original message.

Media outlet behavior is based on a conceptual model of the media outlet production process that has not appeared previously in the literature. MIM incorporates this model to enable exploratory analysis of media influence when media bias effects are of concern.

4.9 Source Lines of Communication

Public relations practice is to analyze, develop, and maintain desired relationships between an organization and its target public segments through sustained communication outreach. Techniques for building trusted communication relationships vary. They continue to be adapted to the modern information environment of media and to the growing need to develop cross-cultural relations. Understanding mechanisms that describe how a source will adopt its statements to achieve audience resonance is potentially an important contribution of exploratory analysis. In this section, we describe one hypothetical model that characterizes the behavior of source agents in a PIE. This conceptual model has not previously appeared in the literature.

MIM represents a trusted relationship between an information source and a public segment as an LOC (see Fig. 9). An LOC source issues statements to media to maintain influence on its target audience. A source strives to maintain positive influence but does not typically control media distribution. Instead, it must rely on distribution through media channels that provide access to its target audience.

Source actor behavior: A source has a relationship with a target audience defined by an LOC. A source actor T issues statements expressing its opinion toward a subject S if either of two conditions hold (Table 5):

1. T holds a strong opinion toward subject S
2. The opinion held by T differs sufficiently from that of its target audience P

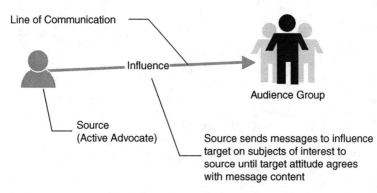

Fig. 9 MIM line of communication

Table 5 MIM source behavior conditions

	Source behavior condition	Model condition
1	Opinion strength of source statement exceeds its latitude of indifference	$\|a(T,S,C)\| \geq d_I$
2	Opinion strength of source statement and target audience sentiment disagree by d_T	$\|a(T,S,C) - a(P,S,C)\| \geq d_T$

5   Illustrative Example

Here, we illustrate how MIM can be used to analyze message influence effects using three cases. The example considers the conditions for successfully influencing attitude within a public segment by issuing a message through a media channel.

In the first case, a source releases a public message statement expressing an opinion toward a subject actor. The message is covered by a media channel that reaches a particular target audience. Let us assume a media outlet distributes content covering this message without distortion or bias to the audience for 21 days (3 weeks).

In the second case, the target audience is also exposed to a dissenting opinion from a second source during the same 3-week period. In both cases, the audience initially holds a neutral (undecided) opinion of confidence in both sources prior to exposure to media coverage. The audience develops or withholds confidence depending on how it interprets and accepts the message influence.

In the third case, we examine the response when the audience holds an initial attitude of distrust toward the source.

MIM can be used to analyze conditions that can lead to accepted influence, attitude change effect, and development of source confidence. The modeler introduces explicit assumptions about the strength of initial audience sentiment, message tone, framing contexts, and the duration of exposure leading to influence. Table 6 summarizes the hypothetical model parameters used in the example to describe the assumed response of a public audience to media information exposure. Table 7 summarizes the audience sentiment and message tone for this example.

Table 6 Assumed parameters.

Symbol	Parameter	Value
d_I	Latitude of indifference	0.4
d_A	Latitude of acceptance	2.4
d_R	Latitude of rejection	3.0
d_α	Latitude of source confidence	0.4
d_ρ	Latitude of distrusted source opposition	−0.2
R_m	Receptivity to message influence	1/6
R_s	Receptivity to source confidence influence	1/12
d_I	Latitude of opinion indifference	0.5

Table 7 Message and public sentiment

	Message sentiment		
	Audience sentiment	Original source	Dissenting source
Framing contexts			
Subject legitimacy	SO	HS	SO
Subject affinity	SS	HO	SS
Subject competency	UD	UD	UD

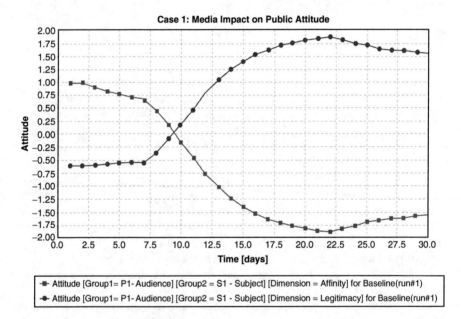

Fig. 10 Case 1 – public attitude impact

Figure 10 illustrates the attitude change in the audience in response to the message exposure in the first case. The simulated response shows that the target loses its conviction toward the subject within 11 days and adopts a new attitude within 21 days under these assumptions. Figure 11 illustrates the change in audience accepted

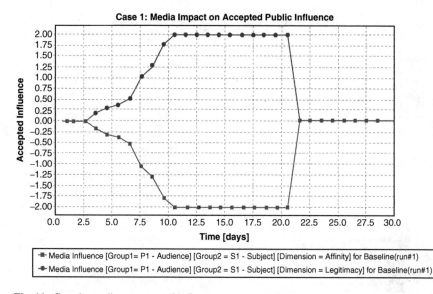

Fig. 11 Case 1 – audience accepted influence

influence as the target is exposed to the message. Within 6 days of exposure, greater than 50% of the target population understands and accepts the message influence. The full influence of the message tone is experienced within the population within 10 days of exposure. In the first case, we assume that the target audience is not exposed to any dissenting opinion, media channel distortion is negligible, and the target confidence in the source is neutral. Hence, the target is poised to accept influence on the subject from this source.

Figure 12 illustrates how the target audience develops confidence in the source as it interprets and accepts the message. In this scenario, media information affects the reversal of public opinion within 21 days of media message exposure.

The second case considers the impact of exposure to dissenting media coverage during the message exposure period. Figure 13 shows that public attitude is affected and opinions are reversed, but the public does not reach full agreement with the tone of the message content before the message coverage terminates. This case illustrates the effect of public dissonance resulting from exposure to dissenting influence. Figure 14 illustrates the level of dissenting content exposure.

The third case considers the impact of initial distrust in the source. Figure 15 illustrates how attitude change is suppressed in this case. Figure 16 plots the fluctuations in accepted influence resulting from source mistrust and the tendency to oppose message opinion. Figure 17 plots the degradation in source confidence that results. Note that under these assumptions, an initial 20% population negative source confidence is sufficient to lead to this failure in reaching the intended effect or even reversing initial sentiment. Table 8 summarizes the qualitative analysis from these three exercises.

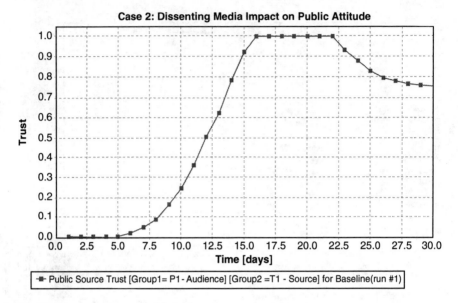

Fig. 12 Case 1 – audience confidence in source

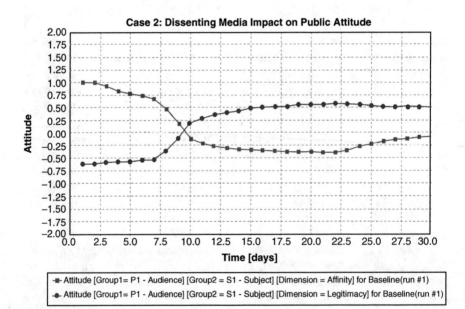

Fig. 13 Case 2 – public attitude impact of dissenting message exposure

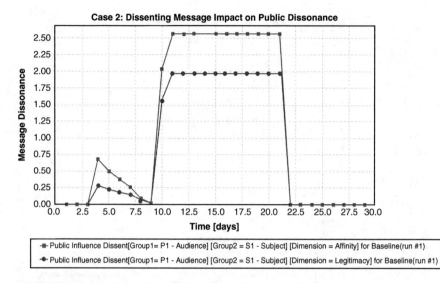

Fig. 14 Case 2 – public exposure to dissonant influence

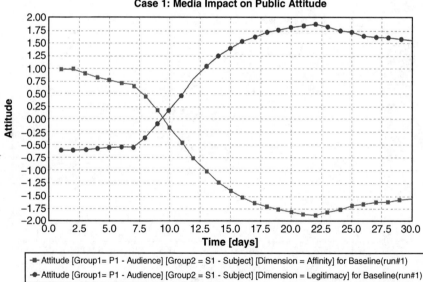

Fig. 15 Case 3 – attitude impact of source distrust

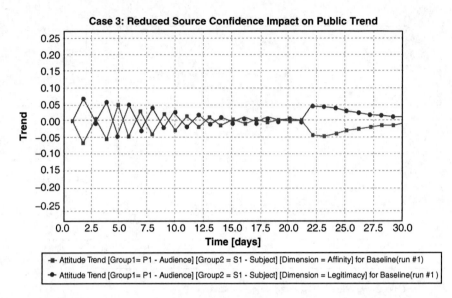

Fig. 16 Case 3 – public trend fluctuations due to source distrust

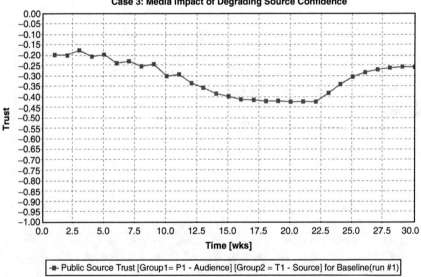

Fig. 17 Case 3 – source confidence degraded

Table 8 Summary assessment

Qualitative state	Case 1	Case 2	Case 3
Target adopts opinion consistent with message tone	Yes, in 21 days	No. Dissonant influence leads to uncertain public opinion	No change in audience opinion
Target reverses opinion toward subject	Yes, in 10 days	Yes, in 10 days	No change in audience opinion
Target interprets and accepts intended message influence	Yes, in 11 days	No. Dissent leads to dissonant influence	No. Influence is inconsistent and counter to message intent
Target adopts 50% trust in source	Yes, in 11 days	No	No. Trust degrades as message intent is rejected

6   Practical Tips

- Begin your modeling project by conducting research to identify the salient themes that resonate in the media within the country or region of interest. Developing an understanding of what will constitute salient media themes for your modeling project will drive much of the modeling process. Your goal at this stage should be to build understanding of the existing public information environment in the client's region of interest. In any country or region having a sophisticated media, the range of themes that may be covered can be vast and change rapidly. Concentrate building your understanding around those themes that are relevant to the client's analytical interest. These may be a blend of political, social, or economic interests that represent regional, national, or district issues.

- Plan to develop an understanding of the client's analytical goals during the early stages of the modeling project. What metrics, level of resolution, and bread of coverage does the client expect to see in the final analytical product. Remember that model will be a tool that will support analytical studies. It may not answer all the analytical questions explicitly, but it should provide insights and derived metrics that enable the client's needs, otherwise the modeling effort will be deemed irrelevant. It is important to set consistent expectations with the client about the time scale of the model validity and to draw out any inconsistencies with the analytical time scale that the client requires. Multiple modeling or model revisions may be necessary to support the client's analytical horizon. Only after you understand the client's analytical goal adequately and the available situational knowledge, can you make an informed value judgment on the modeling fidelity and time scale of validity.

- Plan part of your effort to build a knowledge base to support model instantiation and validation. Place initial emphasis on research to gain an understanding of the breadth of knowledge that is available. The modeling effort should focus on representing the common knowledge to be credible.

Conduct research using credible sources. Be sure to conduct knowledge elicitation from subject matter experts (SME's) who are deemed credible to the client. In many cases, the client will recommend SME's. But be sure to vet any SME's with client before you expend effort to represent their knowledge in the model. Determine the availability of any key SME's and budget your access time wisely.

- Use a combination of broad and shallow analysis to gain confidence with client. Do not push details as single answer analysis. Involve client stakeholders and SME's early in model development and seek to establish an appreciation of the scope and granularity of the modeling process that is consistent with the analytical objectives of the project. Remember that you, the modeler, will soon become a knowledgeable stakeholder in the analysis. Do not neglect your role as analyst. You may become the key SME as you gain insight from model analysis of the domain.

- One key issue in populating any media influence model will involve representing the audience segments and influential media channels to be consistent with available knowledge. You may need to extrapolate from knowledge sources and available data to instantiate an executable model. This baseline model configuration can then form a basis for further elicitation with SME's that will aid in ringing out any inconsistencies or knowledge gaps.

- A substantial aspect of the domain knowledge base will likely be acquired through active elicitation from SME's. Be careful to recognize the knowledge limits of each SME that you engage. The knowledge obtained from any individual SME can often be dated, inaccurate and inconsistent with other SME's. Most SMEs have little appreciation of their own knowledge limitations and have difficulty characterizing the extent of their knowledge uncertainty. Many are uncomfortable with any quantitative representation that may call into question their knowledge. Although the model can often be helpful in reaching consensus on common knowledge, it should not be used as a tool to confront or contest SME's judgments. The model and derived analytic products will be deemed valid if the SME common knowledge is viewed as consistent with the model.

- Analytical studies for clients will most often involve understanding impact of information actions that might contribute to an intervention. Work with the client to understand the nature of information actions that might be taken to shape the media environment in the near to mid-term future. Develop a clear understanding of what types of media shaping actions are to be represented in the analysis. This will aid in selecting the sources and subject that should be represented in the model. Complete this level of representation by conducting research to identify valid knowledge or analytical assumptions about the existing lines of communication from sources to target audiences.

- Conducting research on media channel reach and bias can often require access to data that is proprietary or otherwise restricted. It is important to reach an understanding with the client about the level of model fidelity that can be achieved with available resources. Discuss with the client how the model can be modified or refined to represent conditional assumptions where data is unavailable or highly uncertain.

- Analytical studies involving models often use multiple models to gain broader insights. Be careful not to mix incompatible assumptions and expect compatible results. Media influence effects are best analyzed via ordinal comparison, either to a baseline reference or between pairs of alternative or competing opinions. Keep in mind the limitations of media influence time scale analysis. Media influence theory has evolved as communication technology has evolved. Any model that pretends to predict influence over decades is fundamentally flawed due to lack of understanding of impact of continual evolution of technical communications and social implication of increasing access to media.

7 Summary

The public information media provides information on current events (news), entertainment (programming), and opinions offered by trusted public sources, e.g., business, academic or religious spokespersons, journalists, and government officials. It is a major force in shaping a populace's attitudes toward significant social issues. Theoretical approaches to media influence include Communication Penetration Theory, Source–Message–Channel–Receiver Model, Opinion Leadership Theory, Social Judgment Theory, Media Agenda-Setting Theory, and Elaboration Likelihood Theory of Persuasion. Computational models make use of the theories and include Rational Choice Model, Social Judgment Computational Model, Public Education and Broadcasting Model, and Elaboration Likelihood Model. For example, the Media Influence Model (MIM) employs a hybrid, computational approach that blends multiagent simulation of communications with a system dynamic model of media influence. This approach models the subjects, sources, audience targets, and media channels that comprise a Public Information Environment (PIE), and relates the causal flow of influences from source statements (in a PIE) to changes in attitudes toward subject actors and to degree of source confidence. The model represents the impact of media outlets on message distortion and the dissemination of media through channels that reach certain public audience segments. Media message content is represented as expressing opinion toward a subject. The message contains sentiment and source attribution. Media effects are computed as media channel sentiment on each of several issues or themes, source statement sentiment and public segment attitude. Three cases illustrate MIM analysis of message influence effects (1) the influence of a public message in which the source has expressed an opinion toward an actor, (2) the target audience is exposed to a dissenting opinion from a second source, (3) an audience that initially harbors distrust toward the source. The timing and extent of influences differ significantly in each case. Although the influence of the media on public opinion can be modeled with credible results, current media models should be viewed as exploratory in purpose.

8   Resources

1. Pointers to modeling and simulation tools

   NetLogo an open source library for ABM,
   http://ccl.northwestern.edu/netlogo/

   Repast an open source library for ABM,
   http://repast.sourceforge.net/

   Swarm, an open source library for ABM,
   http://www.swarm.org/

   VenSim, commercially supported stand alone platform for developing SDM,
   http://www.vensim.com/

   SimBLOX, commercially supported development environment for ABM and
      SDM,
   http://www.simblox.com/

   AnyLogic, commercially supported development environment for ABM and
      SDM,
   http://www.xjtek.com/

2. Pointers to data sources for instantiating model of audience segments

   CIA world fact book
   www.cia.gov/library/publications/the-world-factbook

   Conflict Research Consortium
   www.colorado.edu/conflict/peace/treatment/opencomm

   The Pew Research Center for the People & the Press
   http://people-press.org/

3. Pointers to data sources for instantiating models of issues, themes, and public
   opinions

   The Pew Global Attitudes project,
   http://pewglobal.org/

   Center for Strategic and International Studies,
   www.csis.org

   Freedom House,
   www.freedomhouse.org

   United Nations News Centre,
   www.un.org/News/

   InterMedia,
   www.intermedia.org

4. Pointers to data sources to instantiate models of media outlets and channels

   InterNews
   www.internews.org

Internet World Stats
www.internetworldstats.com
ABYZ News Links
www.abyznewslinks.com
Mondo Times
www.mondotimes.com
The Association for International Broadcasting
www.aib.org.uk

References

Ajzen, I., and Fishbein, M. (1980). *Understanding attitudes and predicting social behavior.* Englewood Cliffs, NJ: Prentice-Hall.
Allen, P. D. (2008). Accounting for bias in broadcast media message acceptance. *IO Sphere.* San Antonio, TX: The Joint Information Operations Warfare Command.
Axelrod, R. (1997). The dissemination of culture: a model with local convergence and global polarization. *The Journal of Conflict Resolution*, 41(2), 203–226.
Bennett, W. H. (2009). Media Influence Modeling in support of Conflict Modeling, Planning, and Outcome Experimentation (COMPOEX), Presentation. Presented to the Military Operations Research Society (MORS) Irregular Warfare Analysis Workshop, MacDill AFB. Retrieved from http://www.mors.org/events/09iw.aspx.
Berlo D. K., Lemert, J. B., and Mertz, R. J. (1969). Dimensions for evaluating the acceptability of message sources. *Public Opinion Quarterly*, 33(4) 563–576.
Berlo, D. K. (1960). *The process of communication: an introduction to theory and practice.* New York, NY: Holt, Rinehart and Winston.
Chaiken, S. (1987). The heuristic model of persuasion. In M.P. Zanna, J. M. Olson, and C. P. Herman (Eds.), *Social influence: The Ontario Symposium*, 5, 3–39. Hillsdale, NJ: Lawrence Erlbaum Assoc.
Chong, D., and Druckman, J. N. (2007). Framing theory. *Annual Review of Political Science*, 10, 103–126.
Cover, T. M., and Thomas, J. A. (1991). *Elements of information theory.* New York, NY: John Wiley & Sons.
Goidel, R. K., Shields, T. G., and Peffley, M. (1997). Priming theory and RAS models: toward an integrated perspective of media influence. *American Politics Research*, 25(3), 287–318.
Gonzalez-Avella et al. (2005). Page 19.
Gonzalez-Avella, J. C., Eguiluz, V.M., San Miguel, M., Consenza, M. G., and Klemm, K. (2007). Information feedback and mass media effects in cultural dynamics. *Journal of Artificial Societies and Social Simulation*, 10(3), 9.
Jager, W., and Amblard, F. (2004). A dynamical perspective on attitude change. In *Proceedings of the North American Association for Computational Social and Organizational Science (NAACSOS) Conference*. Pittsburgh, PA.
Johnson (2007). Page 13.
Katz, E., and Lazarsfeld, P. F. (1955). *Personal influence: the part played by people in the flow of mass communication.* New York, NY: Free Press.
Latané, B. (1981). The psychology of social impact. *American Psychologist*, 36, 343–356.
McCombs, M., and Shaw, D. (1972). The agenda-setting function of mass media. *The Public Opinion Quarterly*, 36(2), 176–187.
McQuail, D. (2005). *McQuail's mass communication theory.* London, England: Sage Publications Ltd.

Mosler, H. J., Schwartz, K., Ammann, F., and Gutscher, H. (2001). Computer simulation as a method of further developing a theory: simulating the elaboration likelihood model. *Personality and Social Psychology Review,* 5(3), 201–215.

Perse, E. M. (2001). *Media effects and society.* Mahwah, NJ: Lawrence Erlbaum Assoc.

Petty, R. E., and Cacioppo, J. T. (1986). The elaboration likelihood model of persuasion. *Advances in Experimental Social Psychology,* 19, 123–205.

Rogers, E. M. (2003). *Diffusion of innovations,* (5th ed.). New York, NY: Free Press.

Rucker, D. D., and Petty, R. E. (2006). Increasing the effectiveness of communications to consumers: recommendations based on elaboration likelihood and attitude certainty perspectives. *American Marketing Association,* 25(1), 39–52.

Scheufele, D. A., and Tewksbury, D. (2007). Framing, agenda setting, and priming: the evolution of three media effects models. *Journal of Communication,* 57, 9–20.

Schumann, D., Petty, R., and Clemons, D. (1990). Predicting the effectiveness of different strategies of advertising variation. *Journal of Consumer Research,* 17(2), 192–202.

Sherif, C. W., Sherif, M. S., and Nebergall, R. E. (1965). *Attitude and attitude change: the social judgment-involvement approach.* Philadelphia, PA: W.B. Saunders Co.

Stone, G., Singletary, M., and Richmond, V. (1999). *Clarifying communication theories: a hands-on approach.* Ames, IA: Iowa State University Press.

Waltz, E. (2008). Situation analysis and collaborative planning for complex operations. In *Proceedings of the 13th International Command and Control Research Symposium.* Bellevue, WA: Office of Assistant Secretary of Defense.

Waltz, E. (2009). Modeling the Dynamics of Counterinsurgency (COIN). *In Analytical Tools for Irregular Warfare, Proceedings of NATO System Analysis and Studies Panel Specialists Meeting* (SAS-071/RSM-003), Ottobrunn, Germany: NATO Research and Technology Organization.

Weisbuch, G., Deffuant, G., Amblard, F., and Nadal, J.P. (2002). Meet, discuss and segregate. *Complexity,* 7(3), 55–69.

Chapter 6
Governance and Society

Corey Lofdahl

Countries with collapsed governance – so-called failed states – are common targets of international interventions. Not only do failed states impose enormous suffering on their own populations, often necessitating humanitarian aid and other forms of intervention, but they also constitute a key threat to world security. *The National Security Strategy of the United States of America* states, in fact, that we are now "threatened less by conquering states than we are by failing ones" (NSC 2002).

The most common characteristic of a failed state is the collapse of governance: the government loses physical control of its territory, the monopoly on legitimate use of force, the authority to make decisions, and the ability to provide public services. Equally important, the lack of adequate governance makes it difficult for an intervention of any kind – diplomatic, economic, or military – to succeed. The intervening parties face an unappealing choice between supporting the failed state indefinitely or accepting the failure of their intervention and discontinuing their efforts.

In examining governance, looking to definitions is of little help. Governance implies effective control, and the modern state controls, or at least seeks to control, many social and national features, and this list varies from state to state. However, for our purposes, we can build a working definition of "governance" by looking to places where governance is weak, has failed, or does not exist.

The Fund for Peace (2005) constructed a Failed State Index that lists countries in order of weaker governance. The index is composed of the following 12 metrics:

1. Demographic pressures,
2. Refugees and internally displaced persons,
3. Group grievance,
4. Human flight,
5. Uneven development,

C. Lofdahl (✉)
IGEN Corporation, 84 Hobblebush Lane, Carlisle, MA 01741, USA
e-mail: clofdahl@igencorp.com

A. Kott and G. Citrenbaum (eds.), *Estimating Impact*,
DOI 10.1007/978-1-4419-6235-5_6, © Springer Science+Business Media, LLC 2010

6. Economic decline,
7. Delegitimization of the state,
8. Public services,
9. Human rights,
10. Security apparatus,
11. Factionalized elites, and
12. External intervention.

The metrics break out into three basic groups: social, economic, and political-military indicators. The first of the social indicators is *demographic pressures*. Additional population can lead to additional pressures for food, land, and jobs that when unmet can lead to conflict. The second metric is *refugees and internally displaced persons*: large movements of refugees resulting from intergroup competition, oppression, or war can lead to serious and long-lasting humanitarian emergencies. The third metric is *group grievance*, which can include seeking vengeance for past injustices, institutionalized exclusion of groups, and paranoia. The fourth metric is *human flight*, the result of processes such as "brain drain" in which the professionals, intellectuals, and dissidents leave the country if they can. The fifth metric is an economic indicator, *uneven development*, which is especially problematic if it occurs along group lines that affect child mortality, education, and employment. The sixth indicator is *economic decline*: sudden drops in income for a large number of people that can place great social stress on a country.

The remaining Failed State Index measures are political and military. The seventh metric is *delegitimization of the state* and criminalization of the leadership. Corruption, lack of accountability, and resistance to transparency can all lead to lack of confidence, trust, and legitimacy. The eighth metric is *public services*, a reduction of basic services offered to the population – for example, police, health, education, sanitation, and public transportation – or a narrowing of those services to the country's elites. The ninth metric is *human rights*, the suspension or arbitrary application of rule of law. The tenth metric is a *security apparatus* that operates as a lawless "state within a state," attacks enemies of the state, and serves the interests of a small portion of the country. The 11th metric is *factionalized elites* who fight among themselves, hinder effective governance, and employ group-based rhetoric. The 12th metric is *external intervention* by other states, which could be either military or economic. A failed state can feature one, some, or all of the pathologies described by the 12 metrics, but such states are almost certain to be unstable and unable to sustain themselves and their populations over time.

The failed state metrics give an idea of the components and analytic units of governance: the society, economy, state, and military. Determining the relationship among these units of analysis, how their relationship varies over time, and what goals the government should seek through governance go unanswered by the index. Clearly, delegitimization of the state is to be avoided – a population should trust its government – but determining how best to achieve and sustain this relationship is a nontrivial problem. The tension between the military and economy has a rich history featuring disagreements over which should come first, security or development, and heated political debates between promilitary "hawks" and prodiplomacy "doves."

The failed state metrics are not prescriptive. The index does not tell how to formulate policies that improve governance. Economic policy mistakes can lead to intervention by the International Monetary Fund, and security mistakes can lead to terrorism, domestic insurgencies, or in extreme cases regime change at the hands of foreign troops. Understanding these potent, interrelated forces and formulating policies that manage them effectively is a formidable challenge.

1 Theories of Governance

A review of the governance literature reveals no commonly accepted definitions. Dobbins (2003) examines the American experience of nation-building, including the complexity of the undertaking, the focus on democratization, the initial successes of post-World War II Germany and Japan, and the limited success of more recent nation-building efforts. Fukuyama (2004), in studying governance, asserts a lack of "organizational tradition" in many countries and the need for "institutional transfer" to those countries, as well as the need for capitalist and free-market values. Such observations fall short of a theory that can be used to order data and structure more rigorous and systemically informed analyses.

A working theory of governance, we call the Quest for Viable Peace (QVP) model, is developed here based on Covey et al. (2005), and particularly (Blair et al 2005), who distil lessons learned from the international community's experience in Bosnia and Kosovo during the 1990s. Governance is examined in terms of capable government institutions making decisions regarding both security and economy that result in state stability. Government capacity is expressed in terms of four governance elements: (1) political, (2) economic, (3) military, and (4) rule of law. These elements do not exist separately but instead are interwoven and influence one another.

Figure 1 depicts the organizational structure of a state with ineffective governance, like the failed states described by the Fund for Peace (2005). The *state* and its institutions have been captured by a *criminal political elite*, making the state effectively illegitimate for the *mass of society*. The economy is composed of three separate subeconomies: the *legitimate economy* in which transactions for legal goods occur and taxes are paid to the state, the *gray economy* in which legal goods are traded but taxes are not paid, and the *illegitimate economy* in which illegal goods such as drugs, weapons, and people are traded and taxes are not paid. In a failed state, the illegitimate and gray economies are large relative to the legitimate economy, and the benefits of the economy are directed by the political criminal elite to a favored *client group* rather than the mass of society. Potential economic pathologies from this arrangement include uneven development due to the two-class system, economic decline due to illegal activity, and lack of public services for most of the population. The state's military and legal frameworks are used to

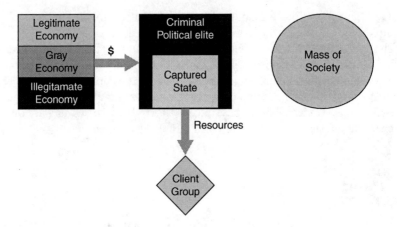

Fig. 1 Ineffective governance results in failed states (Blair et al. 2005, [p. 208])

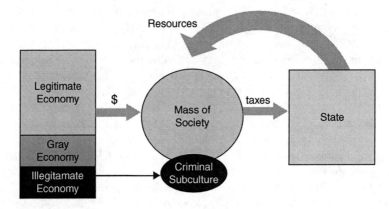

Fig. 2 Effective governance leads to state stability (Blair et al. 2005, [p. 209])

keep the criminal elite in power and oppress the mass of society, a misuse of the security apparatus that can cause group grievance, internally displaced persons, human-rights violations, and human flight.

Figure 2 depicts the organization of a state featuring effective governance. Placed at the center is the mass of society who now receive the benefits of the economy and resources from the state. The legitimate economy benefits primarily the mass of society, while the smaller gray and illegitimate economies pass resources to a *criminal subculture*. The mass of society pays taxes to the state and in return receives benefits and services from the state, and the rule of law is used to keep the gray economy, illegitimate economy, and criminal subculture in check. Figure 2 presents a picture of rightly ordered relationships that provide stability and avoid the failed state pathologies depicted in Fig. 1.

In the twenty-first century, failed states like those depicted in Fig. 1 are more than a national tragedy; they are an international security threat to effectively

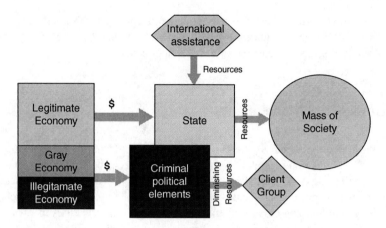

Fig. 3 International assistance seeks to transform failed states (Blair et al. 2005, [p. 221])

governed states through terrorism and weapons of mass destruction (Popp 2005). This then provides an incentive for the international community to transform Fig. 1 states into Fig. 2 states as shown in Fig. 3. The process is accomplished through *international assistance* that must accomplish several goals. First, it must engage militarily the criminal political elements and restrict resource flows to its client group. Second, it must reconstitute a state that provides effective governance and resources to society. Specifically, the state must provide the rule of law, which entails the rehabilitation or creation of noncorrupt police, judiciary, and prison systems. Third, the illegitimate and gray economies must be reduced to lessen the strength of the criminal political elements and to create an incentive for them to reintegrate in society. The legitimate economy must be strengthened, and taxes must be collected to fund the state. Note the interrelationships among these various factors and that they must be changed simultaneously, and hence the complexity and difficulty of international assistance and nation-building.

Hawley and Skocz (2005) outline a planning paradigm to coordinate these efforts, to harmonize civilian governance and security efforts in a failed state intervention. Coordinating these political and security efforts requires a coherent intervention strategy. This strategy must address core, driving issues underlying the state's instability as well as mobilize resources over what could be a considerable length of time because security forces, civilian experts, and aid flows must stay in place long enough to complete the job and not leave too early due to political pressures or cost considerations. Hawley and Skocz (2005) articulate a three-stage intervention to accomplish the nation-building process. Stage 1 imposes stability through an international intervention strong enough to take on war-hardened powers, a process that takes approximately 3 years. In stage 2, with the violent extremists marginalized, international forces are downsized, legitimate institutions are strengthened, and local leaders are trained, which can take 5–7 years. In stage 3, the peace process has become self-sustaining with extremists even further marginalized and effective governance in place. With security at acceptable levels and maintained with domestic forces and institutions, international aid efforts focus primarily on economic development issues. Note that the time frame for this nation-building effort is measured in years if not decades.

2 Governance Modeling

Computer-based modeling is inherently an exercise in *abstraction*; that is, the real world is sufficiently complex that only a few key aspects can be represented. Inherent in the modeling process is the act of deciding which real-world objects, situations, or processes to concentrate on and which, for the time being, to ignore. The choice of modeling methodology may influence the decision as each features its own combination of strengths and weaknesses. Three modeling methodologies are presented in this section – statistics, agent-based modeling, and system dynamics – and within each methodology, two models are contrasted and compared. The purpose of this section is to give the flavor, strengths, and weaknesses of each methodology as it applies to the problems of governance rather than impart the details necessary to construct and implement such a model.

2.1 Statistics

Statistical results tend to stress correlation over causation; that is, the examination of empirical data demonstrates that two measures vary together in a way not explainable by chance. Two statistical models are examined here: the Conflict Assessment System Tool (CAST), which is used to generate the Failed State Index discussed earlier, and the Analyzing Complex Threats for Operations and Readiness (ACTOR) model.

The CAST model (Fund for Peace 2005; OSD 2009, 323–324) is calculated by adding the 12 separate measures shown in (1), each of which ranges from 0 to 10 and results in a maximum score of 120.

$$
\begin{aligned}
\text{Failed State Index} = {} & \text{demographic pressures} + \text{refugees and displaced} \\
& \text{persons} + \text{groupgrievance} + \text{human flight} + \text{uneven development} \\
& + \text{economic decline} + \text{delegitimization of state} + \text{public service} + \text{human} \\
& \text{rights} + \text{securityapparatus} + \text{factionalized elites} + \text{external intervention}
\end{aligned} \tag{1}
$$

The lower the score, the better and more stable the government. Some of the factors, like demographic pressures, are outside the direct realm of governance yet make governance more difficult, while others, like public service, result from decisions made directly by the government. In 2009, the seven highest scores were 114.7 for Somalia, 114 for Zimbabwe, 112.4 for Sudan, 112.2 for Chad, 108.7 for the Democratic Republic of the Congo, 108.6 for Iraq, and 108.2 for Afghanistan (Fund for Peace 2009).

CAST model analysis begins by examining thousands of documents to generate the 12 social, economic, political, and military indicators. Once this initial work is done, the capabilities of five core state institutions that are essential for security are assessed: military, police, civil service, the system of justice, and leadership. Finally, idiosyncratic factors, surprises, and the conflict risk history are identified for each country. Thus, source documents are evaluated by computer to generate initial scores and are then adjusted by experts as needed. It is not necessary to automate every step of the analysis as long as adjustments are consistent, justified, and defensible. The resulting Failed State Index is a number computed from a sample of numbers, a simple form of statistical model.

The ACTOR model takes a different analytical tack by employing regression analysis on (2) (O'Brien 2001).

$$\text{Intensity of Conflict} = \text{F(pct history spent in state of conflict, infant mortality rate, tradeopenness, youth bulge, civil liberties index, life expectancy, politicalrights index, democracy, religious diversity, caloric intake, GDPper capita, ethnic diversity)} \tag{2}$$

Regression analysis entails gathering data for both the terms on the right side of the equation, the *independent* or *explanatory* variables, and the variable on the left side of the equation, the *dependent* or *response* variable. To populate the model, *cross-national* data was collected for approximately 160 countries between the years 1975 and 1998 for each of the 13 variables in (2), almost 30,000 observations. The ACTOR model attempts to predict country instability or conflict level between 2001 and 2015, in that way it is similar to the CAST model. However, the ACTOR model is different because its outputs are a set of parameters that measure how well the independent variables on the right explain, predict, or "fit" the level of intensity of conflict variable on the left. The ACTOR model was a research effort that reported an 80% accuracy level for predicting conflict, and this research continued in other programs (O'Brien 2007).

2.2 Agent-Based Modeling

In agent-based modeling, "agents" that correspond to real objects move and interact within a computer world (Waldrop 1993) according to low-level rules that lead to higher-level, unintended, realistic, and *emergent* behaviors. For example, high-level behaviors that emerge from low-level agent rules have provided insights into complex systems as varied as the flocking birds, automobile traffic, and ant colonies. Time is explicitly represented in agent-based models as the agents move and change state in their computer world, which makes it a simulation. An agent-based simulation's low-level rules are of the type, "If A is true, then the agent does B," so to the extent that A causes B, agent-based models are causal rather than correlative-like statistical models. Two agent-based models are discussed here: the

Table 1 SEAS overview (Prevette and Snyder 2007)

Actions	Effects	Agents	Behaviors	Environment
Diplomatic	Political	Government	Initiate	Roads and bridges
Information	Military	Organizations	Search	Communication lines
Military	Economic	Leaders	Decide	Oil and gas
Economic	Social	Citizens	Execute	Sea ports
	Infrastructure		Communicate	Financial
	Information		Update	Public utilities
			Terminate	Health services
				Security

Synthetic Environment for Analysis and Simulation (SEAS) and the Power Structure Toolkit (PSTK).

The SEAS model is used for applications ranging from government policy analysis to corporate strategy analysis by the pharmaceutical industry (Baard 2007). For the government analysis project (Table 1), governance is analyzed though Diplomatic, Informational, Military and/or Economic (DIME) actions that represent elements of national power and lead to Political, Military, Economic, Social, Infrastructure, and Information (PMESII) effects (JFCOM 2004). SEAS allows these DIME actions and PMESII effects to be specified and developed using Measures of Performance that capture governance efforts and Measures of Effectiveness that capture the consequences of these efforts. As SEAS is an agent-based simulation, agents can represent various levels of group aggregation: for example, agents can represent governments, leaders, citizens, or organizations such as corporations, agencies, and nongovernmental organizations. SEAS seeks to minimize aggregation so that, as much as possible, every person in a scenario is represented one-to-one by an agent.

Each agent features a set of rules that it uses to interpret and interact with the world: agents can be initiated in the world, they can search their world, they can make decisions, they can execute instructions, they can communicate with each other, they can update their information and rules, and they can be terminated. It is here, at the agent level, that theories of social behavior are embodied within the simulation. Environmental factors and infrastructures can also be represented, including transportation, communication, oil, financial flows, utilities, health, and security. With these features – actions, effects, agents, behaviors, and environment – scenarios that explore the logical consequences of shutting off the water supply, increasing unemployment, or experiencing an earthquake or tsunami can be run (Baard 2007). Exploration of possible futures through simulation allows high-level decision-makers to think about governance under varied circumstances.

PSTK is an agent-based simulation largely focused on explaining power structures and the way they change over time (Taylor et al. 2006). PSTK models are driven by four *domains* – political (P), military (M), economic (E), and social (S) – that describe an agent's capabilities. Figure 4 shows three agents, each featuring different PMES levels called *effective capital*. Each agent has a set of explicit goals and makes

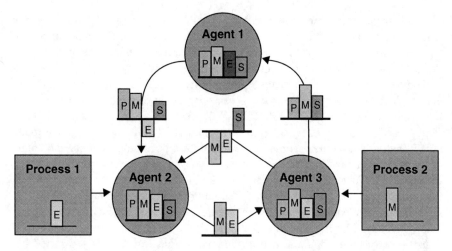

Fig. 4 An example model in the PSTK graphical user interface (Taylor et al. 2006)

decisions about how it will expend its capital to achieve those goals. Processes, like those to the left and right of Fig. 4, are similar to agents but do not make decisions.

For this application, the power structure relationships are examined among three regional rivals as the economic capability of Agent 2 and the military capability of Agent 3 are increased. Political, military, economic, and social capital flows among the three agents according to the relative capabilities and desired goal states of each agent. In this example, political, military, and social capital flows from Agent 1 to 2, while economic capital flows from Agent 2 to 1. Military and economic capital flows from Agent 2 to 3, while social capital flows from Agent 3 to 2. Agent 2 is aided in its ability to give economic aid to both Agents 1 and 3 by the Process 1 economic capital flow. Lastly, political, military, and economic aid flows from Agent 3 to 1. Note that Agent 3 receives additional military aid from Process 2 as well as military capital from Agent 2 and gives military aid to Agent 1.

One of the key insights that emerge from the PSTK model is that a number of well-understood relationships may combine to result in the emergence of unexpected consequences and counterintuitive behaviors. The PMES capital that drives the PSTK model largely corresponds to the DIME actions and PMESII effects described in the SEAS model. Both agent-based models thus address the fact that effective governance requires employing and balancing different types of power to maintain sovereignty, security, and stability.

2.3 System Dynamics

System dynamics (SD) is another modeling methodology that can be used to study governance. SD, like agent-based modeling, allows analysts to study the changing system behavior over time. However, rather than focusing on individual agents, SD features aggregated quantities that are manipulated through differential equations (Sterman 2000). Two governance-related models are examined here: the State Stability Model (SSM) and the Counterinsurgency (COIN) model.

The SSM models and evaluates state stability by balancing the *load* an insurgency places on the regime with the *capacity* of the regime to withstand the load (Choucri et al. 2007; OSD 2009, 334–335). Governance is thus represented in this model by the ability of the regime to handle the insurgency load, which increases as some people in the population become dissidents and some of them become insurgents. The load is lessened when some of the insurgents are removed. These transitions can be seen at the upper part of Fig. 5 in the Population, Regime Opposition, and Reducing Overt Opposition dashed boxes. Four rectangular *stocks* depict these groups, and *flows* depict the transitions among groups. The logic within the recruiting dashed box shows how people transition from the population to dissidents, a transition that is supported by Communication and Mobilization and countered by Regime Resilience, which is one place where the insurgency load comes into direct contact with regime capacity; the Reducing Overt Opposition dashed box is another. Note that Insurgents can return to Dissidents and Dissidents can return to the Population as a function of Government Appeasement.

The COIN model depicted in Fig. 6, like the SSM model, puts the population at the center of the analysis. The COIN model is based on the counterinsurgency theory of U.S. Army (2006), the goal of which is not to kill the enemy but to win over the population, which will in turn help defeat an insurgency. At the center of Fig. 6 are three stocks that show the extent to which the population supports the Host Nation (HN) government or the insurgency. Governance is represented by a capacity that takes as input support from the population and time to develop; governance can be harmed or hindered through insurgent acts of violence.

Governance then drives three effects: it influences economic investment, the population's support of the HN government, and psychological operations effectiveness, which then influences perceived security. Note that although one may disagree, the COIN model is clear about what governance is and is not. Its inputs and outputs are well defined. Pierson et al. (2008) use the COIN model to make their assumptions explicit and open for evaluation, review, and criticism. SD modeling of governance is developed further in the next section.

3 Modeling Governance with System Dynamics

In this section, several governance concepts are explored by representing them in the system dynamics (SD) modeling methodology generally and iThink (Richmond 2005) simulation tool specifically. First, a simple model is presented and analyzed, and then more complex elements are added to the model to develop additional governance and modeling concepts.

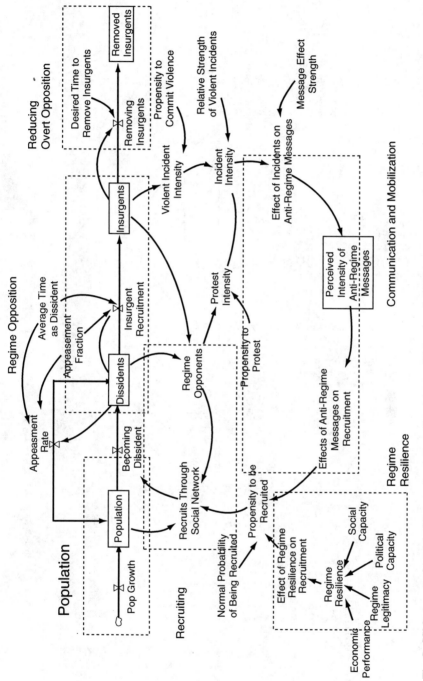

Fig. 5 SSM conceptual model of insurgent activity and recruitment (Choucri et al. 2007)

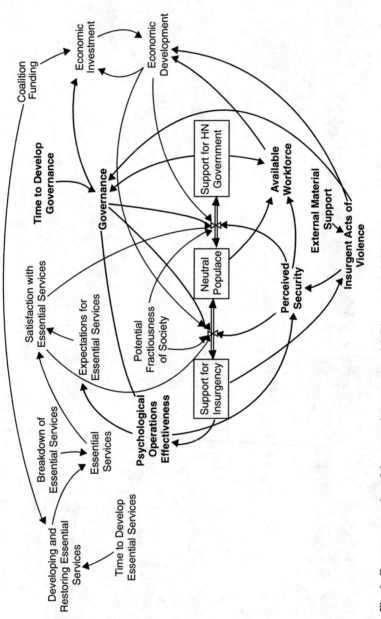

Fig. 6 Governance section of the counterinsurgency model (Pierson et al. 2008)

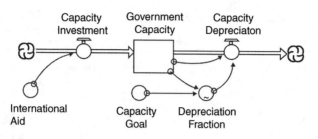

Fig. 7 Government capacity model

Figure 7 shows a simple governance model using the four key SD primitives: stocks, flows, converters, and action connectors. First are the rectangular *stocks* that contain the quantity being measured, in this case Government Capacity. Second are the *flows* that change the level of a stock. Two are defined here: (1) the inflow Capacity Investment and (2) the outflow Capacity Depreciation. Stocks can only change through flows, so the only way to increase Government Capacity is through investment, and the only way to reduce it is through depreciation. Flows contain equations that determine their value. Stocks feature no equations, only initial conditions. Third, the remaining three circles – International Aid, Capacity Goal, and Depreciation Fraction – are *converters* that contain additional equations and constants. Converters can also contain tabular relationships as denoted by the tilde (~) in Depreciation Fraction. Fourth, the final primitive are the arrows that connect the other three, which in iThink are called *action connectors* and show and denote the model variables that are used in the flow and converter equations. For example, as Capacity Depreciation has two connectors coming into it, its defining equation must include those two variables. The equation is in fact this – Government Capacity × Depreciation Fraction – so both variables are included in the definition.

Figure 8 shows the dynamic behavior of the simple Government Capacity model. Looking at the horizontal, X-axis, the simulation runs for 24 months. Looking at the vertical, Y-axis, there are several scales: the first one for Government Capacity between 0 and 80, and the second and third for Capacity Investment and Capacity Depreciation between 0 and 20. The units for these two ranges are different. The first is an aggregate measure of capacity. The Government Capacity variable, a stock, comprises many different components, for instance labor and capital. Government capacity requires trained employees, buildings, and equipment, but in this model these and other components are aggregated into a single capacity variable. The two flows represent changes in government capacity per month. In Fig. 8, the inflow, Government Capacity, remains constant at about 16. Capacity Depreciation starts small but rises until it equals Capacity Investment. Government Capacity also rises until Capacity Investment equals Depreciation, at which point Capacity levels off.

This is to be expected because, when the inflow equals the outflow, there is no change in the overall level, in this case Government Capacity. The interpretation is that the international community has funded a nation-building effort for a failed

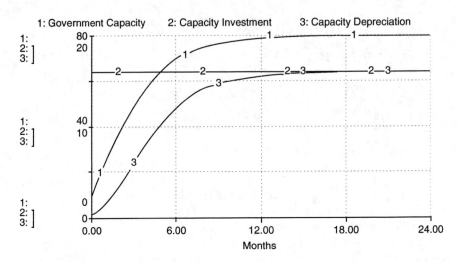

Fig. 8 Dynamic behavior of government capacity model

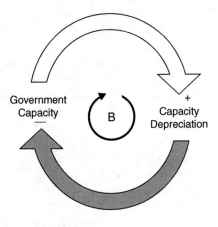

Fig. 9 Negative, balancing feedback

state. Government Capacity increases quickly with the funding, but as employees, buildings, and equipment become part of the government, their recurring mainte- nance costs increase: employees need to be trained, buildings need to be repaired, and equipment needs to be fixed or replaced. That is, the bigger the capacity, the more it costs to maintain that capacity, so the government grows until its mainte- nance costs are equal to its income.

The point at which Investment equals Depreciation is not defined explicitly but is instead determined experimentally through the feedback process shown in Fig. 9. When the simulation is run, increases in Government Capacity cause increases in the Capacity Depreciation flow as denoted by the positive sign next to the light arrowhead, which indicates "change in the same direction." However, increases in Capacity Depreciation cause decreases in Government Capacity as denoted by the negative sign next to the dark arrowhead, which indicates "change in the opposite

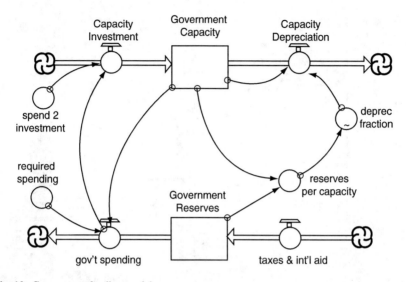

Fig. 10 Government funding model

direction." These two causal relationships combine to form a negative feedback relation, which can be determined by counting the negative causal relationships. An odd number of negative signs denote a negative, balancing, or goal-seeking feedback loop; zero or an even number denotes a positive or reinforcing feedback loop. The "B" at the center of Fig. 7 shows that it is a negative or balancing loop.

Figure 10 introduces some additional complexity to the Government Capacity model by showing how Capacity Investments are funded. A stock of Government Reserves contains the money available to invest in Government Capacity. The Government Spending outflow drains this stock as investments are made, and an inflow from locally collected taxes and international aid increases the stock. The Reserves per Capacity converter to the right of Fig. 10 captures the tendency for people to make do with less as the Government Reserves get low and the money runs out.

Figure 11 shows a sensitivity analysis of the Government Funding model; each behavior over time line is driven by an increasing inflow of taxes and international aid. Note that for every value of taxes and international aid inflow, Government Capacity reaches a maximum and then falls back to a lower level after the Government Reserves have been depleted, although higher flows allow for higher and later capacity peaks. The lesson here is that over the long term, monetary outflows must match inflows regardless of initial reserves.

The initial upward curve in Fig. 11 can be explained by the positive, reinforcing behavior of Fig. 12. More Government Capacity is paid for with more Government Spending, more Government Spending is required for more Capacity Investment, and more Capacity Investment leads to more Government Capacity. There are no negative causal relationships in the loop, which makes it positive or reinforcing. Positive feedback gives Fig. 11 that initial, upward sloping behavior until

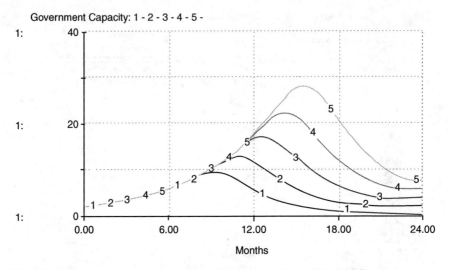

Government Capacity: 1 - 2 - 3 - 4 - 5 -

Fig. 11 Government capacity behavior for various taxes and international aid flows

Fig. 12 Positive, reinforcing feedback

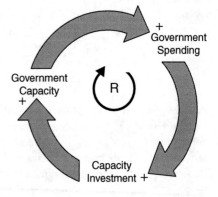

Government Reserves are depleted, and then negative loops take over and capacity falls. This is a general SD lesson: systems and their positive loops cannot grow forever. Eventually, negative loops will assert themselves and the system will return to equilibrium.

The Government Rivalry model (Fig. 13) builds on the previous models in this section in two ways. First, Government Capacity has been duplicated as a Rival Capacity. Second, some population stocks have been added, and the government and rival organizations compete for the loyalty of this population per Pierson et al. (2008). Three of the more interesting causal relationships that emerge from these additions are described. First, the more capable the Government is with respect to its Rival, the more loyalty it commands from the population, and the more it is able to collect taxes from the population. Second, connections from the two Capacity stocks to the other's Depreciation flows allow the organizations to attack each

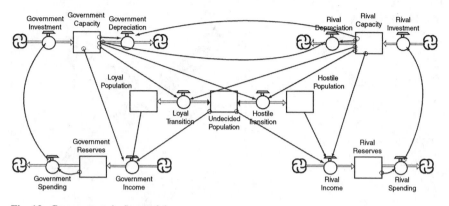

Fig. 13   Government rivalry model

other. Third, even though the structure of the Government and its Rival are similar, the values that depict the strengths and weaknesses for each organization need not be the same, thus allowing for asymmetric competition. For example, Government Reserves may be much larger than the Rival Reserves, but the rival may easily be able to depreciate Government Capacity, while the government may have a comparatively difficult time depreciating Rival Capacity. These competitive themes are explored and developed in the next section.

4   The QVP Model

As an additional example of governance modeling, in this section we translate the Quest for Viable Peace (QVP) governance theory presented earlier in this chapter into a computer model using the SD simulation methodology.

In conjunction with the QVP theory, Covey et al. (2005) develop the notion of X-charts shown in Fig. 14 which depicts the power of obstructionists going down over time and the capacity of legitimate government institutions going up. Here, we apply the QVP theory to provide a simulation that allows decision-makers to experiment with "active" X-charts. That is, users should be able to modify the model inputs or actions, run the simulation, see how the outputs or effects change, and then repeat the process. Accomplishing this requires developing metrics that represent the "power of obstructionists" and "capacity of legitimate institutions."

Recall that Covey et al. (2005) developed four governance elements: the (1) political, (2) military, (3) rule of law, and (4) economic. Figure 15 develops three of them, with the political element represented in the "legitimate government" and "enemy forces" sectors, the military in the "friendly forces" and "enemy forces" sectors, and the economic in the "economic" and "economic distribution" sectors. The population sector represents the "mass of society" as shown in Figs. 1–3, and policy levers available to senior decision-makers are contained in the "data inputs" sector.

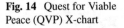

Fig. 14 Quest for Viable
Peace (QVP) X-chart

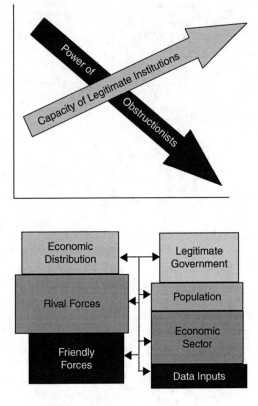

Fig. 15 Quest for Viable
Peace model

Figure 16 shows the model logic in its visual, simulatable form. The large arrow in the center indicates the many connections among model sectors, which are actually handled by the dashed icons called *ghosts*. The model explanation begins in the lower left-hand corner of Fig. 16 as an international assistance intervention arrives (Fig. 3). This causes the criminal political elements to engage the intervention security forces and to transition from an overt to covert posture as the security forces suppress them. The criminal political members desert as they realize that it is pointless to continue fighting, but then they are left to decide between taking employment within the economy and joining the insurgency. A shortage of jobs increases the odds of a deserter joining the insurgency. This tracking of the criminal political population through various stocks matches the core functionality of Choucri et al. (2007) and Pierson et al. (2008). Unlike those models, though, the Population Sector on the right half of the model features an undifferentiated populace.

The Economic sector contains two economies, legitimate and illegitimate, the benefits of which are directed toward either the criminal political elite or the population in the Economic Distribution sector. As the intervening powers take over control of the country, the benefits of the economy will be redirected away from the criminal political elite and toward the mass of society. A small Legitimate

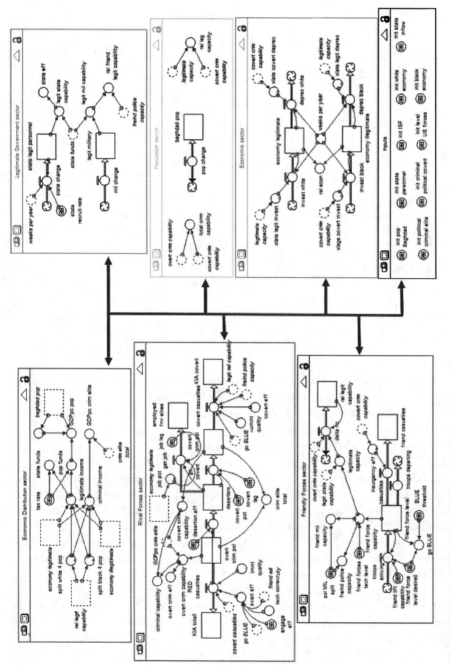

Fig. 16 Quest for Viable Peace (QVP) system dynamics simulation

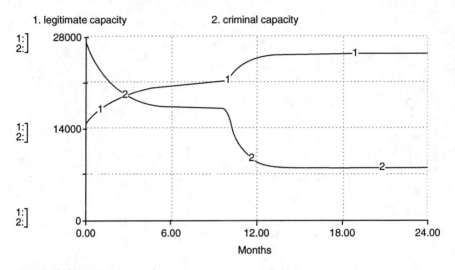

Fig. 17 QVP governance metrics

Government sector that features separate stocks for civilian and military personnel is provided. However, these legitimate government capabilities are built up only over the long term, so this sector serves primarily as a placeholder for future model development.

Figure 17 graphs the behavior of two metrics that track governance and the efficacy of the international intervention: Legitimate Capacity and Criminal Capacity. Legitimate Capacity is the sum of the Legitimate State Capacity and Intervention Force Capacity. These capacities in turn are driven by the product of the number of personnel in the organization and the technology level of the organization. The graph shows Legitimate Capacity increasing initially and Criminal Capacity decreasing. At about month 10, though, the political criminal element gives up fighting and deserts, which causes Criminal Capacity to drop quickly and Legitimate Capacity to rise. The resulting graph shows Legitimate Capacity increasing and Criminal Capacity decreasing like the X-chart in Fig. 14, but Fig. 17 is more specific and quantitative. The Legitimate and Criminal Capacity variables have definitions, measurable values over a set time frame, and dynamic variance over the time frame that can be causally explained.

Figure 18 shows the economic consequences of the international intervention. Legitimate and Criminal Income are both calculated in the Economic Distribution sector. As intervention forces gain control of the country, the economic resources that have previously been misdirected to the country's criminal political elite are redirected to the general population or mass of society. The scale of the Y-axis is in thousands of dollars, so this is reflected in the increase of Legitimate Income from a little over $30 million to a little over $50 million and the decrease of Criminal Income from a little less than $30 million to about $7.5 million. The regions' population is seven million people, which leads to a GNP per capita of about $8,500 per

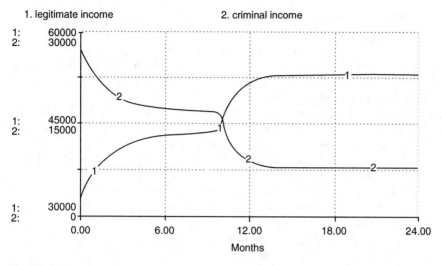

Fig. 18 QVP economic metrics

year. Note that the scales of the two curves are different for Fig. 18, with Legitimate Income ranging from $30 million to 60 million and Criminal Income from $0 to 30 million. Note also that the nonlinearity at time 10 that was encountered in Fig. 17 shows up again in Fig. 18. Lastly, recall that our goal in creating the QVP model was to show that active X-charts could be created using SD simulation. Figs. 16–18 demonstrate that this goal was achieved.

5 Practical Tips

- Expose models and modelers to clients early in the model development process so that when the models are finally presented for use, clients know what to expect. Clients are rarely comfortable with "big bang" exposures, in which the modeler receives the requirements, disappears for a while, and returns with a completed model. If clients have not contributed to a model's construction, they tend to lack a sense of ownership and confidence, and even if a model is correct, they may not embrace it.
- Keep in mind that a simulation can be transformed into a useful collaboration and consensus-building tool. Policy and governance problems of the type discussed here require a "broad and shallow" perspective that encompasses political, military, rule of law, economic, and myriad other concerns (Richmond 2005). Decision-makers appreciate computer-based tools that synthesize such diverse perspectives, and if members of the experts team gain a sense of ownership in the model, it can serve as a consensus and team-building tool.

- Recognize that in the process of building a model, discussing it with experts, and performing sensitivity analysis, it is often found that most variables are uncontroversial; i.e., everybody agrees on the variables' values and recognizes that they are unlikely to change drastically. Consider using the insights gleaned in this development and vetting process to focus your analysis and data collection resources on the 5–10% of variables that are controversial, dynamic, or unknown.
- Distinguish policy levers from model initialization inputs. Although a modeler, who sees both as straightforward numerical inputs, may be inclined to treat the two similarly, policymakers who can change the former (e.g., by means of economic aid) but not the later (e.g., a population's culture) will tend to think of them quite differently.
- Always have available debugging tools and methods that can quickly differentiate between insights and bugs. When decision-makers suggest a set of governance policies, and modelers report to them the results of a simulation, decision-makers will question results that contradict their experiences and expectations. The questioned results may be either legitimate counterintuitive insights or programming errors. Modelers hope for the former, but must be prepared for the latter.

6 Summary

A loss in a government's capacity to govern is accompanied by a decline in its ability to control territory, limit violence, enforce policies and regulations, collect revenue, and provide services to population. Statistics-based approaches study data to predict changes in governance in response to specified conditions. For example, the CAST model calculates a score for each of 12 indicators of governance and employs correlation analysis and expert opinion to derive the Failed State Index. Another statistical model, ACTOR employs regression analysis on two sets of data, a list of PMESII indicators and SME judgments on the country's stability. Agent-based approaches to the study of governance are exemplified by such models as SEAS and PSTK. Agents can represent various levels of group aggregation: for example, agents can represent governments, leaders, citizens, or organizations such as corporations, agencies, and nongovernmental organizations. SEAS seeks to minimize aggregation. The Quest for Viable Peace (QVP) theory expresses governance as a function of four elements – political, economic, military, and rule of law. In the theory, an international intervention contributes to a conflict between "friendly" and "enemy" forces. The criminal subclass joins the conflict but eventually withdraws, and must then choose between joining the economic model as well-behaved actors and joining an insurgency. The international assistance serves to shrink the illegitimate sector of economy and feed the legitimate sector. Based on this conceptual model, a computational QVP model uses system dynamics technique: stock variables; flow variables; converters; and action connectors. In translating the QVP theory into a working system-dynamics computational model, consideration had to be given to partitioning PMESII sectors appropriately (e.g., political and military into friendly and enemy components).

The example simulation depicts the societal development over 24 months and outputs evolution of several indices such as Government Capacity, Capacity Depreciation, Criminal Income, and Criminal Capacity.

7 Resources

1. Foreign policy

   Failed State Index (Fund for Peace 2005):
   At its publisher, Foreign Affairs, www.foreignpolicy.com/failedstates
   At its creator, The Fund for Peace, http://www.fundforpeace.org

   The National Security Council or NSC:
   http://www.whitehouse.gov/administration/eop/nsc

   Foreign Policy: http://www.foreignpolicy.com

   Foreign Affairs: http://www.foreignaffairs.com

   The RAND Corporation: http://www.rand.org

   The Heritage Foundation: http://www.heritage.org

   The Brookings Institute: http://www.brookings.edu

   The Ford Foundation: http://www.fordfound.org

2. Journalistic sources

   The Economist: http://www.economist.com

   The New York Times: http://www.nytimes.com/world

   The Wall Street Journal: http://www.wsj.com/world

   The Washington Post: http://www.washingtonpost.com/world

   The Atlantic: http://politics.theatlantic.com

   The New Republic: http://www.tnr.com

   The National Review: http://www.nationalreview.com

   Frontline: http://www.pbs.org/wgbh/pages/frontline

3. System dynamics

   System dynamics methodology: John Sterman's Business Dynamics (2000)
   http://www.mhhe.com/business/opsci/sterman

   Ithink: http://www.iseesystems.com

   Vensim: http://www.vensim.com

   Powersim: http://www.powersim.com

   AnyLogic: http://www.xjtek.com

4. Agent based simulation

   Repast: http://repast.sourceforge.net
   Swarm: http://www.swarm.org

Mason: http://cs.gmu.edu/~eclab/projects/mason

StarLogo: http://education.mit.edu/starlogo

NetLogo: http://ccl.northwestern.edu/netlogo

5. Statistics

R statistical computing environment: http://www.r-project.org

Matlab: http://www.mathworks.com

SAS: http://www.sas.com

SPSS: http://www.spss.com

Stata: http://www.stata.com

6. Data

CIA World Factbook: https://www.cia.gov/library/publications/the-world-factbook

World Bank: http://www.worldbank.org/data

International Monetary Fund or IMF: http://www.imf.org/external/data.htm

Penn World Tables Purchasing Power Parity or PPP: http://pwt.econ.upenn.edu/php_site/pwt_index.php

Kansas Event Data System (KEDS): http://web.ku.edu/~keds

University of North Texas: http://www.paulhensel.org/data.html

Buffalo: http://cas.buffalo.edu/classes/psc/duchesne/psc328/database.html

Emory: http://einstein.library.emory.edu/intlinks.html

Georgia Tech: http://www.library.gatech.edu/research_help/subject/index.php?/international_affairs/datasets

Tennessee: http://web.utk.edu/~gsops/dataSETs.html

7. Visualization

Edward Tufte: http://www.edwardtufte.com

Environmental Sciences Research Instistute (ESRI): http://www.esri.com

Google Earth: http://earth.google.com

References

Baard, M. (2007). Sentient World: War games on the grandest scale. *The Register*. http://www.theregister.co.uk/2007/06/23/sentient_worlds/.

Beal, R. S. (1985). Decision Making, Crisis Management, Information and Technology. *Program on Information Resources Policy, Technical Report*. Cambridge, Mass: Center for Information Policy Research, Harvard University.

Blair, S. A., Eyre, D., Salome, B. & Wasserstrom, J. (2005). Developing a Legitimate Political Economy. In J. Covey, M. J. Dziedzic & L. R. Hawley (Eds.) In *The Quest for Viable Peace: International intervention and strategies for conflict transformation* (pp. 205–243). Washington, DC: US Institute of Peace.

Choucri, N., Goldsmith, D., Madnick, S. & Mistree, D., Morrison, J. B. & Siegel, M. (2007). Using System Dynamics to Model and Better Understand State Stability. Presented at *International Conference of the System Dynamics Society*, Boston.

Covey, J., Dziedzic, M. & Hawley, L. (Eds.) (2005). *The Quest for Viable Peace: International intervention and strategies for conflict transformation*. Washington, DC: US Institute of Peace.

Davis, P. K. (2001). *Effects-Based Operations: A grand challenge for the analytical community*. Santa Monica, CA: RAND.

Davis, P. K. & Cragin, K. (Eds.) (2009). *Social Science for Counterterrorism: Putting the pieces together*. Santa Monica, CA: RAND.

Dobbins, J. (2003). Nation-Building: The inescapable responsibility of the world's only superpower. *RAND Review*, 27(2), 16–27.

Eberlein, R. (2007). *Vensim User's Guide (version 5)*. Harvard, MA: Ventana Systems.

Effects Based Research (2007). *Verification, Validation, and Accreditation Final Report*. Vienna, VA: Evidence Based Research.

Forrester, J. W. (1961). *Industrial Dynamics*. Cambridge, MA: Productivity Press.

Forrester, J. W. (1971). Counterintutitve Behavior of Social Systems. *Technology Review*, 73(3), 52–68.

Fukuyama, F. (2004). *State-Building: Governance and world order in the 21st century*. Ithaca, NY: Cornell.

Fund for Peace. (2005). The Failed State Index. *Foreign Policy*. This survey has been updated yearly since 2005 http://www.foreignpolicy.com/articles/2009/06/22/the_2009_failed_states_index.

Fund for Peace. (2009). The Failed State Index: Frequently Asked Questions (FAQ) and methodology. *Foreign Policy*. http://www.foreignpolicy.com/articles/2009/06/22/2009_failed_states_index_faq_methodology.

Hawley, L. R. & Skocz, D. (2005). Advance Political-Military Planning: Laying the foundation for a viable peace. In J. Covey, M. J. Dziedzic, & L. R. Hawley (Eds.) *The Quest for Viable Peace: International intervention and strategies for conflict transformation*, (pp. 37–76). Washington, DC: US Institute of Peace.

JFCOM (2004). *Operational Net Assessment: A concept paper for joint experimentation*. Norfolk, VA: US Joint Forces Command.

Jones, S. G. (2008). *Counterinsurgency in Afghanistan*. Santa Monica, CA: RAND.

Kristoff, N. D. (1999). Global Contagion Series. *New York Times*. Four part series: Part 1, "Who sank, or swam, in choppy currents of a world cash ocean." with E. Wyatt (Feb. 15); Part 2, "How US wooed Asia to let cash flow in." with D. E. Sanger (Feb. 16); Part 3, "World's markets, none of them an island." with S. WuDunn (Feb. 17); Part 4, "The world's ills may be obvious, but their cure is not." with S. WuDunn (Feb. 18).

Morecroft, J. D. W. (1983). System Dynamics: Portraying bounded rationality. *International Journal of Management Science*, 11(2), 131–142.

MSCO (2006). *Verification, Validation, and Accreditation (VV&A) Recommended Practices Guide (RPG)*. Alexandria, VA: Modeling and Simulation Coordination Office. http://vva.msco.mil/.

NSC (2002). *The National Security Strategy of the United States of America*. National Security Council (NSC), Presidential Report.

O'Brien, S. (2001). *Analyzing Complex Threats for Operations and Readiness*. Technical Report, Center for Army Analysis (CAA-R-01-59), Ft. Belvoir, VA.

O'Brien, S. (2007) *Integrated Crisis Early Warning System (ICEWS) Program Description*. Arlington, VA: Defense Advanced Research Projects Agency (DARPA). http://www.darpa.mil/ipto/Programs/icews/icews.asp.

OSD (2009). *Requirements for a Government Owned DIME/PMESII Modeling Suite*. Arlington, VA: Office of the Secretary of Defense (OSD) Modeling and Simulation Steering Committee.

Perry, W.L. & Gordon IV, J. (2008). *Analytic Support to Intelligence in Counterinsurgencies*. Santa Monica, CA: RAND.

Pierson, B., Barge, W. & Crane, C. (2008). The Hairball that Stabilized Iraq: Modeling FM3-24. *Presented at the Human, Social, Cultural, and Behavioral (HSCB) Modeling Conference*, Center for Technology and Security Policy, National Defense University, Washington, DC.

Popp, R. (2005). *Utilizing Social Science Technology to Understand and Counter the 21st Century Strategic Threat*. Arlington, VA: DARPA DARPATech.

Prevette, M. & Snyder, D. (2007). Synthetic Environment for Analysis and Simulation (SEAS). Technical brief, United States Joint Forces Command (USJFCOM)/J9, Norfolk, VA for the *CCRP Information Age Metrics Working Group (IAMWG)*.

Randers, J. (1980). Guidelines for Model Conceptualization. In J. Randers (Ed.) *Elements of the System Dynamics Method* (pp. 117–139). Cambridge, MA: Productivity Press.

Reddy, R. (1996). The Challenge of Artificial Intelligence. *Computer* 29(10), 86–98.

Richmond, B. (2005). *An Introduction to Systems Thinking*. Lebanon, NH: ISEE Systems. First published in 1992.

Sterman, J. (2000). *Business Dynamics: Systems thinking and modeling for a complex world*. New York: Prentice Hall.

Taylor, G., Bechtel, R., Morgan, G. & Waltz, E. (2006). A Framework for Modeling Social Power Structures. Presented at the *14th Annual Conference for the North American Association for Computational Social and Organizational Sciences*.

Tufte, E. (2001). *The Visual Display of Quantitative Information* (2nd ed.). Cheshire, CT: Graphics Press.

US Army (2006). *Counterinsurgency* (FM3-24). Field Manual. Headquarters, Department of the Army, Washington, DC.

Waldrop, M. (1993). *Complexity: The emerging science at the edge of order and chaos*. New York: Simon & Schuster.

Yergin, D. & Stanislaw, J. (2002). *The Commanding Heights: The battle for the world economy*. New York: Simon & Schuster.

Chapter 7
Groups and Violence

Ravi Bhavnani, Dan Miodownik, and Rick Riolo

Violence can take place along a multitude of cleavages, e.g., (1) between political groups like the Kach Movement, pitting West Bank settlers against Israeli governments supporting the land-for-peace agenda; (2) between religious groups, such as Christians and Muslims in the Nigerian cities of Jos and Kaduna; (3) along class lines, as in India between Dalits and members of the Brahminical upper castes, upwardly mobile intermediate castes, and even other backward castes such as the Thevars; and (4) between ethnic groups such as the Hutu and Tutsi, both within and across state boundaries in Rwanda and neighboring Burundi.

Currently, conflict between ethnic groups is understood best; therefore, we focus here on modeling ethnic violence, noting that similar approaches apply to other forms of intergroup violence. And while we do not directly address the effect of intervention on ethnic violence, we discuss how a commonly cited set of factors may all affect the incidence, intensity, and duration of violence between ethnic rivals. These factors include uncertainty about a rival's resolve, the speed at which divergent beliefs converge, direct intervention versus the use of side payments, neutral versus biased interveners, the role of spoilers, and neighborhood effects.

Specifically, the literature suggests that wars involving international intervention are typically more violent, more atrocious, and longer than those without (Regan 2002). The implication for ethnic wars often among the most difficult to resolve, is that intervention by a neutral party may in fact prolong the fighting as opponents maneuver to strike a better deal at the negotiation table, whereas biased intervention favoring the opposition may reduce conflict duration (Collier et al. 2004).

Others argue that successful intervention depends upon the ability of the intervening party to deploy forces to monitor and enforce peace agreements, while at the same time fulfilling local civic and economic functions (Doyle and Sambanis 2000, 2006). Still others suggest that success is more closely tied to the following factors: (1) mediation style – manipulation proving more effective for formal agreements

R. Bhavnani (✉)
Department of Political Science, East Lansing,
Board of Trustees, MI 48824, USA
e-mail: rvibhav@gmail.com

A. Kott and G. Citrenbaum (eds.), *Estimating Impact*,
DOI 10.1007/978-1-4419-6235-5_7, © Springer Science+Business Media, LLC 2010

and crisis abatement, and facilitation for diffusing tension (Beardsley et al. 2006); (2) preferences – with mediation efforts being more likely to build trust if the mediator cares about the issue at stake, has a moderate ideal point, and does not consider the cost of conflict to be prohibitive (Kydd 2006); and (3) time horizon – with the ability to help rivals identify and reach satisfactory outcomes waning in the long run, as demands also begin to change (Beardsley 2008).

This chapter is organized as follows: It begins with a review of prominent theoretical and methodological approaches to the study of ethnic violence. Next, it discusses why agent-based models (ABM) lend themselves particularly well to the study of violence between members of nominally rival ethnic groups using some illustrative examples. Then, it turns to a discussion of REsCape – an agent-based computational framework for studying the relationship between natural resources, ethnicity, and civil war. It concludes with some thoughts about the distinction between exploratory and consolidative modeling, and the appropriateness of each for studying ethnic violence.

1 Approaches to Studying Ethnic Violence

In this section, we briefly describe the main alternative methodological approaches to ABM. Each approach has its own strengths and weaknesses, and as a result each focuses on particular questions, units of analysis, and modes of understanding. We view these approaches as complementary – each is useful in different situations and several of them can be used in combination to yield a deeper understanding of the causes and nature of ethnic violence.

1.1 Theoretical Approaches

Several theories attempt to explain the incidence of violence between ethnic rivals, varying in their degree of formality, emphasis on micro- versus macrolevel factors, use of statistical evidence versus case-based reasoning, and focus on underlying processes versus equilibrium outcomes. In political science, theories of ethnic violence are commonly divided into the following categories: (1) *primordial*; (2) *institutional*; (3) *instrumental*; and (4) *constructivist*.

Widely discredited in the literature, but with a fair amount of currency in the policy community, primordial approaches highlight the *intrinsic nature of ethnic antagonisms* – as purported to exist between Hutu and Tutsi in both Rwanda and Burundi dating back to the period preceding colonial rule (Braeckman 1994; Chrétien 1995; Mamdani 2001; Uvin 1996). One variant of the identity-based explanation emphasizes the particularity of culture. In the Rwandan case, the argument suggests that a culture of conformity and unquestioning obedience to authority existed among the Hutu, facilitating mass participation in the genocide (Gourevitch 1998; Khan 2001; Prunier 1995; Scherrer 2002).

Colonial intervention often factors prominently in this explanation, given the role of colonial administrators in reinforcing or solidifying ethnic stereotypes and antagonisms through preferential treatment or divide-and-rule policies, with the decades old Hindu–Muslim conflict in India being one prominent example. A related explanation emphasizes individual and group psychology and suggests that the heightened salience of social identity, together with the revival of strong childhood emotions and membership in religious groups, constitute a volatile brew that foments ethnic or sectarian tension. The heightened salience of social identity, together with the revival of strong childhood emotions and membership in religious groups, come together to make a volatile brew that leads to ethnic or sectarian tension. According to this view, cultural identity is an unconscious human trait which assumes salience when threatened (Kakar 1996). The result is that fear, anxiety, and panic all serve to heighten the salience of group identity, making a crowd more susceptible to manipulation. Rumors, propaganda, or other catalysts can then set the crowd off, resulting in "collective bloodlust" (Scherrer 2002:113). Yet, the large number of willing participants in episodes of mass violence ostensibly undermines theories of deviant behavior, given that for the majority of perpetrators, participation, far from being considered deviant behavior, more often tends to be sanctioned by the state.

A second explanation has its roots in *institutional theory*, suggesting that in areas prone to ethnic violence, a nexus of politicians, criminals, informants, and thugs exist (Brass 1997). As Paul Brass, its key proponent, puts it, "There are, then, a whole series of specialized roles that are occupied in larger riots, including provocateurs, monitors, informers, 'riot captains and thugs,' provisioners of transport and liquor, criminals, bomb manufacturers, journalists and pamphleteers, graffiti writers, and distributors and plasterers of scurrilous posters" (p. 16). In addition, fire tenders "… maintain communal, racial and other ethnic relations in a state of tension, of readiness for riots" and conversion specialists "… know how to convert a moment of tension into a grander, riotous event" (Brass 1997). Brass refers to this collection of actors and the informational network connecting them as an *institutionalized riot system*.

Other variants of the institutional argument analyze the role of the state in fostering violence. Turning more specifically to sectarian violence, Kalyvas and Kocher (2007) argue that the second conflict in Iraq has been misportrayed as an "ethnic" civil war, and that such an assessment is misleading. Instead, they suggest that the sectarian conflict between Shia and Sunni militias that began in 2003 and is still continuing, albeit to varying degrees of intensity, throughout the war is not simply the outcome of cleavages in Iraqi society but to an important extent is a legacy of the U.S. occupation that started in 2003. As such, they argue that while some conflicts evolve into sectarian wars, others develop dynamics virtually indistinguishable from ideological civil wars, with the difference hinging critically upon the role of the state. However, even the state's role is often open to question. In the context of the Rwandan genocide, popular accounts suggest that the frequent bouts of interethnic violence were attributable to a weak or fragmented state structure, typical of many failing African regimes (Thorning 2005). Others argue to the contrary, suggesting that the highly rigid and hierarchical organization of Rwandan society into *préfectures*, *communes*, *collines*, *secteurs*, and *cellules* (Des Forges 1999; Prunier 1995;

Verwimp 2006) exemplified a high level of state capacity and penetration – key for successfully coordinating mass violence against Tutsi in 1994.

A third explanation points to the *instrumental use of violence*, highlighting the role of compulsion by ethnic entrepreneurs for whom violence is a means to maintain or increase power. During La Violencia, a period of human rights abuses against the Mayan Indians in Guatemala from 1960 to the mid 1990s, the "guilty" as well as those who failed to obey orders were killed, all in a purposive effort to engender collective silence and compliance (Zur 1994). During the war in the early 1990s, the Serbian population in Croatia was terrorized into submission and ethnically mobilized by Serbian guerillas, police, and army units, while in Serbia the SPS (Socialist Party Serbia) accused those who questioned the war of treason, sent reservists from opposition strongholds to the front first, and tortured or killed Serbs identified as "disloyal" to the Serbian cause (Gagnon 1995; Kuran 1998b; Rieff 1995; Vulliamy 1994). In each instance, conformity and participation increased as a result of compulsion, with grave consequences for those who failed to comply. This is illustrated most vividly in the case of the Rwandan genocide, which took place in 1994 and claimed upward of 800,000 lives, in which uncooperative officials were rapidly eliminated. Hutu who sheltered or hid Tutsi were punished, with sanctions that ranged from fines and beatings to rape and death at the hands of fellow Hutu. The reluctant were taught how to kill and compelled to participate in the killing (Mamdani 2001).

Finally, *constructivists* suggest that the salience individuals attach to ethnicity – one of multiple identities individuals may possess – varies as a function of incentives or strategic manipulation (Chandra 2001, 2006). To be sure, constructivism, which has discredited the primordialist approach by showing that ethnicity is fluid and endogenous to a set of social, economic, and political processes, is better suited to framing questions rather than providing answers to some important but nonetheless complicated questions: How large are individual identity repertoires? How easily can repertoires change? And do changes in repertoire size and salience affect larger political processes? Examples of more or less standard positions within this approach are Aronoff (1998), Brass (1980), Laitin (1998), and Nagel (1994).

Building upon this approach, associational or social capital arguments suggest that interethnic engagement contains violence. According to this view, trust based on inter- rather than intraethnic networks is critical, and preexisting local networks of civic engagement "stand out as the single most proximate cause explaining the difference between peace and violence" (Varshney 2003). These networks – which assume the form of business associations, trade unions, cadre-based parties, professional organizations, festival organizations, reading clubs, film clubs, and sports clubs – cut across ethnic groups and are distinct from communal organizations that foster trust among members of a single group. Thus, in contrast to civic ties that exist between groups, communal or group-specific organizations are often incapable of preventing violence and may even intensify violence. Where these networks of interethnic engagement exist, they effectively contain or dampen violence, whereas their absence leads to widespread violence. Furthermore, these networks can be divided into more formal and organized engagement and informal or everyday engagement. Whereas both forms of engagement promote peace, the argument suggests that formal interethnic associations are necessary to promote peace in large urban settings.

In short, members of an ethnic group share affinities with and commitments to one another that should facilitate collective action. Yet, while group consciousness – abetted by fear, frustration, deprivation, customary obligation, and, arguably, a culture of obedience – is *necessary* for collective action, it proves *insufficient* when the individual costs of participation are high.

Ascribing mass participation to a deeply ingrained animosity between Hutu and Tutsi fails to discern that anti-Tutsi sentiment has never been constant among Rwandan Hutu, and that even in 1994 – when as a group they had the most to fear from the impending Tutsi invasion – the country's Hutu population was not monolithic in its propensity to engage in violence against Tutsi. In a similar vein, characterizing Rwandan culture as one of conformity and obedience fails to account for selective Hutu defiance of the genocidal regime and the need for advance planning, propaganda, and persuasion to generate mass participation among the Hutu. Moreover, while there is little doubt that structural factors that pertain to economic and social conditions have important implications for participation, the conventional preference for tracking structural factors – which either tend to remain constant or are replicated to some degree in most episodes of conflict – appears to be overstated. Thus, the contention that Hutu stood to gain economically from a mass Tutsi exodus fails to explain why participation extended beyond criminal or disadvantaged classes (Mamdani 2001: 202) or why Hutu in Butare – one of the poorest and most overpopulated préfectures in Rwanda – were the last to participate in the killing (Des Forges 1999: 262; Straus 2004: 381).

Likewise, trust based on formal interethnic associations may, under the right set of conditions (active policing, representative membership in associations), prevent violence, but suggesting that the absence of these associations leads to violence reveals precious little about the causes and underlying mechanisms generating violent encounters.

1.2 Statistical Models and Empirical Claims

A second set of approaches focuses on empirical relationships of static snapshots to extract hypothetical causal relations from correlations. These approaches tend to assess the significance of aggregate explanatory factors and generally shy away from specifying the mechanisms that generate violence.

Take scholarship that seeks to assess the role of ethnicity in civil war. Few scholars of civil war would go so far as to argue that ethnicity is not an important factor in civil violence. Yet, contrary to expectation, the bulk of the statistically-oriented literature fails to establish a clear association between ethnicity and civil war onset, even in wars that are commonly identified as "ethnic" wars. Using the Ethno-Linguistic Fractionalization (ELF) Index (Roeder 2001; Alesina et al. 2003; Fearon 2003), an aggregate measure of ethnic heterogeneity, these studies make one of the following arguments: (1) ethnic heterogeneity decreases the likelihood of civil war as coordination for rebellion becomes harder in more diverse societies (Collier and Hoeffler 2004); (2) ethnic heterogeneity increases the likelihood of internal armed conflict, and to a lesser degree of civil war (Sambanis 2004; Hegre and Sambanis 2006); (3) ethnic heterogeneity increases the likelihood of civil war, yet the effect is indirect (Blimes 2006); (4) ethnic heterogeneity has a nonmonotonic association with the outbreak of civil war (Horowitz 1985; Elbadawi and Sambanis 2002) – low when the population is ethnically homogeneous and extremely diverse, and high when divided into a few prominent ethnic groups; or (5) ethnic heterogeneity has no significant relationship with the outbreak of civil war (Fearon and Laitin 2003; Fearon et al. 2007). Given the indeterminacy of these findings, it might be plausible to suggest that the focus on outcomes needs to be complemented by a focus on the *process* by which ethnicity becomes salient in civil war.

A similar problem plagues scholarship on the link between natural resources and civil war. The puzzle concerns why countries with high levels of risk for civil war – identified as low per capita income, a large population, rough terrain, petroleum, and political instability – often fail to generate conflict, while countries with ostensibly low risk levels experience conflict. The inability of models driven by aggregate, system-level variables to explain this empirical puzzle (see Table 1 for a summary of conflicting results) may also be traced to the absence of attention to mechanisms linking natural resources to civil war – how, for example, these mechanisms may influence or be influenced by the *type* of actors and *method* of resource extraction. Ross's (2004) summary of the findings from 13 cross-national studies suggests little apparent consensus on the resource-conflict relationship across five dimensions: conflict onset, conflict duration, the type of civil war, the type(s) of resource, and the underlying causal mechanisms. The only identifiable regularity is that oil dependence affects civil war initiation, not duration, with the reverse holding true for gemstones, coca, opium, and cannabis. Hegre and Sambanis's (2006) global sensitivity analysis of 88 variables finds the oil-exports-to-GDP variable to be a marginally significant predictor of low-intensity armed conflict, not civil war, while other commonly used measures of resource dependence are insignificant (Hegre and Sambanis 2006). And in a more recent study, Ross (2006) finds exogenous measures of oil, gas, and diamond wealth to be robustly correlated with the onset of civil war, with the caveat that these findings are based on a small number of cases and sensitive to certain assumptions.

1.3   Game-Theoretical and Negotiation-Based Approaches

Game-theoretical and negotiation-based approaches to the study of ethnic violence typically specify a few well-defined mechanisms, focus on interactions between representative agents (leaders, individuals, and groups), and draw conclusions about equilibrium outcomes, devoting less attention to underlying dynamics.

Table 1 Statistical research on resources and civil war

	Coverage	Resource measure	Dependent variable	Finding
Buhaug and Gates (2002)	262 conflicts, 1946–2000	Mineral resources in conflict zone (dummy)	Size of the conflict zone	Increases size of conflict zone
Collier and Hoeffler (1998)	27 wars, 1960–1992	Primary exports/GDP	War onset	Increases likelihood of war (curvilinear)
Collier and Hoeffler (2002)	52 wars, 1960–1999	Primary exports/GDP	War onset	Increases likelihood of war (curvilinear)
Collier and Hoeffler (2004)	48 wars, 1960–1999	Primary exports/GDP	War onset	Increases likelihood of separatist wars only (curvilinear)
Collier et al. (2004)	52 wars, 1960–1999	Primary exports/GDP	War duration	No significant effect
Elbadawi and Sambanis (2002)	108 wars, 1960–1999	Primary exports/GDP	War onset	Weak or no effect
Elbadawi and Sambanis (2002)	108 wars, 1960–1999	Primary exports/GDP	War prevalence	Weak or no effect
de Soysa (2002a)	76 states, 1989–2000	Natural resource stocks/capita	Conflict onset (>25 deaths)	No significant effect
de Soysa (2002a)	76 states, 1989–2000	Mineral stocks/capita	Conflict onset (>25 deaths)	Decreases likelihood of conflict (curvilinear)
de Soysa (2002b)	138 states, 1989–1999	Oil exporter (dummy)	Conflict onset (>25 deaths)	Increases likelihood of conflict
Doyle and Sambanis (2000)	124 wars, 1945–1997	Primary exports/GDP	Peacebuilding success	Harms successful peace building

(continued)

Table 1 (continued)

	Coverage	Resource measure	Dependent variable	Finding
Fearon (2004)	122 wars, 1945–1999	Contraband (drugs, gems)	War duration	Increases duration of war
Fearon and Laitin (2003)	97 wars, 1960–1999	Primary exports/GDP	War onset	No significant effect
Fearon and Laitin (2003)	122 wars, 1945–1999	Oil exporter (dummy)	War onset	Increases likelihood of war
Hegre (2002)	50 wars, 1960–2000	Mineral exports/total exports	War onset	No significant effect
Hegre (2002)	50 wars, 1960–2000	Primary exports/GDP	War onset	Increases likelihood of war (curvilinear)
Humphreys (2005)	122 wars, 1945–1999	Oil reserves	War onset	No significant effect
Humphreys (2005)	122 wars, 1945–1999	Diamond production	War onset	No significant effect
Humphreys (2005)	122 wars, 1945–1999	Diamond production	War duration	Reduces war duration
Humphreys (2005)	122 wars, 1945–1999	Oil production	War onset	Increases likelihood of war
Reynal-Querol (2002)	91 wars, 1960–1995	Primary exports/GDP	War onset and prevalence	Increases likelihood of nonethnic wars only

A citation from Brubaker and Laitin (1998) (reprinted in Brubaker 2004) effectively summarizes how game-theoretical models have been used to study ethnic violence: "There is no unitary or complete game theory of ethnic violence. Rather, game theorists have identified certain general mechanisms that help account for particular aspects of the problem of ethnic violence" (1998, p. 438; 2004, p. 105). In their extensive review of the literature on ethnicity and violence, Brubaker and Laitin identify three general mechanisms that game theorists have identified as significant for understanding violence: (1) commitment problems, (2) asymmetric information, and (3) intragroup dynamics.

The "commitment problem" was initially applied to interethnic conflict by James Fearon (1994). Fearon's basic model consists of at least two groups: an ethnic majority in control of a newly formed state, and a subordinate yet powerful ethnic minority. In the absence of a credible commitment on the part of the ethnic majority – a guarantee that it will not renege on its commitment to protect the security of the minority group – the minority group chooses to fight for independence from a nascent or weak majority-controlled state, under the calculation that this constitutes a superior alternative. Needless to say, the ethnic war that ensues leaves both majority and minority worse off than if the majority could make a credible commitment not to abuse the minority in the new state.

According to Fearon, key factors that make ethnic war more or less likely include the following: (1) the size of the expected change in relative military power between groups that would result from formation of a new state; (2) the relative size of the ethnic minority; (3) whether the costs of fighting of majority and minority groups are low, as may occur if they are more rural than urban and if they are not strongly interdependent in economic terms; and (4) whether institutions can be created that give minority groups political power that is at least proportional to their numbers.

Commitment problems also affect the likelihood of a peaceful settlement, given fear on the part of rivals that any agreement on disarmament will be violated by the opponent (Walter 1997); fear on the part of the minority that the state will not respect negotiated settlements that guarantee partial autonomy, especially when the state has a vested interest in controlling the periphery (where the minority resides); and fear, again on the part of the minority, that powerful international states, agencies, and organizations will be unable, in the long run, to effectively enforce negotiated settlements (Fearon 1998). Taken together, these factors explain why intrastate ethnic wars tend to last much longer than nonethnic civil wars (Fearon 2004).

The issue of asymmetric information and its effect on ethnic violence was put forth by Fearon and Laitin (1996), who developed a social matching game model to specify the conditions under which intergroup cooperation prevails, even during episodes of ethnic violence. The main premise, they argue, is that cooperation is contingent upon the availability of information on the intentions of nominal rivals. High levels of communication may produce ethnic solidarity, and repeated interactions between rivals may engender cooperation. The problem, however, lies with the fact that intergroup interactions are typically characterized by a dearth of information about the behavior of nominal rivals. Given asymmetric information, any violent incident involving members of the two ethnic groups may spiral into large-scale violence.

Fearon and Laitin nonetheless suggest that the spiral outcome is not inevitable and that cooperation is possible and may be sustained with effective "in-group policing." Groups are better able to monitor and police the behavior of their own members, and if members ignore offenses by rivals under the expectation that the offender will be held accountable and punished by coethnics, then the mechanism serves to contain or dampen violence. Challenging the expectation that in-group policing is required to prevent violence from spiraling, Arfi (2000) demonstrates that interethnic cooperation can emerge spontaneously even among ethnic groups whose members do not fear punishment by coethnics. The emergence of stable patterns of intergroup cooperation is attributable to the fact that, while individuals are calculative and can act strategically, they also act interdependently and learn from past interactions.

Lastly, game-theoretical models help identify the microfoundations of intra-group processes such as policing, the instigation and intensification of violence, ethnic outbidding, and ethnic recruitment. Abandoning the assumption that members of an ethnic group are uniform in their goals and interests permits game theorists to specify conditions that generate the processes listed above. Kuran (1998a, b), for instance, suggests that individuals hold unique preferences for consuming "ethnic" versus "other" goods. Their exposure to ethnic entrepreneurs, individuals with a vested interest in activating ethnic identities, leads to an "ethnifcation" cascade under the right conditions. In a similar vein, Laitin (1995, 1998) and Petersen (2001) explore how contextual factors such as tactics of humiliation, institutional incentives, symbolic values, and norms may affect the behavior of ethnic entrepreneurs and occasionally trigger the decision to instigate violence.

Overall, game-theoretical models have helped explore some of the theoretical foundations for explanations of ethnic violence. Many of these models include empirical tests of their implications (e.g., Fearon 2004), while others remain purely theoretical.

1.4 System-Dynamics Approaches

System-dynamics models, as the name suggests, focus on the dynamics of an entire system, by representing putative causal relationships as "flows" between key aggregate variables ("stocks"). This high-level approach, representing mean properties of populations rather than individuals, makes it possible to include many factors and processes in a model, focusing on the complex nonlinear dynamics that result from various direct and indirect feedbacks (positive and negative) built between model components. While this approach is well suited for examining model dynamics, if the models reach equilibria those can be studied as well. That said, if system-dynamics models primarily aimed at studying ethnic violence are relatively rare, there are some system-dynamics models that are relevant. We briefly describe a few typical models in this section.

First, Penzar and Srbljinovi (2004) model "social conflicts" in the territory of the former Yugoslavia in the early 1990s. That is, they explicitly built the model to reflect prevailing hypotheses about the preconditions, factors, and processes that contributed to ethnic, sectarian, and other social violence in that area in the period between 1991 and 1995.

The model includes four entities: two social groups and two organizations – a government and a mobilized out-group. The groups are represented by four state variables (stocks) reflecting population mean level of grievance and ethnocentrism, and the level of support from the group for the government and for the out-group. The variables change over time based on assumed causal relationships, e.g., (1) grievance increases with violence toward the group and as a result of government policies that decrease group privileges; (2) ethnocentrism increases with grievances and propaganda (from government, out-group, and external sources); and (3) support depends on the differences between a group's ethnocentrism and the organization's (government or out-group) policy position. The organizations are represented by four state variables: strength, political position, violent posture, and propaganda toward each of the groups. Again, the state variables change over time based on causal relationships: (1) strength increases with support; (2) policy is affected by ethnocentrism of groups, violence, etc.; (3) violence depends on position, strength, degree of threat, etc.; and (4) propaganda depends on strength and differences between policy and ethnocentrism.

Because the model has 68 parameters and 12 initial values to set, with little empirical or theoretical support for setting parameters, the authors focus on three cases that represent observed situations in the 1991–1995 conflicts in the former Yugoslavia and then conduct some sensitivity analysis around those cases. Some of the system behavior they observe is as expected (i.e., passes a face-validity test); e.g., increasing the cost of violent behavior reduces overall violence. On the other hand, they do not observe behavior reflecting positive feedbacks as often as expected. Sensitivity analysis indicates that the model behavior is not particularly sensitive to initial conditions, but it is to some parameters. Overall, the authors conclude that the models can be useful tools for exploration and "policy gaming," to increase experts' appreciation of the complex dynamics and nonintuitive effects of some interventions.

Second, Akcam and Asal (2005) propose a system-dynamics model of ethnic "terrorism" (conflict) based on a theoretical framework sketched by Gurr (2000), in turn based on analysis done as part of the Minorities at Risk Project (2008), with data on nearly 300 politically active ethnic groups from 1945 to the present, including information on their location and activities. The Akcam and Asal model includes various key feedback loops described by Gurr, representing hypothetical causal relations between aggregate variables. For example, state variables represent levels of: (1) government repression; (2) leadership for rebellion; (3) incentives for rebellion; (4) disadvantage of ethnic group; (5) government's understanding of rebellion; (6) group capacity for rebellion; and (7) group rebellion (conflict) itself, as well as other variables. While the authors do not report any results, they do show how constructing a system-dynamics model can represent an informal model in a relatively concise and clear form, which can serve to help guide theory construction itself and which could be implemented to gain further insights into the resulting dynamics the system can generate.

Third, Choucri et al. (2006) describe a system-dynamics model designed to increase our understanding of state stability; thus, while it is not specifically about ethnic violence, most or all of the components and processes are common to both modeling goals. The core model considers state stability a function of state capacity and resilience, decreased by insurgent activities. These high-level factors are disaggregated into a number of other factors and processes representing causal relationships between them, e.g., economic performance, capacity for and use of violence, social mobilization, civic capacity, and liberties.

In this particular model, they focus only on some aspects of the overall system, for reasons both theoretical (e.g., the time scale of those components was short) and methodological (the model would be easier to understand and thus show the value of the system-dynamics modeling approach). In particular, they focus on factors and processes contributing to the dynamics of dissident recruitment in response to state resilience and capacity to carry out activities meant to control insurgents.

The resulting "proof of concept" model was developed by reviewing the theoretical literature, consulting with experts, and using empirical work to identify key components, factors, and causal relationships. The model consists of about 140 equations, so even though it is a simplified representation of only one part of the overall system they identified, it is fairly complex. In any case, the authors believe this approach is useful both for consolidating a diverse set of theoretical, empirical, and expert claims, and for formalizing the model so that computational experiments can be carried out and compared to real-world situations and expert judgments. In this work, they focus on two example uses of the model: (1) to show some "tipping points," i.e., large changes in system behavior that occur under some conditions that cause parts of the regime to be overwhelmed, and (2) to compare the outcomes and particular various side effects of policy alternatives, i.e., the effects of "removing" insurgents (which reduces antiregime messages in the short term but has other self-defeating effects in the long term), versus reducing antiregime messages by other means.

In sum, system-dynamics models can be a useful way to create formal models that include a fairly complete set of components, factors, and processes that are believed to be related to the occurrence of ethnic violence. At the least, creating such models encourages the systematic description and integration of various hypothetical causal relationships; the models can also help researchers and policymakers increase their understanding of the nonlinear dynamics that such systems can generate and help them to hone their intuitions about how changes (e.g., from policy) may affect the overall dynamics and outcome.

2 Using ABM to Study Ethnic Violence

A growing body of scholarship across academic disciplines utilizes ABM. Within this body of work, an emerging literature analyzes puzzles in the area of ethnic violence (see Table 2 for a summary). Examples of scholarship that combines ABM with a substantive focus in this area include work on the evolution of ethnocentric

Table 2 A typology of agent-based models

	General models	Models of violence
Complex	Synthetic ACtors (COGNET, Pat-Net)	
	Tac-Air-Soar (architectures for human cognition)	
	BDI based on NDM (SWARM, dMARS)	
	Urban-Sprawl models (UrbanSim, Brown et al. 2002)	Civil violence (Dibble and Feldman 2004)
	Sugarscape (Epstein and Axtell 1996)	Civil violence (Epstein et al. 2001)
	Adaptive parties and fitness landscapes (Kollman et al. 1992)	Secessionism (Lustick et al. 2004)
	Dynamics of corruption (Hammond 2000)	REsCape (Bhavnani et al. 2008b)
	Endogenous borders (Cederman 2002)	Rumor dynamics (Bhavnani et al. 2009)
	Landscape model of alliances (Axelrod and Bennett 1993)	Interethnic rivalry (Bhavnani and Backer 2000)
	Tag-tolerance model of cooperation (Riolo et al. 2001)	Ethnic norms and violence (Bhavnani 2006)
Simple	Culture model (Axelrod 1997)	Ethnocentric behavior (Hammond and Axelrod 2006)

behavior (Hammond and Axelrod 2006), on variation in individual characteristics to explain localized conflict and genocide (Bhavnani and Backer 2000), on the role of rumors in ethnic violence (Bhavnani et al. 2009), on the relationship between ethnicity and civil war (Bhavnani et al. 2008b; Bhavnani and Miodownik 2009), on the endogenization of borders (Cederman 2002), on a comparison of GeoSim and FEARLUS (Cioffi-Revilla and Gotts 2003), on Geograph3D applications to civil violence (Dibble and Feldman 2004), on containing civil violence (Epstein et al. 2001), on the survival of disagreement (Johnson 2001), on secessionism (Lustick et al. 2004), and on cultural violence (Lim et al. 2007).

ABM lends itself well to modeling problems characterized by numerous, hetero-geneous, and adaptive agents in which patterns of agent interaction – the interaction topology – also matter. In addition, ABM is well suited for modeling agents in a dynamic world that is rarely in equilibrium, which makes understanding the general *processes* being modeled equally if not more interesting and useful than predicting particular outcomes. Finally, and perhaps most importantly, ABM makes it possible to study dynamics that are inherently nonlinear, in which small changes in individual characteristics, heuristics, or interaction patterns can generate large changes in collective behavior. Our discussion of why ABM lends itself well to the study of ethnic and sectarian violence covers the following: (1) agent heterogeneity, (2) agent adaptation, (3) interaction typology, (4) specifying mechanisms, and (5) emergence of nonlinearities, structure, and dynamics.

As with all models, ABMs (also known as individual-based models) are constructed by specifying simplified representations of the entities and processes of interest to the modeler. Their distinguishing feature is that they are constructed in a "bottom-up" manner. That is, ABMs are defined in terms of entities and *mechanisms* at a microlevel, at the level of individual actors – their characteristics, behavioral rules, and interactions with each other and with their environment. Thus, when using an ABM approach, the focus is often on finding particular assumptions about agents (traits and behavior) and their environment that *generate* macropatterns (e.g., of ethnic violence) like those we see in the real world. While showing that particular micromechanisms and conditions lead to the patterns we see in the world does not "prove" that those micromechanisms are necessarily what is happening in the world, it does provide support for their plausibility relative to other nongenerative explanations (Epstein 2006).

Generally, an ABM is comprised of one or more types of agents, as well as a nonagent environment in which the agents are embedded. Agents in an ABM can represent individuals or institutions. This flexibility – also known as agent granularity – makes it possible to study systems at many scales and to integrate parts that are specified at different scales into a coherent whole. The profile, or state, of an agent can include various characteristics and preferences, as well as particular social connections (i.e., identities, memberships, and networks) and a memory of recent interactions and events. In addition to individual characteristics, agents are defined by their decision-making heuristics and capabilities to act in response to inputs from other agents and from the environment. Agents may also possess adaptive mechanisms (learning or evolutionary) that lead them to change their heuristics based on their own experiences.

The environment can encompass any variables external to the agents that are relevant to agent behavior – whether as stimuli, factors or constraints, or targets. These can range from physical features such as geography or topography to things comprising states of the world such as political, military, economic, or social conditions. An environment, therefore, can be specified in terms of various entities or dimensions, each with an associated "state." The environmental entities in a model usually have their own dynamics, describing how they change over time independent of agent behavior. These changes can reflect natural processes, according to logical rules. They can also involve uncertainty or noise. In addition, they can represent the effects of shocks or "triggers" such as sudden economic collapse, the mobilization of ethnic rivals, or a rebel incursion.

The model dynamics are studied by implementing the agents and environment as a computer program. One then runs the program to model the behavior of the agents – including their interaction with each other and the environment – and the dynamics of the environment independent of the actions of the agents. When an ABM is run on a computer, agent behavior is generated as agents determine which other agents to interact with, what to do when they interact, and how to interact with the environment. Each agent's behavior affects other agents as well as the environment. The environment, in turn, changes not only in response to agent behaviors but also in following its own dynamical rules. Thus, ABMs *generate* elaborate interlaced feedback relationships, leading to the nonlinear, path-dependent dynamics that are characteristic of complex adaptive systems.

The output from model runs consists of both the microlevel behavior of agents and changes in the environment, as well as the emergent macrolevel structures, patterns, relationships, and dynamics that result from the aggregation of this microlevel activity and in turn affect behavior at the microlevel. In principle, the model can be run hundreds or thousands of times – with various tracking measures or outcome variables summarized across runs – to study the variations in and sensitivity of results.

2.1   Agent Heterogeneity

ABMs are distinct in that they are constructed in a "bottom-up" manner – specified at the level of individual agents and their interactions with each other and their environment. As such, ABMs are capable of providing insight into how the diverse characteristics and behavioral rules of individual agents lead to the system-level patterns in space (correlations and structure) and time (dynamics), and one can accommodate myriad differences in agent characteristics and decision-making heuristics within the same model. In short, ABMs lend themselves well to modeling individuals as heterogeneous actors because the actors are represented as a population of distinct individuals, with a resulting explicit representation of the distributions of traits across as many dimensions as required. Thus, ABMs can represent heterogeneity as richly as needed, compared to the homogeneity of agents in game-theoretical models or the limited heterogeneity in system-dynamics models that track a population mean (and sometimes a standard deviation for an assumed normal distribution).

With respect to ethnic violence, for instance, members of an ethnic group are *not* homogeneous in their antipathy for nominal ethnic rivals. As noted by Brubaker and Laitin (1998), violence is not a natural outgrowth of conflict; interethnic hostility does not translate directly into group solidarity, which is then manifested in group violence. Rather, the tension between individual and group interests peaks when solidarity is costly and entails risking one's life (Gould 1999). It follows that individuals vary in their propensity to engage in (or refrain from) violence against nominal rivals – what in effect may be conditional upon a host of additional factors at the individual and group levels (intragroup monitoring and sanctioning or information flows and interaction patterns, to name but a few possibilities).

2.2   Agent Adaptation

Agents in ABM are capable of adapting their behavior, characteristics, and behavior rules as a result of their interactions with other agents and with the nonagent environment in which these agents are embedded.

Note that it is often prudent to introduce adaptation into the model only after having carried out many experiments to explore and understand the dynamics that are generated by models using nonadaptive agents for two main reasons. First, many situations to be modeled may involve relatively short time scales so that the people and institutions involved do not have time to adapt their fundamental characteristics or decision rules. Thus, when modeling these situations, there is no need to include adaptation of agent features or decision rules (note that the lack of adaptive processes as defined here does not imply that the agents' behaviors will not change, since their choices depend on the order of inputs they receive from other agents and from the environment, both of which generally will be changing over time.) Second, even without adaptive processes, ABMs generate a wide variety of very complex dynamics simply as a result of changing model parameters, sensitivity to initial conditions and the emergence of complex feedbacks. Thus, it is crucial to have a good understanding of the basic dynamics that result from such nonadaptive models before trying to understand the additional complexities that may result when adaptation is introduced.

In order to adapt, i.e., respond differently to the same situation, adaptive agents can change traits or behavioral rules immediately in response to a situation, or they can retain some history of actions and results and use this history to shape their behavior in a variety of ways.

In their analysis of turnout in popular rebellions, Bhavnani and Ross (2003) treat the credibility of the government and opposition as endogenous: agents in the model compare government and opposition announcements with subsequent events and devise credibility scores for both. The model thus generates its own history, and agents update their beliefs about the credibility of official information, as well as their own information, and act on updated beliefs based upon this information. This enables the model to capture the emergent, path-dependent properties of popular rebellions, since its dynamics are driven endogenously rather than by exogenous events or shocks to the system. In addition, the simulation can be run repeatedly – with various tracking measures or outcome variables summarized across runs – to study the variations in and sensitivity of results.

This direct or indirect dependence on a history of interactions and events is one key source of the sensitivity to initial conditions and path-dependence that is prevalent in the dynamics of complex adaptive systems.

Usually, adaptive processes are defined as either learning processes or evolutionary processes. In the "bottom-up" spirit of the agent-based approach, both of these adaptive processes are generally defined at the level of agents. For example, learning may consist simply of agents imitating the characteristics or decision rules of other successful agents. Learning may be modeled by more complicated algorithms such as those found in neural network models or in classifier systems that use reinforcement, associative, and various rule-discovery algorithms, e.g., inductive or genetic algorithms (Lanzi and Riolo 2003). Similarly, evolutionary adaptive processes are modeled by various forms of "survival of the fittest agent" algorithms, in which agents with new characteristics or individual decision rules result from "mutation" and "recombination" of the characteristics and decision rules of existing successful agents (Holland 1995). However, also note that evolutionary algorithms are used to represent various kinds of individual and social learning (Reschke 2001; Fisher 2003).

Once again, with respect to ethnic violence, a relatively basic assumption is that members of an ethnic group react to the threat of violence from rivals in a variety of ways: some initiate violence preemptively; others respond in kind; still others participate in collective violence only when there is safety in numbers; the majority, in most cases, refrain from participating at all, although under compulsion, even mass behavior is susceptible to change. Using an ABM that consists of agents who vary in their disposition to engage in violence against nominal ethnic rivals, their propensity to punish coethnics for failing to behave accordingly, the strength of punishments administered to coethnics, and the particular update rules used to adapt their behavior in response to punishment, Bhavnani (2006) finds that under compulsion, behavioral conformity within an ethnic group increases significantly.

Thus, whereas *heterogeneity* may capture variation in the initial behavioral response of artificial agents in a computational model, *adaptation* makes it possible to model how the traits and behavioral rules of these agents change over time in response to the behavior of ethnic rivals, coethnics, and the nonagent environment. In contrast, employing a mean-field approach to describing trajectories and variances for reasons of analytical tractability can be misleading precisely because the heterogeneity and adaptability of agents lead to sensitive, path-dependent dynamics that are not adequately captured by the mean trajectory or even by a simple distribution over such trajectories.

2.3   Interaction Topology

With ABM, it is relatively easy to embed agents in both physical and social spaces in the same model. For example, agents can move in a two-dimensional spatial topology. The resulting spatial relationships can bias the agents which are more likely to interact with other agents, and explicit representation space allows agents to move in and interact with heterogeneous environments. ABM can also include social networks of various kinds, each defining interaction topologies based on who talksto, observes, influences, or ultimately sanctions or rewards whom.

In the context of ethnic violence, networks that connect group members to one another are therefore instrumental in determining behavioral trajectories and the outcome of efforts to achieve collective compliance. Social movement theorists (Curtis and Zurcher 1973; Finifter 1974; McAdam and Paulsen 1993; Oliver 1984; Oberschall 1973; Opp and Gern 1993; Tilly 1978) regard networks as important for recruiting participants for protest or rebellion. Despite their prominence in this literature, social networks have received limited attention in the context of ethnic violence. For instance, Brass (1997) notes that all riot-prone towns have – to a greater or lesser degree – informal organizational networks that serve to mobilize members. He does not, however, distinguish between different types of networks. Likewise, Varshney (2003) bases his argument on the existence of interethnic networks that promote civic engagement and reduce conflict but does not specify the structure of these networks – whether and how these networks are likely to differ across contexts.

Group networks can, in effect, determine how and how often "like-minded" agents observe and sanction the behavior of agents with contrasting or opposing views (Granovetter 1976; Morikawa et al. 1995). For instance, one may differentiate an environment in which interaction is unrestricted – influenced by either encounters with other individuals selected at random or by widely disseminated information – from an environment in which interaction is spatially bounded and influenced by local perceptions of appropriate behavior. Likewise, one may differentiate "ethnic entrepreneurs" – individuals with high connectivity or centrality – from other group members, or, in the case of popular rebellions, examine how the level of technological sophistication on the part of individuals – the structure of social networks – affects the dynamics of rebellion (the use of "texting" in the Philippines at the beginning of the twenty-first century versus more rudimentary forms of communication in Indonesia during the same era).

Take the study of cultural violence by Lim et al. (2007) which utilizes an ABM framework – a landscape comprising grid cells and agents who migrate – based upon the key assumption that violence arises due to the structure of boundaries between ethnic groups rather than inherent conflicts between groups themselves and that the spatial population structure, as opposed to measures of ethnic diversity such as fragmentation, increases the propensity for violence. Underlying this assumption is the notion that spatial heterogeneity itself serves as a predictor of violence and that modeling violence at the individual level is both unnecessary and impractical. Yet, capturing the spatial distribution of ethnic groups largely fails to explain when and why some and not other members of these very groups mobilize along ethnic lines (Bhavnani and Miodownik 2009), which effectively requires careful specification at the microlevel, that is at the level of individual characteristics and interaction, to explain how violence emerges from the "bottom up." We turn, next, to the specification of mechanisms.

2.4 Specifying Mechanisms

ABM is a mechanistic (as opposed to phenomenological) approach: with ABM, it is necessary to specify the causal micromechanisms that determine agent choices and behavior. The explicit representation of the microbehavior of all agents over time makes it possible to deepen our understanding of the system by examining not only how behavioral trajectories (of individual agents or groups of agents) differ across various parameter settings, but also how changing the micromechanisms themselves affects the behavior of the system as a whole. Additional understanding of how the individual agent-level behaviors, in response to each other and a nonagent environment, generate macropatterns in space and over time is possible by examining individual histories of behavior and looking for correlations and patterns at that level of analysis.

For example, an empirically-grounded ABM of the conditions and mechanisms that contribute to political stability or instability in resource-rich countries over time could go some way in explaining why lootable resources are associated alternatively with stability and instability during different phases of a country's history as well as across countries over time. By making it easier to explore the effects of different basic microlevel causal mechanisms and various parameters (resource distributions, agent distributions, institutional structures), as well as by examining specific counterfactual scenarios, this approach could help build our intuition about the dynamics that can emerge when salient actors and institutions are viewed and modeled as complex adaptive systems (Bhavnani et al. 2008b).

One way to conceive of the relationship between the field-research component of such a project and model development is as an iterative "two-way street." The first round of field research could provide the initial empirical inputs for the ABM. This would help generate valid specifications of agents' characteristics, their behavioral rules, the interaction topology, adaptive processes, and a set of characteristic environments that capture key similarities and differences across the set of countries of interest. In turn, initial models could be used to guide and sharpen the second round of field research, to gather microlevel data to guide selection of values for model parameters shown by sensitivity analysis to be critical in determining model behavior, to better understand the mechanisms and processes at work, and to gather the kind of aggregate data needed to validate results generated by the first round of simulations. This second round of ABM-informed data collection could, in turn, be used to help structure computer experiments by deepening and refining the specification of agents and environments. As such, it is possible to envision multiple iterations of field research followed by modeling, followed by a new round of "ABM-informed" field research, and followed by a new round of "field-research informed" modeling. In this way, it is possible to tie the "real world" of political phenomena to ABM while simultaneously using ABM to guide data collection and future theory development.

3   REsCape

REsCape (Bhavnani et al. 2008b) is an example of a specific agent-based computational framework for studying the relationship between natural resources, ethnicity and civil war. REsCape was developed in an effort to explicitly specify a plausible set of underlying mechanisms and processes, given the wide range of theories about causal relations, the range of claims in the quantitative, empirical literature, and the data limitations at both the micro- and aggregate levels on this topic. For example, some of the many claims include (1) resources significantly increase the likelihood of war but the effect is curvilinear; (2) resources have a weak or no significant effect on civil war; (3) resources reduce war duration; (4) resources increase the likelihood only of nonethnic civil wars; and (5) resources have no effect on war initiation but increase war duration. This focus is reflected in the second generation of work on "new" civil wars in the post-Cold War era, characterized as distinctly criminal, depoliticized, and predatory – driven or motivated by greed and loot, by a lack of popular support and by gratuitous violence. "Old" civil wars, in contrast, were more likely to be ideological, fought over collectively articulated grievances, and characterized by broad popular support and controlled violence. For an analysis of the distinction between old and new civil wars, see Kalyvas (2001).) These findings are summarized in Table 1.

By permitting the user to specify (1) different resource profiles ranging from a purely agrarian economy to one based on the artisanal or industrial extraction of lootable resources (i.e., alluvial or kimberlite diamonds); (2) different patterns of ethnic domination, ethnic polarization, and varying degrees of ethnic salience; and (3) specific modes of action for key agents, REsCape may be used to assess the effects of key variables – whether taken in isolation or in various combinations – on the onset and duration of civil war.

3.1 Model Description

Figure 1 presents a summary of key model components (1–9), mechanisms (a–d), and feedback loops (i–n). To begin with, we define a landscape (1) as a discrete cellular grid with fixed borders and a capital city (C) located in the center. The size and shape of this grid are alterable by the user. In the 2008 implementation, each of the 441 (21×21) cells may contain any number of agents, divided into members (peasants) and leaders of two rival ethnic groups (2), and house production, which falls into one of four economic sectors (3). Sectoral and spatial spending decisions (4) by leaders of each ethnic group determine the amount of revenue (5) available to garner peasant support (6). Where such support is weak, peasants may relocate or migrate (7) to cells populated and controlled by members of their own ethnic group. Leaders also use revenue to control territory (8), and territorial control is important in this framework, given that control is a necessary condition for spending on investment, revenue generation, and popular support. All control is cell-specific, as is the breakdown of economic sectors, spending decisions, and peasant support. Conflict (9), also cell-specific, emerges when group leaders seek to control the same territory or cell.

Specific model mechanisms include the following: (a) robbery leads to a decline in economic growth, undermining peasant support, and weakening the state, making it more vulnerable to capture rebels over time; (b) spending on social welfare increases popular support but remains economically unviable in the long term; (c) spending on coercive power alters support and is essential for territorial control; and (d) investment in the economy serves to increase the flow of revenue over time and has a robust effect on peasant support.

Key feedback loops in the framework include the following: (i) changes in revenue (relative to the revenue of nominal rivals) increase (or decrease) the salience of ethnicity – the weight individuals place on ethnicity as a defining or core identity; (j) ethnic salience affects peasant support; (k) high levels of peasant support decrease the cost of control, and control has a nonmonotonic effect on support (excessive control lowers

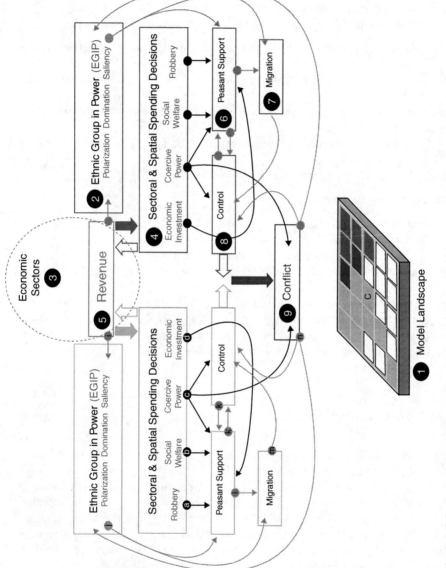

Fig. 1 REsCape – summary of key model components and mechanisms

support, as does weak or insecure control); (*l*) when peasant support for the leader in control of a cell is weak, peasants may exercise the option to migrate to ethnic enclaves in an effort to find safety in numbers; (*m*) migration changes the calculus of control and thus affects spending, investment, and support for leaders; (*n*) conflict, which arises when leaders seek to control the same territory, alters the control of individual cells and may ultimately alter control of the state (the ethnic group in power, or EGIP).

The basic sequence of steps in the modeling process is as follows:

- Determine the resource base and spatial distribution of resources in the economy.
- Determine the strategy defining spending and investment decisions on the part of group leaders.
- Determine the degree of ethnic polarization by specifying the population share of rival ethnic groups.
- Determine the structure of ethnic domination by specifying the EGIP and by default the EGOP.
- Determine whether ethnic salience is fixed or variable.

In each time step of a model run:

- Group leaders make sectoral and spatial spending decisions
- Spending decisions generate revenue for leaders and peasants
- Future spending and investment is, in turn, constrained by revenue
- Peasants determine their level of support for leaders as a function of revenue, security, and ethnicity
- If support is low, peasants migrate to ethnic enclaves
- Conflict emerges when group leaders seek to control the same cell
- Conflict determines new patterns of territorial control
- Change in control of the capital city effectively changes the EGIP.

3.2   Demonstration Run

Now, let us walk through a demonstration run of the REsCape model in Repast (Recursive Porous Agent Simulation Toolkit), a free and open source set of tools for creating ABMs using the Java language. Specifically, we analyze a case in which members of the EGIP (group *A*) constitute 85% of the population and members of the EGOP (group *B*) constitute 15% of the population, with the dominant ethnic majority *A* exercising political control and the minority group *B* excluded from power. We also permit the salience of ethnicity to vary across peasants, who periodically shift their location on the landscape. The government (*A*) plays a benevolent strategy for 149 time steps, after which it engages in robbery, with the switch in strategy implemented exogenously, in an effort to demonstrate the effects of a change in leadership on spending and investment. In contrast, group *B*'s leadership (*B*) plays the benevolent strategy for the duration of the model run. We utilize a landscape characterized by high population density in the center, moderate density in the NW, NE, and SE corner regions, and low density in the remaining areas, with alluvial diamond deposits located in a ring around the capital city. A screenshot from the demonstration is presented in Fig. 2.

Priming the Model (time steps 0–149). The first 149 time steps are characterized by high levels of agricultural revenue and the absence of violence, given that A plays a benevolent strategy during this period. Migration levels are moderate, with peasants moving to cells characterized by greater numbers of ethnic kin, where control is exercised by leaders from their own ethnic group. For members of minority group B, migration results in the formation of enclaves toward the edges of the landscape, given that rival control is weakest here. In contrast, members of majority group A begin to cluster around the central mining region. Despite the ethnic clustering, overall levels of sympathy favor A.

Switching A's Strategy: From "Benevolent" to "Robbery" *(time steps 150–200)*. The exogenously determined change in A's strategy at time step 149 generates a shift in peasant sympathy, which now begins to favor B in ethnically heterogeneous cells in which the mode of production is predominantly agricultural. In the central, diamond-rich region, however, sympathy for A remains high. Of note is that the newly formed minority enclaves tend to be located at such great distance from the majority-controlled resource-rich regions that comparisons of per capita income fail to generate grievances on the part of the minority, resulting in levels of sympathy that are largely neutral, insofar as they favor neither A nor B.

The Growth of Conflict (time steps 201–540). By time step 400, conflict begins to occur in ethnic enclaves within "per capita range" (see our explanation in the Appendix) of diamond deposits, where the income differential between nominal rivals increases the salience of ethnicity, undermines support for A, and results in flight by the more privileged members of group A and the subsequent occupation of that abandoned territory by members of group B.

The Diminution of Conflict (time steps 541–1720). By time step 1,000, a large minority enclave has formed along the southern edge of the landscape. As this enclave grows in size, it becomes home to nearly all members of B, pushing members of A out of "per capita range" and weakening A's control. As ethnic homogeneity increases in this enclave, both ethnic antagonism and violence subside.

The Reassertion of Government Control (time steps 1,721–2,000). Ethnic clustering stabilizes by time step 2,000, with unrest limited to isolated pockets of majority-group members in the minority-dominated southern enclave and isolated pockets of minority-group members in the majority-dominated northern enclave. Over time, even these isolated pockets disband, as peasants migrate to find safety in numbers. Moreover, with its control of diamond deposits secure, A's ability to quell unrest in the north remains high, resulting in the eventual elimination of conflict and the almost complete segregation of peasants along ethnic lines.

By focusing upon (1) government and rebel allocation of revenue; (2) peasant support for the government or rebels; (3) ethnicity and its salience; and (4) the nature of the physical landscape – the type, size, and location of resource deposits – research with REsCape has identified the conditions under which changes in the behavior of key agents, the resource base of the economy, and the salience agents place on ethnicity as a defining or core identity generate violence. REsCape may

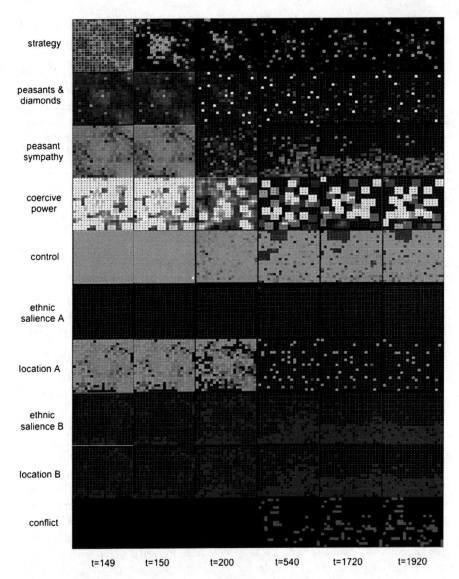

Fig. 2 Demonstration run screenshots *Note*: The demonstration was run with the following parameters settings: government strategy ($t \leq 150$) = *benevolent*; government strategy ($t > 150$) = *robbery*; rebel strategy = *benevolent*; resource base = *alluvial*; resource location = *point source*; $n_A = 0.85$; e_i^A, e_i^B defined on the unit interval with "per-capita range" = 3; ethnic group in power = A; migration period $x_i(min, max) = (25, 200)$. The text that follows describes the screenshots in the figure. ROW 1 (*white, green, black*): the brighter the color of a cell, the higher the *priority* accorded to the cell by the government; white cells reflect a high priority, followed by deepening shades of green (decreasing priority) which merge into black (a lack of interest in the cell). ROW 2 (*gray scale*): the brighter the cell, the higher the *population density* of the cell; white cells are heavily populated, Gray cells are moderately populated, whereas black cells are not populated; magenta dots indicate the presence of *alluvial diamond deposits* within a cell. ROW 3 (*green, blue, black*): green cells denote *peasant*

be modified to capture the characteristics of sectarian violence, although this would require reconsideration of the number of political actors, the issue of revenue allocation between or among actors, agent identities and the depth of various identity-based cleavages, and the nature of support given to the current emphasis on rural peasants, and the observation that most sectarian violence presumably occurs in urban settings.

3.3 Modeling Intervention in REsCape

Third-party interventions, be they diplomatic, economic, humanitarian, or military, have the ability to either terminate or prolong ethnic conflict, the outcome contingent upon the strategies used by interveners and the nature of the conflict itself. In the literature on intervention, factors that influence the incidence, intensity, and duration of violence between ethnic rivals include (1) uncertainty about a rival's resolve, (2) the speed at which divergent beliefs converge, (3) direct intervention versus the use of side payments, (4) neutral versus biased interveners, (5) the role of spoilers, and (6) neighborhood effects. We note, however, that Regan's (2002) analysis of intervention strategies – including (1) early intervention, (2) early intervention with force, (3) taking the side of one party to the conflict, (4) economic intervention, and (5) neutral multilateral intervention – finds that only biased intervention reduces the length of conflict; that on the whole, intervention does more harm than good. With this caveat in mind, we explain how it may be possible to model a few forms of intervention in REsCape.

Mediation: In the same vein, as commitment problems (Fearon 1994) lies uncertainty over a rival's resolve to continue fighting. As fighting continues, resolve may be partially inferred by an opponent. However, the continuation of fighting alone cannot provide insight into the willingness of a rival group to risk casualties until either decisive victory or defeat. By endowing each warring group with the ability to privately reveal its resolve for fighting to a neutral mediator, the latter may successfully

---

sympathy in favor of the government; blue cells denote peasant sympathy in favor of the rebels; black cells denote neutrality. ROW 4 (*gray scale*): the brighter the cell, the higher the level of government coercive power in the cell; white cells indicate a significant troop presence; Gray cells a moderate presence; and black cells the absence of any government troops. ROW 5 (*green, blue, black, red*): green cells denote government control; blue cells denote rebel control; the brighter the color, the greater the extent of control; red indicates that conflict over control of the cell has erupted. ROW 6 (*green, black*): green cells denote ethnic salience for members of group A; the brighter the color, the greater the salience of ethnicity. ROW 7 (*green, black*): green cells denote the location of members of group A; the brighter the color, the greater the density of A's. ROW 8 (*blue, black*): blue cells denote ethnic salience for the members of group B; the brighter the color, the greater the salience of ethnicity. ROW 9 (*blue, black*): blue cells denote the location of members of group B; the brighter the color, the greater the density of B's. ROW 10 (*black, red*): a cell colored red at timestep t^* indicates that there has been at least one conflict in the cell at time $t < t^*$

present the option for a cease-fire without a loss of face for one or both groups. Such an extension would be relatively simple to implement in REsCape.

Biased Intervention: Multiparty intervention tends to be fairly neutral, favoring neither warring group while unilateral intervention more often takes the side of one group over another (Regan 2002). By taking a neutral position, multiparty intervention may be viewed as fair by both groups, encouraging peaceful settlement through concessions the rivals would otherwise be unwilling to make on their own. On the other hand, neutral intervention may lack teeth, given an unwillingness to let either group obtain a preponderance of its demands. By extension, a biased intervener may be able to provide one group with sufficient resources (be they arms, funds, or tactical support) to overpower its rival, effectively shortening the conflict. By exogenously enhancing the revenue available to one group to supplement its coercive capacity, one can begin to capture the effect of biased intervention in REsCape.

Neighborhood effects: In regions characterized by transborder ethnic kin, that is, a similar configuration of ethnic groups in a set of neighboring states, the type of neighborhood a state is located in – peaceful or violent – has a significant effect on domestic events. For instance, refugee flows from neighboring states may serve to destabilize already tense ethnic relations in the host state, especially when refugees have a vested interest in stirring up ethnic antagonism. Likewise, transnational rebels (Salehyan 2007) crossing from one neighboring state to another may effectively serve as biased interveners, tipping the balance of power in favor of ethnic kin. By analyzing two or more interlinked landscapes in REsCape, it becomes possible to model neighboring countries with transborder ethnic kin and thus analyze the "neighborhood effects" of otherwise domestic conflicts.

4 Practical Tips

- Choose carefully between exploratory and consolidative modeling (Bankes 1994; Casti 1997). Take into account the following considerations: consolidative modeling employs measurable physical characteristics and components, and usually requires a detailed specification of inputs, a long period of knowledge transfer, and significant feedback between modelers and subjects. Thus, this approach is most useful in cases in which there are extensive opportunities for validation and controlled experimentation. Exploratory modeling, on the other hand, stops short of offering precise and detailed forecasts. Consequently, it requires less input, less accurate input, less rigorous validation, and less modeler-subject feedback.
  Ethnic violence is characterized by significant information uncertainties, practical barriers to validation, and in most cases, limited modeler-subject interaction. Thus, exploratory models, despite their lack of rigor, are often the best choice for violence models, especially for preliminary and "scoping" analyses.

- To understand the output of a model of group violence, look at distributions of model-generated histories vis-à-vis single histories. This concern is dictated by the uncertainty and chaotic behavior often present in the models of complex adaptive systems. When possible, consider the developing sets of alternative virtual histories to predict which aspects of the histories are persistent across outcomes vis-à-vis "accidental." (Brown et al. 2005).
- Beware of trying to "calibrate" a model of group violence. Overzealous calibration can result in an "over-fitting" of model mechanisms, parameter values, and initial states, and can degrade a model's usefulness outside the regime for which it was calibrated.
- Recognize that using empirical data to validate a model is often difficult or even infeasible. While establishing face validity (ensuring that a model behaves reasonably both within single runs and as parameters are changed) is onerous, establishing microvalidity (confirming that agents behave as expected) is even more difficult, particularly in situations in which a modeler has little or no access to the "real world" and must make do with whatever data is available.
- Make use of possibilities to "dock" model outcomes to models of other types between ABM and system-dynamics models, for instance (Axtell et al. 1996), and to compare model behavior to empirical data. Consider, for instance, seeding a computational model with empirical data and running the model to assess the extent to which "virtual histories" generated by the model conform to real events. Or, consider constructing and running a model to test predictions generated by empirical analyses.
- Ensure that users of a model understand that the quality of models should not be gauged by their ability to point-predict particular events and outcomes. Instead, models should be treated as frameworks capable of producing what Epstein (2008) refers to as generative explanations – i.e., explanations "in which macroscopic explananda, or large-scale regularities, emerge in populations of heterogeneous software individual agents interacting locally under plausible behavioral rules." In particular, models of ethnic and other group violence should be judged by their ability to provide *insight* into the processes and mechanisms underlying violence, and not by an ability to make predictions.

5 Summary

Alternative methodological approaches to modeling group violence can be complementary. The first covers four prominent theoretical approaches in political science: primordial, which highlights the intrinsic nature of ethnic antagonisms; institutional, variants of which emphasize the existence of "riot systems" comprising criminals, informants, and thugs capable of stoking the flames of violence; instrumental, which stresses the role played by ethnic entrepreneurs who use violence as a means to maintain or increase power; and constructivist, which emphasizes the construction and contingency of ethnicity as a core or defining identity. A second approach focuses on the statistical analysis of empirical data to assess the significance of aggregate explanatory factors and generally shies away from specifying the mechanisms that generate violence. A third approach, game-theoretic and negotiation-based, focuses on interactions between representative agents (leaders, individuals, groups), and draws conclusions about equilibrium outcomes, devoting less attention to underlying dynamics. And a fourth approach focuses on the dynamics of an entire system, by representing putative causal relationships as "flows" between key aggregate variables ("stocks"). This high-level approach, representing mean properties of populations rather than individuals, makes it possible to include many factors and processes in a model. ABM is a particularly effective approach for analyzing how the aggregation of microlevel agent behavior effects and is effected by environmental changes, emergent macrolevel structures, patterns, relationships, and dynamics. REsCape is a concrete example of how ABM can be used to understand ethnic violence. This framework, which captures the relationship between natural resources, ethnicity, and civil war, takes into consideration: resource profiles that range from a purely agrarian economy to one based on the artisanal or industrial extraction of lootable resources; different patterns of ethnic domination, ethnic polarization, and varying degrees of ethnic salience; and specific modes of action for key agents. In applying such approaches, it is important to examine distributions of histories from model runs rather than drawing conclusions from single histories.

6   Resources

1. Pointers to special-purpose tools or specific systems for modeling/analysis of ethnic violence:
   International Conflict Research, ETH Zurich:
   GROWLab (Luc Girardin and Nils Weidmann);http://www.icr.ethz.ch/research/growlab
   WarViews: Visualizing and Animating Geographic Data on Conflict (Nils B. Weidmann and Doreen Kuse): http://www.icr.ethz.ch/research/warviews
   GeoContest – Simulating Strategies of Conquest: http://www.icr.ethz.ch/research/geocontest
   Lim, Metzler, Bar-Yam:
   Dynamic GIS Model of Ethnic Violence: http://www.sciencemag.org/cgi/content/full/317/5844/1540?ijkey=S.Kb5wAK45Q5.&keytype=ref&siteid=sci
2. Pointers to collections of data that can be used to initialize a model:
   Center for the study of civil wars, PRIO:
   Geographical and Resource Datasets: http://www.prio.no/CSCW/Datasets/Geographical-and-Resource/
   Economic and Socio-Demographic Data: http://www.prio.no/CSCW/Datasets/Economic-and-Socio-Demographic/

Data on Governance: http://www.prio.no/CSCW/Datasets/Governance/
James Fearon, Stanford University:
Ethnic and Cultural Diversity by Country: http://www.stanford.edu/~jfearon/
data/egroupsrepdata.zip
Minorities at Risk, University of Maryland: http://www.cidcm.umd.edu/mar/

References

Akcam, B. & Asal, V. (2005). The Dynamics of Ethnic Terrorism. Paper presented at *The 23rd International Conference of the System Dynamics Society*, Boston.

Alesina, A., Devleeschauwer, A., Easterly, W., Kurlat, S. & Wacziarg, R. (2003). Fractionalization. *Journal of Economic Growth*, 8(2), 155–194.

Arfi, B. (2000). Spontaneous Interethnic Order: The Emergence of Collective, Path-Dependent Cooperation. *International Studies Quarterly*, 44(4), 563–590.

Aronoff, M. (1998). The Politics of Collective Identity. *Reviews in Anthropology*, 27(1), 71–85.

Axelrod, R. (1997). *The Complexity of Cooperation*. Princeton, NJ: Princeton University Press.

Axelrod, R. & Bennett, D.S. (1993). A Landscape Theory of Aggregation. *British Journal of Political Science*, 23(2), 211–233.

Axtell, R., Axelrod, R., Epstein, J. & Cohen, M. (1996). Aligning Simulation Models: A Case Study and Results. *Computational and Mathematical Organization Theory*, 1, 123–141.

Bankes, S. (1994). Exploratory Modeling for Policy Analysis. *Operations Research*, 41(3), 435–449.

Bankes, S. (2002). Tools and Techniques for Developing Policies for Complex and Uncertain Systems. *Proceedings of the National Academy of Science*, 99(Suppl. 3), 7263–66.

Beardsley, K. (2008). Agreement without Peace? International Mediation and Time Inconsistency Problems. *American Journal of Political Science*, 52(4), 723–740.

Beardsley, K., Quinn, D., Biswas, B. & Wilkenfeld, J. (2008). Mediation Style and Crisis Outcomes. *Journal of Conflict Resolution*, 50(1), 58–86.

Bhavnani, R. (2006). Ethnic Norms and Interethnic Violence: Accounting for Mass Participation in the Rwandan Genocide. *Journal of Peace Research*, 43(6), 651–669.

Bhavnani, R. & Backer, D. (2000). Localized Ethnic Conflict and Genocide: Accounting for Differences in Rwanda and Burundi. *Journal of Conflict Resolution*, 44(3), 283–307.

Bhavnani, R., Backer, D. & Riolo, R. (2008). Simulating Closed Regimes with Agent-Based Models. *Complexity*, 14(1), 36–44.

Bhavnani, R., Findley, M. & Kuklinski, J. (2009). Rumors Dynamics in Ethnic Violence. *Journal of Politics*, 71(3) (July), 876–892.

Bhavnani, R. & Miodownik, D. (2009). Ethnic Polarization, Ethnic Salience, and Civil War. *Journal of Conflict Resolution*, 53(1), 30–49.

Bhavnani, R., Miodownik, D. & Nart, J. (2008). REsCape: An Agent-Based Framework for Modeling Resources, Ethnicity, and Conflict. *Journal of Artificial Societies and Social Simulation*, 11(2). http://jasss.soc.surrey.ac.uk/11/2/7.html.

Bhavnani, R. & Ross, M. (2003). Announcement, Credibility and Turnout in Democratic Rebellions. *Journal of Conflict Resolution*, 47(3), 340–366.

Blimes, R. (2006). The Indirect Effect of Ethnic Heterogeneity on the Likelihood of Civil War Onset. *Journal of Conflict Resolution*, 50(4), 536–547.

Braeckman, C. (1994). *Rwanda: Histoire d'un Génocide [Rwanda: History of a Genocide]*. Paris: Fayard.

Brass, P. (1980). Ethnic Groups and Nationalities: The Formation, Persistence, and Transformation of Ethnic Identities Over Time. In P. Sugar (Ed.) *Ethnic Diversity and Conflict in Eastern Europe* (pp. 1–68). Santa Barbara, CA: ABC Clio.

Brass, P. (1997). *Theft of an Idol: Text and Context in the Representation of Collective Violence*. Princeton, NJ: Princeton University Press.

Brown, D., Page, S., Riolo, R. & Rand W. (2002). Modeling the Effects of Greenbelts at the Urban-Rural Fringe. *iEMSs 2002*, June 24–27, Lugano, Switzerland.

Brown, D., Page, S., Riolo, R., Zellner, M. & Rand, W. (2005). Path dependence and the Validation of Agent-Based Spatial Models of Land Use. *International Journal of Geographical Information Science*, 19(2), 137–174.

Brubaker, R. & Laitin, D. (1998). Ethnic and Nationalist Violence. *Annual Review of Sociology* 24, 423–452.

Brubaker, R. (2004). *Ethnicity without Groups*. Cambridge, MA: Harvard University Press.

Buhaug, H. & Gates, S. (2002). The Geography of Civil War. *Journal of Peace Research*, 39(4), 417–33.

Carley, K. (2002). Simulation Modeling Practice and Theory. *Computational Organizational Science and Organizational Engineering*, 10(5–7), 253–269.

Casti, J. (1997). *Would-Be Worlds: How Simulation is Changing the Frontiers of Science*. New York: Wiley.

Cederman, L. (2002). Endogenizing Geo-Political Boundaries with Agent-Based Modeling. *Proceedings of the National Academy of Science USA*, 99 (May), 7296–7303.

Chandra, K. (2001) . Symposium: Cumulative Findings in the Study of Ethnic Politics. *APSA-CP Newsletter* (Winter), 7–11.

Chandra, K. (2006). What is Ethnic Identity and Does It Matter? *Annual Review of Political Science* 9, 397–424.

Choucri, N., Electris, C., Goldsmith, D., Mistree, D., Madnick, S., Morrison, J., Siegel, M. & Sweitzer-Hamilton, M. (2006). "Understanding & Modeling State Stability: Exploiting System Dynamics." *Proc. of 2006 Institute of Electrical and Electronics Engineers Aerospace Conference*. Big Sky, MT: IEEE, 2006.

Chrétien, J. (1995). Rwanda, les Médias du Génocide [The Media of the Genocide]. *Collection 'Hommes et Sociétés'*. Paris: Karthala.

Cioffi-Revilla, C. & Gotts, N. (2003). Comparative Analysis of Agent-Based Social Simulations: GeoSim and FEARLUS Models. *Journal of Artificial Societies and Social Simulation* 6(4). http://jasss.soc.surrey.ac.uk/6/4/10.html.

Collier, P. & Hoeffler, A. (1998). On Economic Causes of Civil War. *Oxford Economic Papers*, 50(4), 563–573.

Collier, P. & Hoeffler, A. (2002). The Political Economy of Secession.http://users.ox.ac. uk/~ball0144/self-det.pdf

Collier, P. & Hoeffler, A. (2004). Greed and Grievance in Civil War. *Oxford Economic Papers*, 56(4), 563–595.

Collier, P., Hoeffler, A. & Söderbom, M. (2004). On the Duration of Civil War. *Journal of Peace Research*, 41(3), 253–273.

Curtis, R. & Zurcher, L. (1973). Stable Resources of Protest Movements: The Multi-Organizational Field. *Social Forces*, 52, 53–61.

de Soya, I. (2002a). Ecoviolence: Shrinking Pie, or Honey Pot? *Global Environmental Economics*, 2(4), 1–36.

de Soysa, I. (2002b). Paradise Is a Bazaar? Greed, Creed, and Governance in Civil War, 1989–99. *Journal of Peace Research*, 39(4), 395–416.

Des Forges, A. (1999). *Leave None To Tell the Story: Genocide in Rwanda*. New York: Human Rights Watch.

Dibble, C. & Feldman, P. (2004). The GeoGraph 3D Computational Laboratory: Network and Terrain Landscapes for RePast. *Journal of Artificial Societies and Social Simulation*, 7(1). http://jasss.soc.surrey.ac.uk/7/1/7.html.

Doyle, M. & Sambanis, N. (2006). *Making War and Building Peace: United Nations Peace Operations*. Princeton, NJ: Princeton University Press.

Doyle, M. & Sambanis., N. (2000). International Peacebuilding: A Theoretical and Quantitative Analysis. *American Political Science Review*, 94(4), 779–801.

Elbadawi, I. & Sambanis, N. (2002). How Much War Will We See? Explaining the Prevalence of Civil War. *Journal of Conflict Resolution*, 46(3), 307–334.

Epstein J. & Axtell, R. (1996). *Growing Artificial Societies: Social Science from the Bottom Up.* Cambridge, MA: MIT Press.

Epstein, J., Steinbruner, J. & Parker, M. (2001). Modeling Civil Violence: An Agent-Based Computational Approach. *CSED Working Paper* No. 20 (January).

Epstein, J. (2006). *Generative Social Science: Studies in Agent-Based Computational Modeling.* Princeton: Princeton University Press.

Epstein J. (2008). Why Model? *Journal of Artificial Societies and Social Simulation*, 11(4). http://jasss.soc.surrey.ac.uk/11/4/12.html

Fearon J. (1994). Ethnic war as a commitment problem. Paper presented at the *Annual Meeting of the American Political Science Association*, New York.

Fearon, J. (1998). Commitment Problems and the Spread of Ethnic Conflict. In *The International Spread of Ethnic Conflict: Fear Diffusion, and Escalation* ed. D. Lake and D. Rothchild. Princeton: Princeton University Press. 107–126.

Fearon, J. (2003). Ethnic and Cultural Diversity by Country. *Journal of Economic Growth*, 8(2), 195–222.

Fearon, J. (2004) .Why Do Some Civil Wars Last So Much Longer Than Others? *Journal of Peace Research,* 41(3), 275–301.

Fearon, J., Kasara, K. & Laitin, D. (2007). Ethnic Minority Rule and Civil War Onset. *American Political Science Review*, 101(1), 187–193.

Fearon, J. & Laitin, D. (1996). Explaining Interethnic Cooperation. *American Political Science Review,* 90(4), 715–35.

Fearon, J. & Laitin, D. (2003). Ethnicity, Insurgency, and Civil War. *American Political Science Review,* 97(1), 75–90.

Finifter, A. (1974). The Friendship Group as a Protective Environment for Political Deviants. *American Political Science Review*, 68(2), 607–625.

Gagnon, V. P. (1995). Ethnic Nationalism and International Conflict: The Case of Serbia. *International Security*, 19(3), 130–166.

Gleditsch, K.S. (2007). Transnational Dimensions of Civil War. *Journal of Peace Research*, 44(3), 293–309.

Gould, R. (1999). Collective Violence and Group Solidarity: Evidence from a Feuding Society. *American Sociological Review*, 64(3), 356–380.

Gourevitch, P. (1998). *We Wish To Inform You that Tomorrow We Will Be Killed with Our Families: Stories from Rwanda.* New York: Farrar, Straus and Giroux.

Granovetter, M. (1976). Network Sampling: Some First Steps. *American Journal of Sociology*, 81(6), 1287–1303.

Gurr, T. (2000). *People vs. States.* Washington, DC: United States Institute of Peace.

Hammond R. & Axelrod, R. (2006). The Evolution of Ethnocentrism. *Journal of Conflict Resolution*, 50(6), 926–936.

Hammond, R. (2000). Endogenous Transition Dynamics in Corruption: An Agent-Based Computer Model. *Center on Social and Economic Dynamics Working Paper* No. 19. The Brookings Institution.

Hegre, H. (2002). Some Social Requisites of a Democratic Civil Peace: Democracy, Development, and Armed Conflict. Presented at the *Annual Meeting of the American Political Science Association, Boston.*

Hegre, H. & Sambanis, N. (2006). Sensitivity Analysis of the Empirical Literature on Civil War Onset. *Journal of Conflict Resolution*, 50(4), 508–535.

Holland, J. (1995). *Hidden Order: How Adaptation Builds Complexity.* Addison-Wesley.

Horowitz, D. (1985). *Ethnic Groups in Conflict.* Berkeley: University of California Press.

Humphreys, M. (2005). Natural Resources, Conflict, and Conflict Resolution. *Journal of Conflict Resolution,* 49(4), 508–537.

Johnson, P. (2001). Persuasion and Political Heterogeneity within Networks of Political Communication: Agent-Based Explanations for the Survival of Disagreement. Presented at the *Annual Meeting of the American Political Science Association*, San Francisco.

Kakar, S. (1996). *The Colors of Violence: Cultural Identities, Religion, and Conflict*. Chicago: University of Chicago Press.

Kalyvas, S. (2001) 'New' And 'Old' Civil Wars: A Valid Distinction? *World Politics*, 54(1), 99–118.

Kalyvas, S. & Kocher, M. (2007). Ethnic Cleavages and Irregular War: Iraq and Vietnam. *Politics & Society*. 35(2), 183–223.

Khan, S. (2001). *The Shallow Graves of Rwanda*. London: I.B. Tauris.

Kollman, K., Miller, J. & Page, S. (1992). Adaptive Parties and Spatial Elections. *American Political Science Review*, 86(4), 929–937.

Kuran T. (1998a). Ethnic Dissimilation and its International Diffusion. In *Ethnic Conflict: Fear, Diffusion, and Escalation*, ed. D.A. Lake, D Rothchild, Princeton, NJ: Princeton University Press, 35–60.

Kuran T. (1998b). Ethnic Norms and their Transformation through Reputational Cascades. *Journal of Legal Studies*, 27(2), 623–659.

Kydd, A. (2006). When Can Mediators Build Trust? *American Political Science Review*, 100(3), 449–462.

Laitin, D. (1995) .National Revivals and Violence. *Archives Europeennes de Sociologie*, 36(1), 3–43.

Laitin, D. (1998). *Identity in Formation: The Russian-Speaking Populations in the Near Abroad*. Ithaca, NY: Cornell University Press.

Lanzi, P. & Riolo, R. (2003). Recent Trends in Learning Classifier Systems Research. In A. Ghosh & S. Tsutsu (Eds.) *Advances in Evolutionary Computing: Theory and Applications* (pp. 955–988). Berlin: Spriger-Verlag, 955–988.

Lim, M., Metzler, R. & Bar-Yam, Y. (2007). Global Pattern Formation and Ethnic/Cultural Violence. *Science*, 317(5844), 1540–1544.

Lustick, I., Miodownik, D. & Eidelson, R. (2004). Secessionism in Multicultural States: does Power Sharing Prevent or Encourage it? *American Political Science Review*, 98(2), 209–229.

Mamdani, M. (2001). *When Victims Become Killers: Colonialism, Nativism, and the Genocide in Rwanda*. Princeton, NJ: Princeton University Press.

McAdam, D. & Paulsen, R. (1993). Specifying the Relationship between Social Ties and Activism. *American Journal of Sociology*, 99(3), 640–667.

Minorities at Risk Project. (2008). Minorities at Risk Dataset. College Park, MD: Center for International Development and Conflict Management. Retrieved on 4 June 2009 from http://www.cidcm.umd.edu/mar

Morikawa, T., Orbell, J. & Runde, A. (1995). The Advantage of Being Moderately Cooperative. *American Political Science Review*, 89(3), 601–611.

Nagel, J. (1994). Constructing Ethnicity: Creating and Recreating Ethnic Identity and Culture. *Social Problems*, 41(1), 152–76.

Oberschall, A. (1973). *Social Conflict and Social Movements*. Englewood Cliffs, NJ: Prentice Hall.

Oliver, P. (1984). Rewards and Punishments as Selective Incentives: An Apex Game. *Journal of Conflict Resolution*, 28(1), 123–148.

Opp, K. & Gern, C. (1993). Dissident Groups, Personal Networks, and Spontaneous Cooperation: The East German Revolution of 1989. *American Sociological Review*, 58(5), 659–680.

Penzar D. & Srbljinovi , A. (2004). Dynamic Modeling of Ethnic Conflicts. *International Transactions in Operational Research*, 11(1), 63 – 76.

Petersen, R. (2001). *Resistance and Rebellion: Lessons from Eastern Europe*. Cambridge: Cambridge University Press.

Prunier, G. (1995). *The Rwandan Crisis: History of a Genocide*. New York: Columbia University Press.

Rand, W., Zellner, M., Page, S. E., Riolo, R. L., Brown, D. G. & Fernandez, L. E. (2002). The Complex Interaction of Agents and Environments: An Example in Urban Sprawl. *Agent 2002*, October 11–12, Chicago, IL, USA

Regan, P. (2002). Third-Party Intervention and the Duration of Intrastate Conflicts. *Journal of Conflict Resolution,* 46(1), 55–73.

Reschke C. (2001). Evolutionary perspectives on simulations of social systems, *Journal of Artificial Societies and Social Simulation,* 4(4) http://jasss.soc.surrey.ac.uk/4/4/8.html

Reynal-Querol, M. (2002). Ethnicity, Political Systems, and Civil Wars. *Journal of Conflict Resolution,* 46(1), 29–54.

Rieff, D. (1995). *Slaughterhouse: Bosnia and the Failure of the West.* New York: Simon & Schuster.

Rimassa, G., Greenwood, D. & Kernland, M. E. (2006). The Living Systems Technology Suite: An Autonomous Middleware for Autonomic Computing. *International Conference on Autonomic and Autonomous Systems* (ICAS).

Riolo, R., Cohen, M. & Axelrod, R. (2001). Evolution of Cooperation without Reciprocity. *Nature,* 414, 441–443.

Roeder, P. (2001). Ethnolinguistic Fractionalization (ELF) Indices, 1961 and 1985. http://weber.ucsd.edu/~proeder/elf.htm.

Ross, M. (2004). What Do We Know About Natural Resources and Civil War? *Journal of Peace Research,* 41(3), 337–356.

Ross, M. (2006). A Closer Look and Oil, Diamonds, and Civil War. *Annual Review of Political Science,* 9, 265–300.

Salehyan, I. (2007). Transnational Rebels: Neighboring States as Sanctuary for Rebel Groups. *World Politics,* 59(2), 217–242.

Sambanis, N. (2004). What is Civil War? Conceptual and Empirical Complexities of an Operational Definition. *Journal of Conflict Resolution,* 48(6), 814–858.

Scherrer, C. (2002). *Genocide and Crisis in Central Africa: Conflict Roots, Mass Violence, and Regional War.* Westport, CT: Praeger.

Schmidt, J. R. (2008). Can Outsiders Bring Democracy to Post-Conflict States? *Orbis,* 52(1), 107–122.

Srbljinovic, A., Penzar, D., Rodik, P. & Kardov, K. (2003). An Agent-Based Model of Ethnic Mobilization. *Journal of Artificial Societies and Social Simulation,* 6(1). .http://jasss.soc.surrey.ac.uk/6/1/1.html.

Staub, E. (1989). *The Roots of Evil: The Origins of Genocide and Other Group Violence.* Cambridge: Cambridge University Press.

Straus, S. (2004). *The Order of Genocide: Race, Power, and War in Rwanda.* Doctoral dissertation, Department of Political Science, University of California, Berkeley.

Thorning, R. (2005). Civil Militias: Indonesia and Nigeria in Comparative Perspective. In D.J. Francis (Ed.) *Civil Militia: Africa's Intractable Security Menace?* (pp. 89–116). London: Ashgate Publishing.

Tilly, C. (1978). *From Mobilization to Revolution.* Reading, MA: Addison-Wesley.

Uvin, P. (1996). Tragedy in Rwanda. *Environment,* 38(3), 6–29.

Varshney, A. (2003). *Ethnic Conflict and Civic Life: Hindus and Muslims in India.* New Haven, CT: Yale University Press.

Verwimp, P. (2006). Machetes and Firearms: The Organization of Massacres in Rwanda. *Journal of Peace Research,* 43(1), 5–22.

Vulliamy, E. (1994). *Seasons in Hell: Understanding Bosnia's War.* London: Simon & Schuster.

Walter, B. (1997). The Critical Barrier to Civil War Settlement. *International Organization,* 51(3), 335–364.

Wooldridge, M. (2000). *Reasoning About Rational Agents.* Cambridge, MA: MIT Press.

Zur, J. (1994). The Psychological Impact of Impunity. *Anthropology Today,* 10(3), 12–17.

Chapter 8
Insurgency and Security

Alexander Kott and Bruce Skarin

More writings about insurgency appeared in the last few years than in the preceding 100 years (Kilcullen 2006). The explosion of interest in the subject has much to do with international interventions: insurgency is among the most difficult challenges that an intervention – military or nonmilitary – may face.

For the purposes of this chapter, we define insurgency as an organized movement that uses armed violence to overthrow a country's government while often hiding within the civilian population and using civilians to perform combat support functions. The use of the civilian population differentiates insurgency from regular warfare, in which such exploitation of civilians would constitute a war crime. Similarly, a rebellion in which antigovernment forces do not disguise themselves as civilians and fight as a regular, identifiable military is different from insurgency as we treat it in this chapter. Although our definition, like any other (e.g., U.S. DoD 2007), leaves room for gray areas, it serves to emphasize the key feature of insurgency – its reliance upon and exploitation of the civilian population. Because the literature on modeling, simulation, and analysis of regular warfare is vast and readily available, and because insurgencies are often associated with interventions, in this chapter we limit our discussion to insurgencies.

Insurgency forces may include a combination of the following:

- An ideology-based movement that fights to overthrow the current form of the country's government and to establish a different regime;
- A personality-based movement driven to install its leader as the ruler of the country;
- A religious movement that wishes to defend its religious freedoms or to establish a religion-based regime in the country;
- An ethnic minority demanding greater rights or independence;

A. Kott (✉)
Army Research Laboratory, 2800 Powder Mill Road, Adelphi, MD 20783, USA
e-mail: alexander.kott1@us.army.mil

A. Kott and G. Citrenbaum (eds.), *Estimating Impact*,
DOI 10.1007/978-1-4419-6235-5_8, © Springer Science+Business Media, LLC 2010

- A regional movement demanding secession or a greater share of the country's resources;
- An ethnic majority fighting against the rule of an ethnic minority or a colonial power.

Counterinsurgency forces also take a variety of forms:

- A democratic state that enjoys the support of a majority of the population;
- A dictatorship that relies on coercion to maintain its rule;
- A colonial government that represents a foreign power;
- A state that receives limited support of a foreign power but is independent in its actions and could conceivably survive on its own;
- A state largely reliant on resources and support of a foreign power.

Regardless of the forces on the insurgent and counterinsurgent sides, the importance and effectiveness of insurgencies have grown since World War II for numerous reasons. These include the reluctance of Western or Western-supported governments to apply the brutal methods common in prior centuries; the effectiveness, low cost, and ease of use of modern small arms like the Kalashnikov rifle (Singer 2006); and the easy availability of arms from a range of state and nonstate supporters through channels of modern commerce (Anderson 2007).

1 Influences and Models

An international intervention can be a response to an insurgency, either in support of the insurgent side, e.g., African Union peacekeeping in Darfur since 2004, or in support of the counterinsurgents (e.g., the U.S. support to the Colombian government fighting the FARC insurgents (Marcella 2003)). On the other hand, an international intervention can be a cause of an insurgency or a major factor in changing the insurgency's intensity or character. Thus, a change in insurgency can be both a cause and an effect of an intervention.

For example, a diplomatic intervention may induce a third party to discontinue its support to an insurgency or compel a counterinsurgency-fighting government to conciliate with insurgents. An international famine aid or economic development assistance may reduce populations' grievances and its support to insurgents, but it may also increase the resources available to insurgents through protection racket (Baker 2009).

Similarly, an international informational campaign that condemns an oppressive government may fan the flames of insurgency against the government; yet a campaign in support of a government may convince a part of the population that the government is an illegitimate foreign puppet. Finally, a military or law-enforcement intervention is likely to cause popular resentment at foreign meddling or drive a segment of population to insurgency by depriving them of their prior privileges and wealth.

In turn, insurgency affects other phenomena we discuss in this book. Economic development suffers, and the illicit economy flourishes. Political dynamics shift toward the competing positions on the issue of how to fight or to accommodate the insurgency. Information channels become key tools – and casualties – of insurgents and counterinsurgents. Crime and corruption multiply as all sides may resort to bribes, death threats, protection racket, drug revenue, ransom, and extortions. Ethnic, social, and religious divisions are exploited and magnified in an insurgency.

1.1 Qualitative Theories and Models

Theorists and practitioners of insurgency and counterinsurgency have outlined a number of key factors that affect the strengths of insurgency. Lenin (1975) stressed the importance of economic and social discontent of masses as a precondition to successful insurgency, as well as the presence of a well-organized core of revolutionaries able to mobilize and guide the insurgency. Lawrence (1935) emphasized the need for an insurgent base inaccessible to the counterinsurgents' forces, with protective terrain, adequate supplies of munitions, and at least a passively supportive population – a safe haven where insurgents can hide and regroup. He also noted that insurgents benefit when counterinsurgent forces rely on a vulnerable technology, such as a railroad.

Galula (1964) wrote about the critical role of a civilian population that tends to be largely neutral in the conflict and shifts its support to insurgents or counterinsurgents depending on the perceived benefits and outcomes of such support. The population's support also depends on the actions, such as assistance or violent reprisals, taken by either side toward the population. Malayan insurgency (Nagl 2002) offered the evidence that insurgency loses its strength when the population is physically separated and protected against the insurgents and when counterinsurgents offer economic benefits, security from violence, and political conciliation to the population. In addition, counterinsurgents benefit when they are able to attract a large fraction of population by exploiting ethnic and other differences. Indigenous counterinsurgent forces are more effective than foreign counterinsurgency forces in gaining a population's loyalty.

Leites and Wolf (1970) point out that insurgency declines when deprived of resource inflows (such as munitions, supplies, and finances) and when its organizational structure and competency are disrupted by counterinsurgents. Respect and fear of government and its forces are important to dissuade a population from supporting insurgents (Peters 2006). War weariness and antiwar sentiments among the counterinsurgent population and government may encourage and strengthen the insurgency (Iyengar and Monten 2008; Anderson 2007). Amnesty, financial rewards, and offers of government and military positions can induce insurgents to switch sides (Kahl 2007).

While the aforementioned factors are the most common drivers of insurgency, many other phenomena are important in specific situations. For example, a large pool of displaced persons or refugees can become a highly productive recruiting ground for insurgents as well as offering an opportunity to skim the foreign food aid (Cuny and Hill 1999). Large-scale international economic aid programs can become the primary financing mechanism for an insurgency, through protection racket (Baker 2009).

In an attempt to integrate a range of theoretical findings and practical observations, the U.S. military produced a counterinsurgency manual (US Army 2006), which is in part a comprehensive qualitative model of insurgency. The widely cited manual identifies multiple factors that encourage and discourage insurgency, stresses that application of force can be a major factor in increasing population's resentment of counterinsurgency, and highlights a population's security, good governance, and essential services as key factors that diminish the population support to insurgency.

Unfortunately, empirical support for qualitative theories of insurgency tends to be anecdotal rather than scientifically rigorous. The work by Iyengar and Monten (2008) is a relatively uncommon example of a model-based, quantitative examination of a qualitative theory. These authors test the argument that antiwar sentiments in the United States embolden the anti-U.S. insurgents in Iraq and influence them to increase the rate of attacks on the U.S. forces. Iyengar and Monten construct a theoretical model that relates the behavior of Iraqi insurgents, specifically the rate of attacks on U.S. forces and Iraqi government forces, to their perception of antiwar sentiments in the United States. In this model, insurgents are rational, strategic actors who attempt to optimize the distribution of their attacks over time in such a manner that the insurgents preserve their resources while maximizing the antiwar opinions in the United States. The authors compute the differences in predictions of the model for different areas of Iraq – some with greater access to information about U.S. public opinion than others – and compare these estimates with the reported insurgent attacks. They find that in periods immediately after the U.S. media reports a spike in antiwar sentiments, the level of insurgent attacks increases.

Also unfortunately, interpretation and application of qualitative models to practical decision-making is an imprecise art. When in 2007–2008 the U.S. decision-makers pondered whether to increase or to decrease the number of U.S. troops in Iraq – the so-called surge decision (Woodward 2008) – the qualitative theory was hardly in question. Most likely, all participants in the debate agreed that increasing the number of U.S. troops fighting the Iraqi insurgency may improve the security for a fraction of the Iraqi population; it may also increase the population's anger at foreign occupation; it may give the Iraqi government additional time to strengthen its political and military posture; or it may also lull the government into complacent reliance on U.S. protection. However, the decision-makers and consultants disagreed strongly on the relative quantitative magnitudes of these potential qualitative effects and on the resulting balance.

The overwhelming majority of U.S. senior military leaders believed that on balance the surge – a rapid, temporary injection of additional U.S. troops into the counterinsurgency efforts – would be counterproductive because it would merely encourage the Iraqi government to continue its complacent dependency on the United States (Woodward 2008, pp. 224–281). A small group of civilian theoreticians and retired generals believed otherwise and urged President George W. Bush to accept the surge plan.

In the event, President Bush decided to execute the surge, and a major reduction of insurgency followed a few months later, in the middle of 2008. Opinions still differ on whether the surge worked as its proponents expected or whether other, unrelated mechanisms caused the reduction in insurgency (Woodward 2008, pp. 380–384; Pierson 2008). Qualitative models are insufficient to answer such questions; they require quantitative models with the corresponding quantitative metrics, variables, and relationships.

1.2 Quantitative Measures of Insurgency

To construct a quantitative model of a complex phenomenon, such as insurgency, one needs ways to measure attributes and dynamics pertaining to that phenomenon. Formulating meaningful metrics of insurgency, however, is a significant challenge. Insurgencies are largely about human perceptions, which are contextual. For example, public opinion about the quality of the current situation in a country is highly dependent on past historical experiences and the availability of alternatives. Thus, interpretation of a metric's magnitude or event trend is dependent on other, often intangible variables (Campbell et al. 2009).

Most commonly used metrics of insurgency measure the level of violence, e.g., the number of insurgent attacks per month; quality of government institutions, e.g., public opinion polls regarding the level of corruption; and strengths of security forces, e.g., the number of counterinsurgent troops and their degree of readiness. For example, the Brookings Institution offers comprehensive data sets of metrics (O'Hanlon and Campbell 2007; Campbell et al. 2009) for insurgencies in Iraq (since March 2003) and Afghanistan (since October 2001). These data sets include several dozen metrics such as fatalities and counterinsurgent troops, number of insurgent attacks of different types, strength of counterinsurgent troops, strength of anti-insurgent militia, unemployment, electricity generation, inflation, GDP, and public opinion polls.

Others begin to explore more comprehensive processes of measuring insurgency, with a special focus on the insurgency's less-tangible aspects such as population attitudes and perceptions. For example, the MPICE program (Dziedzic et al. 2008) has developed a broad-ranging recommendation for gathering a variety of in-depth metrics with computer tools. These would include semiautomated analysis of a country's media content to gauge popular and elite impressions of insurgency-related issues; creation of a panel of experts to assess issues of interest (e.g., the capacity of law-enforcement agencies to perform essential functions); and specially constructed public opinion surveys.

There is no shortage of complaints about metrics being potentially meaningless and even misleading. For example, Clancy and Crossett (2007) describe the history of several insurgencies and find that metrics used in those insurgencies were highly misleading. Still, critics of metrics agree that analysts and decision-makers must look for insightful metrics and for better means to interpret their meaning. Quantitative models can help do exactly that.

1.3 Influence Diagrams

Also called a causal loop diagram, an influence diagram occupies the middle ground between a qualitative model and a quantitative model. Like a qualitative model, an influence diagram describes key aspects of insurgency phenomena and the influences between them. In addition, however, an influence diagram offers features that make it a steppingstone toward a quantitative model: the diagram names specific quantitative variables, identifies dependent variables for each variable, and specifies whether an increase in a variable causes an increase or decrease in its dependent variable.

In Fig. 1, the influence diagram shows key variables and their relations that describe the insurgency of the Anglo-Irish War (also known as the Irish War of Independence) of 1916–1923 (Anderson 2007). Let us begin with the variable called number of insurgents which reflects the number of active anti-British insurgents operating in Ireland. An increase in the number of insurgents leads to an increase in insurgent attacks – note that the two variables are connected with an arrow and marked with the plus sign (increase leads to increase).

The growing number of attacks in turn leads to an increase in British Public War-Weariness (again, the plus sign indicates that increase leads to an increase) and

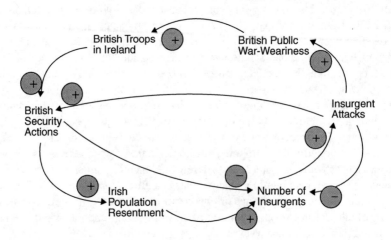

Fig. 1 An influence diagram shows variables and their relations

prompts the British forces in Ireland to energize their British Security Measures, which leads to greater Irish Population Resentment and also more arrests that deplete the number of insurgents. Note that the last arrow (from British Security Measures to number of insurgents) is marked with a minus sign because here increase leads to decrease. And so on.

The diagram points to potentially very complex nonlinear dynamics of insurgency. Even in this simple model, the number of insurgent attacks, for example, affects the number of insurgents in five different ways – there are five distinct paths from insurgent attacks to number of insurgents.

The benefits of constructing such a diagram include the following:

- The modeler or analyst elucidates and formalizes her thinking about the insurgency phenomena;
- Specific variables and their relations are identified and documented;
- Qualitative nature (e.g., increase in A leads to decrease in B) of the variable dependencies are determined and documented;
- Complex feedback loops and side effects become clearly visible;
- Subject matter experts and other analysts and modelers can review and confirm or question the visual representation of the model.

Ideally, the modeler derives the influence diagram directly from relevant qualitative theories. For example, Pierson faithfully followed a single qualitative model – the Counterinsurgency Manual (US Army 2006) – to build an influence diagram of the Iraqi insurgency (started in 2003), with a large number of variables and influence lines (Pierson 2008).

Choucri et al. (2006) formulate their influence diagram of insurgency while rigorously documenting social science theoretical literature in support of each of their model's influences. For example, instead of merely asserting as self-evident the influence "More Insurgents Lead to More Regime Opponents," they cite literature that supports the existence of such an influence.

They also attempt to justify the validity of variables they introduce into their model. For example, they introduce a variable called State Resiliency and justify it by comparing the State Resiliency to the determinants of civil war of Hegre et al. (2001).

1.4 System-Dynamics Models of Insurgency

The modeler may continue to develop the model of Fig. 1 by specifying equations that relate each variable to the variables that influence it, e.g., the equation that computes the number of insurgents as a function of British Security Measures and of Irish Population Resentment. The resulting system of equations (typically a system of coupled nonlinear differential equations) can be solved, for example, by numerical simulation. The solution will show how each variable evolves over time.

System dynamics (Sterman 2001) is a technique that simplifies specifying and solving such systems of equations. A variable is represented as a "stock" of goods. Inflows and outflows represent temporal changes to the variable. A "valve" that opens and closes as a function of other variables controls the rate of a flow.

Figure 2 depicts a fragment of a system-dynamics model that elaborates the influence diagram of Fig. 1. Here, the number of insurgents is a stock, or a level of liquid in a reservoir. The incoming pipe carries the flow of new recruits; the valve opens wider when the Irish Population Resentment is greater. One outgoing pipe represents the depletion of number of insurgents due to arrests. The valve opening on that pipe depends on the British Security Measures. The second outgoing pipe represents the number of insurgents lost in action, and the valve is controlled by the number of insurgent attacks. The modeler must specify an equation for each valve. A computerized system-dynamics tool such as described in isee systems (2009) helps to specify the model and then solves it automatically.

System dynamics is arguably the most popular technique of insurgency modeling. For example, Fig. 1 is partially adapted from Anderson (2007), who constructed a system-dynamics model of the Anglo-Irish War, possibly the first modern urban insurgency. Anderson used only a few causal loops: closed paths through a set of variables. One loop represents insurgency suppression and creation: coercive acts of British forces increase interference in civil life, which increases population resentment, increases the number of insurgents and their anti-British attacks, and leads to an increase in British coercive acts. Another critical loop reflects the impact of British war weariness: as insurgent violence increases, the British public war weariness increases, leading to the public pressure to remove British troops from Ireland.

If Anderson (2007) exemplifies a high-level model capturing overall dynamics of the entire insurgency, the work of Grynkewich and Reifel (2009) is an example of a detailed model of a particular subfeature of insurgency. They model the finan-

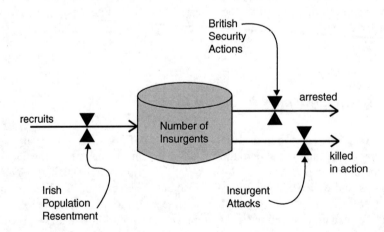

Fig. 2 A fragment of a system-dynamics model

cial operations and organizational behavior of what was then called the Salafist Group for Preaching and Combat (known by its French initials, GSPC). The model relates intensity of insurgent combat operations, expenses to support the operations, and influence of combat operations on a population's willingness to support insurgents financially.

The key stock in this model is the pool of finances available to GSPC; key inflows include extortion from population, voluntary donations, smuggling operations, kidnappings, and ransoms. Outflows include organizational overhead and operational costs. Authors use limited published data and educated guesses to derive values of equation parameters such as the fraction of a population willing to donate to insurgents and the amount of donations. The authors report that the model's simulations agree qualitatively with the available information regarding the GSPC's operations and finances.

While few would claim that a system-dynamics model of insurgency provides a reliable prediction of an insurgency's future evolution, there are other significant benefits in constructing and simulating such a model:

- The model helps analysts and decision-makers see unanticipated side effects, particularly those due to feedback loops;
- The modelers and analysts document systematically a formal model that includes a rich, integrated set of factors, processes, and quantitative dependencies;
- The simulation of the model illustrates the complexity of the nonlinear temporal dynamics of the insurgency system;
- Sensitivity analysis aids analysts and decision-makers in forming insights and intuition toward formulation of intervention plans and policies.

1.5 Agent-Based Modeling of Insurgency

Let us return to Fig. 1. Even in this simple diagram, we see several distinct actors, each with its own set of goals, actions, culture, resources, and relations: insurgent organization, general Irish public, British forces in Ireland, British public, and British government. We may conclude that this set of actors (agents) is too simplistic: after all, Irish insurgents included various movements with different tactics and leadership, the British public included prowar and antiwar segments, and the British government included parties with different views on the Irish question, and so on.

If we wish to model the dynamic relations and mutual influences of all these agents, we should add multiple variables associated with each of the agents and influence lines between the additional variables. The model becomes unwieldy.

An alternative is to use the agent-based modeling paradigm. An agent-based model consists of agents: software representations of individuals or groups of individuals. Groups can be represented at different scales of abstraction: organizations, segments of populations, ethnic or religious groups, social classes, political parties, or movements, even whole countries.

A model may also include elements of the agents' environment. For example, we may want to represent significant geographic areas where the insurgency is unfolded: Dublin, Southern Ireland, Northern Ireland, etc.

An agent has attributes, e.g., an insurgent organization may be characterized by the number of members, amount of munitions, funds available, level of combat training, political objectives, and organizational competency. An agent has relations to other agents, e.g., an insurgent group may be an ally of another insurgent group.

An agent has means by which to make decisions about the actions it will take. Computationally, agents can be implemented, for example, as objects. In that case, an agent has methods by which it makes decision, i.e., choices among available actions. Such a method may include rules or decision-making algorithms, stochastic or deterministic.

An agent has a set of actions it can take, e.g., bombing the barracks of counterinsurgent forces, moving itself to another country, or adopting a more positive attitude toward a rival insurgent group. When executed, an action affects attributes and relations of this and other agents. For example, a bombing attack by an insurgent group reduces the strength and tolerance level of counterinsurgent agents, reduces the amount of explosives available to the insurgent group, and increases its morale and reputation for effectiveness.

To construct a complete model, the modeler identifies the appropriate set of agents (possibly starting with an influence diagram like Fig. 1), assigns attributes and relations for agents, and codes the methods for agent decision-making and for action impacts. Prior to executing the simulation of the model, the modeler also assigns initial values (i.e., the values at the start of the simulation) of the attributes and relations.

The simulation of the model usually proceeds in time steps. A time step for modeling insurgency is often a month or a week. At the beginning of the simulation, at week 1, each agent uses its decision-making method to select one or more actions. Then, each agent executes the selected actions, and each action modifies values of appropriate attributes. This completes the first time step, and the process repeats for the next time step, week 2, and so on. Because attributes and relations of agents change with time, in each time step agents may select different actions (or no action).

The simulation ends when agents reach the last time step. Usually, the analyst who uses the model specifies the number of time steps. For example, if the analyst studies the potential insurgency effects of an international intervention effort that is to last 5 years, she may specify 60 time steps, with each step corresponding to a month.

At the end of the simulation, the analyst reviews the history of the simulated agents: changes in their attributes and relations over time. For example, an analyst may observe that an agent representing an insurgent group rapidly increases its strength between months 1 and 15, then begins to lose support of the local population between months 15 and 20, rapidly depletes its strengths between months 20 and 23, and finally merges with another insurgent group at month 27.

Depending on the tool or model used, an agent may have a memory; it may accumulate experiences, learn new rules, and change its beliefs. For example, the CORES system (Kowalchuck et al. 2004) models an agent's belief in its own actions. When an action does not succeed, the agent's belief in the worth of the action diminishes. Thus, a counterinsurgency agent may gradually come to conclude that harsh retributions are not effective.

The Nexus (Duong et al. 2007) agent-modeling tool pays even greater attention to the cognitive nature of its agents. A Nexus agent has a degree of historical consciousness; it assigns and reassigns blame for past actions of agents, changes beliefs in the trustworthiness of other agents, judges their ideology and looks for friendship with its enemy's enemy. Nexus played a major role in a large-scale, real-world study by a U.S. government agency in a situation that involved potential international intervention and insurgency. Chapter 9 discusses Nexus in more detail.

Senturion's agents possess a complex decision-making mechanism that comprises a set of algorithms drawn from game theory, decision theory, spatial bargaining, and microeconomics. Together, they model how agents interact in a political process. This tool has produced multiple real-world predictions of insurgency-driven situations, such as those in Iraq, the Palestinian Territories, and Darfur in 2004–2009 (Abdollahian et al. 2006; Sentia 2008). Chapter 3 offers more details on Senturion.

2 Other Modeling Methods and Tools

Broad overviews of tools relevant to insurgency modeling are found in Hartley (2008) and Benedict and Dean Simmons (2007). Virtually, all tools able to generate an anticipatory estimate of an intervention's impact on insurgency fall into one of the two categories we already discussed: system-dynamics modeling or agent-based modeling. However, one finds a few exceptions that fall into two other categories: human-driven war gaming and statistical correlations.

2.1 Human-in-the-Loop War Gaming

The PSOM model (Parkman 2005) is an example of a war-gaming-based approach – a computerized, time-stepped war game in which human players decide the actions and moves of insurgent and counterinsurgent forces. In PSOM, the geographic area of operations (the war-game board) is divided into 50-km squares. Each square has attributes such as its degree of urbanization, nature of terrain, population density, quality of infrastructure, cultural values, population's perception of security, and support to the government.

Human players operate the insurgency and counterinsurgency forces. At the beginning of the war game, the players allocate their respective force units to selected squares of the war-game board. Players assign particular missions to these force units: enforce, stabilize, disrupt, and others. During each time step, the computer determines the outcome of each force unit's mission based on the current condition in the square and on actions and strengths of the opponents' forces in the square. The outcome then leads to changes in the square's attribute values. For example, if the counterinsurgent force unit deems successful in its security-enhancement mission, the value of the security attribute in the square increases. Then, the game proceeds to the next time step, and so on.

There is a certain similarity to agent-based modeling, except that in PSOM the human players select the agents' actions (missions and moves), while in the agent-based paradigm agents make decisions without human intervention.

2.2 Statistical Correlations

Application of statistical techniques to historical data on insurgencies yields valuable correlations. Some regularity in data is noticeable even without a formal analysis. For example, Quinlivan (1995) offers a compelling visual correlation between the success of an intervention and the number of security personnel (military plus police) deployed per thousand of the country's inhabitants. Historically, successful suppression of an insurgency requires about ten or more security personnel per thousand of population.

Elbadawi and Sambanis (2000) offer a rigorous quantitative analysis of factors affecting the duration of civil wars (including insurgencies). They find that an external intervention tends to prolong a conflict. They also find a strong U-shaped correlation between the duration of an ethnically based conflict and the ethic fractionalization index. Conflict lasts longer in countries with two or a few large ethnic groups than in those with many small groups or a single dominant group.

3 Initialization, Calibration, and Validation

An insurgency model requires initial values of variables and values of constant parameters or coefficients in equations or rules. The modeler also needs a validation process that shows an acceptable degree of agreement between the model's outputs and data or trends observed in the real insurgency.

Often, modelers have to make educated guesses based on very limited data, e.g., Grynkewich and Reifel (2009) are compelled to use a single newspaper quote of an unnamed "Hezbollah operative" to assign a cost to an insurgent operation.

When no insurgency-specific data are available, modelers resort to the use of data from domains partially similar to insurgency. For example, Robbins (2005) presents a system-dynamics model for reconstruction and stabilization. The model includes an insurgency submodule that accounts for factors such as ethnic fractionalization, effect of unemployment and urbanization. Lacking insurgency-related data, Robbins uses correlations obtained from studies of crime dynamics.

Others derive quantitative parameters using rather sophisticated models, an extensive collection of real-world data and comprehensive statistical analysis. For example, Iyengar and Monten use such a formidable arsenal of tools to quantify the degree of influence that the apparent lack of resolve among the U.S. public has on the intensity of insurgent attacks.

In many cases, the modeler obtains a model's parameters by calibration, i.e., by changing the values until the model's outputs match the available data. For example, Leweling and Sieber (2006) calibrate their model of human resources of an insurgent organization against data derived from publicly available news reports, such as numbers of insurgents arrested. They adjust both the structure and parameters of the model in order to obtain satisfactory agreement between the model's output and the data.

Although hardly the best practice, some modelers consider calibration identical with validation. The modeler calibrates the parameters of his model, shows a reasonable agreement between the model's output and the available real-world data, and then declares the model valid. Ideally, he should calibrate with one set of data and validate with respect to another set data. Often, unfortunately, only one set of data is available.

For example, Anderson (2007) validates his system-dynamics model by comparing the model's results with the data describing the Anglo-Irish War of 1916–1923. He uses this particular civil war for validation purposes because it is the first modern urban insurgency, and because it is a rare case of a well-documented insurgency. The model was able to replicate approximately the dynamic behavior of the Anglo-Irish War, suggesting a degree of model validity. Anderson lists numerous parameters and values of these parameters without explaining how he obtained the values. One has to presume that he calibrated the values in a way that maximized the agreement between the model results and the real-world data.

We discuss validation in detail in Chap. 11.

4   Case Study: Northern Ireland, 1966–1998

The Republic of Ireland occupies about 80% of the island of Ireland. The remaining northeastern area of the island, Northern Ireland, is a part of the United Kingdom. Between 1966 and 1997, Northern Ireland experienced an armed conflict known as the Troubles (Tonge 2006). Several irregular combatant groups rooted in the Irish Catholic population, notably the Provisional Irish Republican Army (Provisional IRA) and the Official Irish Republican Army (Official IRA), fought for unification of Northern Ireland with the Republic of Ireland. Their opponents, irregular combatant groups of Protestant origins, such as the Ulster Volunteer Force (UVF) and Ulster Defense Association (UDA), fought to maintain Northern Ireland as a part of the United Kingdom. All irregular combatant groups were in conflict with the government of the United Kingdom and its armed forces.

Because the irregular combatants disguised themselves as civilians and relied on widespread popular support, the conflict can be characterized as an insurgency according to the definition we discussed in the beginning of this chapter. It may be less obvious whether this insurgency had a connection to an international intervention. However, consider that prounification forces perceived the United Kingdom as a foreign invader, and that there were major influences – ranging from moral support and political pressure to financing and weapon supplies – of governments and communities of both the Republic of Ireland and of the United States.

Political parties in Northern Ireland were strongly polarized along the lines of ethnoreligious affiliation. Pro-British Protestant parties included the Ulster Unionist Party (UUP) and the Democratic Unionist Party (DUP). Supporters of these parties were more likely to sympathize with the UVF and UDA combatants. The Irish Catholic population tended to support parties such as the Social Democratic and Labor Party (SDLP) and Sinn Fein. The last one is often described as the political arm of IRA, and Sinn Fein's supporters were likely to assist IRA combatants (Silke 1999).

Let us review in detail a model (Grier et al. 2008) that focuses on the Troubles in Northern Ireland starting in 1968. The model is agent-based and uses a modeling tool called Simulation of Cultural Identities for Prediction of Reactions (SCIPR). Our objective in this modeling effort is to predict trends in the degree of the population's support of parties in this conflict. In effect, we ask the following question: if we were to have a model like this in 1969, could we predict trends in the population's sympathies to political movements like DUP and Sinn Fein? Arguably, insurgents on both sides draw their strength from population segments that identify with extremes of the political spectrum. If we can predict trends in extreme political opinions, we are better prepared to anticipate changes in the strength of an insurgency.

Another important question is how much data we need to construct and simulate such a model. Models that require less data are less expensive to construct and easier to understand. In this case study, we use a rather simple model that requires little data. We find, encouragingly, that the simple model with a modest amount of readily available data produces potentially useful predictions of trends.

4.1 Agents

Using SCIPR, we construct about 5,000 agents that represent the entire population of Northern Ireland. Each agent represents a group of approximately 300 individuals with approximately similar identities, residing near the same locale.

An agent has several attributes including:

- The district of residence (one of the 26 districts),
- Ethnoreligious affiliation (Catholic or Protestant),

- A number between 0.0 and 1.0 representing the agent's opinion on the issue of Northern Ireland affiliation (0.0 means Northern Ireland must remain British, 1.0 means Northern Ireland must unite with the Republic of Ireland), and
- The political party that the agent supports.

An agent has social links to an average of ten other agents, of which 90% are of the same religion. These networks are fixed and are formed under the assumption that individuals in Northern Ireland have around ten people that they are in regular contact with to discuss political issues, and that they are generally of a similar identity. Using these links, an agent can communicate its political opinion to other agents.

There is no particular theoretical basis for using 5,000 agents and not 500 or 50,000, but for simulations of less than 1,000 agents we find the statistical variance of the generated agent population from the input distributions is significant. With simulations of greater than 5,000, there is no significant difference in either the initialization or outcome. Therefore, the number 5,000 is a modeling assumption that works well for our purposes in this model.

Note what we do not attempt to model: we do not model political movements explicitly, nor political leaders, nor other influential countries like the Eire or the United States, nor the rest of Great Britain. Neither do we represent explicitly the insurgency groups, counterinsurgency forces, economics, nor many other significant factors.

4.2   Agent Actions

An agent can perform several actions:

- Communicate its current political opinion to another agent through an existing link we mentioned earlier,
- Change its political opinion on the question of Northern Ireland affiliation in response to receiving an opinion from another agent,
- Change its political opinion on the question of Northern Ireland affiliation in response to the news of a latest sectarian killing, and
- Change its party affiliation.

An agent changes its political opinion in the manner depicted in Fig. 3. An agent's opinion is characterized by the opinion number, e.g., 0.5, and the opinion's confidence bounds, e.g., (0.25; 0.75). When the agent receives an opinion, e.g., 0.7, from another agent within his social network (i.e., connected by an existing link), the receiver modifies its opinion partially, in the direction of the sender's opinion. When the sender's opinion is outside the receiver's confidence bounds, the receiver ignores the sender's opinion. This model and the opinion scaling equations follow largely Friedkin (1999) and Hegselmann and Krause (2002).

An agent also changes its political opinion in response to events, in this case, the latest episode of sectarian killings. Responses to the event are specified by identities as a distribution of reactions to an opinion. After an event, agents within the region of the

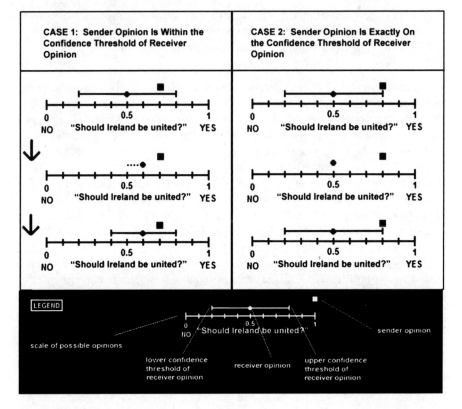

Fig. 3 The model of change in an agent's opinion

event sample a value from the distribution of reactions for their identities. This value is then used to scale the maximum opinion-change parameter and added to an agent's current opinion. This new opinion is then evaluated using the same bounded confidence procedure described above. In this model, when for example a Catholic is killed, agents of the same religion increase the strength of their opinion in favor of the united Ireland.

An agent switches to support of another party when the agent finds its opinion more closely aligned with those of the members of another party than with the members of the agent's current party.

4.3 Model Initialization

We initialize our model to resemble the Northern Ireland of 1968 by assigning each agent the values of its attributes. We pick the values stochastically, but our distributions are such that the total numbers of Protestant and Catholic agents, as well as fractions of supporters of each party in each country, correspond to the demographic and voting data of 1968 (Fig. 4).

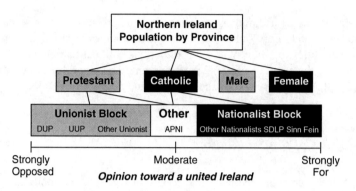

Fig. 4 Population groups

We also create a social network of agents: we assign a link to a randomly chosen pair of agents in such a manner that each agent has on average about ten links, with about 45% of the links connecting to agents within its immediate vicinity, 30% to other agents residing in the same district, and the rest to agents in other locales of Northern Ireland. We also ensure that about 90% of the links are between agents of the same religion. These parameters – 10, 45%, and 30% – are merely modeling assumptions, without a substantive empirical basis. A more rigorous modeling effort should consider obtaining empirical data to support and improve these assumptions.

We also initialize a table of sectarian killings: for each day, how many (if any) Catholics and Protestants perish in intercommunal violence. Our model does not predict such data; they have to come from a separate source.

Now that we have the initial attributes of agents and their links fully defined, the model is ready for simulation.

4.4 Simulation Process

The simulation algorithm begins its process on the first day of the year 1969, in simulated time. The algorithm randomly picks a number of pairs of linked agents. In each pair, the algorithm randomly designates one of the agents to be the sender and another as the receiver of the political opinion. The number of pairs selected on each day is such that every agent, on average, acts as a sender approximately every 3 days. This parameter – three days – is merely a modeling assumption; a more rigorous model should test the empirical validity of this assumption.

When selected by the algorithm, the sender communicates its current political opinion to the receiver. The receiver then either ignores the received opinion or shifts its own opinion partially toward the sender's opinion as we discussed above.

The simulation algorithm also notifies each agent of sectarian killings that occurred on that day. Agents adjust their political opinions correspondingly.

The simulation algorithm then performs the same process on the next day of the simulated time, and so on. At the end of each year, each agent reevaluates the alignment of its political opinion with the members of political parties. Depending on the alignment, the agent changes its party support. The algorithm stops at the end of the year 2005, as instructed by the modeler, and outputs the results.

4.5 Results of the Simulation

Figures 5 and 6 compare the historical results of elections and polls in Northern Ireland with the results of our simulation.

On the one hand, there is a notable similarity in general trends. For example, the simulated growth trends in the levels of support for Sinn Fein post-1981 and for UUP in the years 1973–1993 are broadly consistent with actual historical data. The relatively constant levels of simulated support for DUP and SDL are also comparable to real history.

On the other hand, the simulation clearly arrives at a steady state after 1994 and seems unable to project any further changes. In our experience, this is a general limitation of the Bound Confidence model, which tends to converge to an artificial steady state after a period of simulation.

The simulation also strongly overestimates the support to moderate parties, those between UUP and SDLP. Still, it is encouraging to see that an admittedly

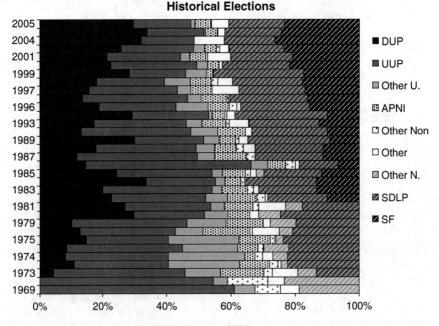

Fig. 5 Actual election data (CAIN Web Service, 2006)

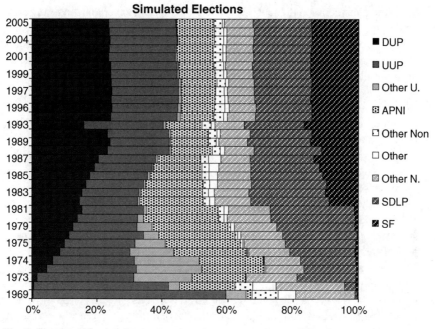

Fig. 6 Simulated election data

simple model shows the ability to predict trends in a population's sympathies nearly 15 years later. It is even more remarkable considering that the modelers used a very small amount of readily available information: election results and basic demographics.

As of this writing, extended models of this type are used to study practical problems of modern insurgencies.

5   Practical Tips

- Recognize that insurgencies and insurgent environments are diverse and that no off-the-shelf model can represent all the features of interest to a particular client. Thus, when evaluating candidate models, take into consideration their ability to be extended or modified.
- Stay cognizant of the many simple yet insightful models of insurgency which have been published by academic researchers such as the Center for Contemporary Conflict of the Naval Postgraduate School (http://www.nps.edu/Academics/ centers/ccc/). Consider using one of these models as a baseline for analysis and experimentation. As the case of interest becomes better understood, the model can be gradually refined and extended. This approach is often much more efficient than starting from scratch.

- Take into consideration the fact that modelers with little experience in insurgency modeling will find system-dynamics models easier to construct, understand, and debug. Recognize, on the other hand, that many insurgencies, and many strategies for managing insurgencies, cannot be properly modeled without representing multiple players, individuals, and groups in the insurgency. In such cases, agent-based models tend to be most appropriate.
- Keep in mind that the ability to trace the chain of influences in analyzing results of a simulation is critically important. System-dynamics models tend to be better in this respect.
- Always model international actors' influences on insurgents and counterinsurgents. Few insurgencies unfold without a significant influence by one or more international participant. The influences can be subtle, diverse, and range from international sympathy to military aid.
- Before constructing a model, ask clients about the actions their organizations consider taking with respect to the insurgency or within an insurgency-plagued region. Make sure the model is capable of representing those actions and their effects.
- For every action the client plans to take, include modeling of undesirable side effects. For example, if the client plans to provide food aid to refugees, include in the model a possible diversion of food by insurgents.
- In selecting variables and attributes for the model, give preference to the most tangible and measurable ones. Include the metrics that the customer expects to use or to affect. Also, consider including available historical or current data for calibration and validation.
- Allocate adequate time and resource to review and refine a model with subject matter experts (SMEs). In preparation for this, develop visualizations specifically designed for SMEs.
- Insurgencies are emotional topics, and SMEs may hold strong, passionate opinions. Therefore, consider collaboration with several SMEs and welcome widely divergent views. Try to find SMEs who have experience on both sides of an insurgency.
- Early in the modeling project, work with insurgency SMEs to create a set of test cases. Ask several SMEs, independently, to produce their estimates of probable evolution of the insurgency in each case, and expect to receive widely divergent estimates.

6   Summary

Qualitative models of insurgency identify factors that encourage and discourage insurgency, e.g., application of force can be a major factor in increasing population's resentment of counterinsurgency; a population's security, good governance, and essential services are key factors that diminish the population support to insurgency. Qualitative models, however, are insufficient to answer practical questions; they require quantitative models with the corresponding quantitative metrics, variables, and relationships. To construct a quantitative model of insurgency, one needs ways to measure its attributes and dynamics. Most commonly used metrics of insurgency measure the level of violence, e.g., the number of insurgent attacks per month; quality of government institutions, e.g., public opinion polls regarding the level of corruption; and strengths of security forces, e.g., the number of counterinsurgent troops and their degree of readiness. System dynamics is arguably the most popular technique of insurgency modeling. An alternative is to use the agent-based modeling paradigm. An agent-based model consists of agents: software representations of individuals or groups of individuals. An insurgency model requires initial values of variables and values of constant parameters or coefficients in equations or rules. In many cases, the modeler obtains a model's parameters by calibration, i.e., by changing the values until the model's outputs match the available data. To illustrate the construction and use of an agent-based model of insurgency, the chapter presents a case study of a model of the Northern Ireland insurgency. The model consists of 5,000 agents, each representing a group of individuals. At each time step, agents pass their attitudes to others within their social network. Key influences under consideration include, for instance, sectarian killings. Predicted changes in issue position among the various groups roughly track the actual growth and decline in support among the political parties of Northern Ireland.

7 Resources

The American Political Science Association, Task Force on Political Violence and Terrorism
Pointers to multiple depositories of datasets related to political violence, including insurgencies
http://www.apsanet.org/content_29436.cfm
Armed Conflict and Intervention (ACI) Datasets
http://www.systemicpeace.org/inscr/inscr.htm
Center for Computational Analysis of Social and Organizational Systems (CASOS) at Carnegie Mellon University
Listings and/or repositories of software tools and libraries
http://www.casos.cs.cmu.edu/computational_tools/tools.html
Conflict Analysis Resource Center
Pointers to multiple depositories of datasets related to armed conflicts, including insurgencies
http://www.cerac.org.co/datasets.htm
Correlates of War Project (COW)
http://www.umich.edu/~cowproj/dataset.html
Ethnicity, Insurgency, and Civil War Project
Data relating ethnic fractionalization and insurgency
http://www.stanford.edu/group/ethnic/publicdata/publicdata.html
Genocide and Politicide project
Data describing cases of genocide, many related to insurgencies
http://globalpolicy.gmu.edu/genocide/
Gompert, D. C. and Gordon IV, J. (2008) War by Other means. RAND Publication
Appendices A and B provide data and analysis of outcomes and correlations for 89 insurgencies, and data on counterinsurgency capabilities of world states and organizations
http://www.rand.org/pubs/monographs/MG595.2/
International Network for Social Network Analysis (INSNA)
Listings and/or repositories of software tools and libraries
http://www.insna.org/software/index.html
Minorities at Risk Data

Datasets characterizing multiple minorities, their conditions and potential risks, including potential
or ongoing insurgency
http://www.cidcm.umd.edu/mar/data.asp
Multiple datasets on state instability and conflict
http://www.systemicpeace.org/
Network Workbench (NWB) Tool
https://nwb.slis.indiana.edu/community/?n=Main.NWBTool
Political Instability Task Force
Multiple datasets related to internal wars, including insurgencies
http://globalpolicy.gmu.edu/pitf/pitfdata.htm
http://globalpolicy.gmu.edu/pitf/pitfpset.htm
War and Health Website
Civilian victims in an asymmetrical conflict
http://warandhealth.com/civilian-victims-in-an-asymmetrical-conflict-data/
World Bank Datasets
Data on civil wars post-WWII
http://econ.worldbank.org/WBSITE/EXTERNAL/EXTDEC/EXTRESEARCH/EXTPROGRAMS/
EXTCONFLICT/0,,contentMDK:20336174~menuPK:637270~pagePK:64168182~piPK:641680
60~theSitePK:477960,00.html

References

Abdollahian, M., Baranick, M., Efird, B. & Kugler, J. (2006). *Senturion A Predictive Political Simulation Model*. Center for Technology and National Security Policy, National Defense University http://www.ndu.edu/ctnsp/Def_Tech/DTP%2032%20Senturion.pdf.

Anderson Jr, E. G. (2007). A Proof-of-Concept Model for Evaluating Insurgency Management Policies Using the System Dynamics Methodology. *Strategic Insights*. 6(5).

Baker, A. (2009). How the Taliban Thrives. *Time*, 174(9), 46–53.

Benedict, J. & Dean Simmons, L. (2007). Enterprise-Wide Opportunities for, Advancing Irregular Warfare Analyses. Brief at *MORS Workshop*, 13 December 2007.

Campbell, J., O'Hanlon, M. & Shapiro, J. (2009). Assessing Counterinsurgency and Stabilization Missions. Foreign Policy Paper Series, Number 14, The Brookings Institution.

Choucri, N., Electris, C., Goldsmith, D., Mistree, D., Madnick, S., Morrison, J., Siegel, M. & Sweitzer-Hamilton, M. (2006). Understanding & Modeling State Stability: Exploiting System Dynamics. Proceedings of *2006 Institute of Electrical and Electronics Engineers Aerospace Conference*. Big Sky, MT: IEEE.

Clancy, J. & Crossett, C. (2007). Measuring Effectiveness in Irregular Warfare. *Parameters*, 37(Summer 2007), 88–100.

Cuny, F. C. & Hill, R. B. (1999). *Famine, Conflict and Response: a Basic Guide*. West Hartford, Connecticut: Kumarian Press. pp. 117–126.

Duong, D., Marling, R., Murphy, L., Johnson, J., Otterberg, M., Sheldon, B. & Stephens, S. (2007). Nexus: An Intelligence Agent Model of Support Between Social Groups. in Proceedings of the *Agent 2007 Conference on Complex Interaction and Social Emergence*, Evanston, Illinois, November 15–17, 2007, 241–246.

Dziedzic, M., Sotirin, B. & Agoglia, J. (2008). *Measuring Progress in Conflict Environment (MPICE)*. Report ADA488249, US Army Corps of Engineers.http://www.usip.org/files/resources/MPICE%20Aug%2008.pdf.

Elbadawi, I. & Sambanis, N. (2000). *External Interventions and the Duration of Civil Wars*. Working Paper 2433, The World Bank, September 2000.

Friedkin, N. E. (1999). Choice Shift and Group Polarization. *American Sociological Review*, 64(6), 856–875.

Galula, D. (1964). *Counterinsurgency Warfare: Theory and Practice*. Westport, CT: Praeger.

Grier, R., Skarin, B., Lubyansky, A. & Wolpert, L. (2008). SCIPR: A Computational Model to Simulate Cultural Identities for Predicting Reactions to Events. Proceedings of the *Second International Conference on Computational Cultural Dynamics* (pp. 32–37), College Park, MD.

Grynkewich, A. USAF & Reifel, C. USAF (2009). Modeling Jihad: A System Dynamics Model of the Salafist Group for Preaching and Combat Financial Subsystem. *Strategic Insights*, 5(8) (November 2006).

Hartley, D. (2008). *DIME/PMESII Tools: Past, Present and Future, Workshop on Analyzing the Impact of Emerging Societies on National Security*. Argonne National Laboratory, Argonne, IL, 14–18 April 2008, http://www.mors.org/events/08es.aspx.

Hegre, H., Ellingsen, T., Gates, S. & Gleditsch, N. (2001). Toward a Democratic Civil Peace? Democracy, Political Change, and Civil War, 1816–1992. *American Political Science Review*, 95(1), 33–48. http://www.worldbank.org/research/conflict/papers/CivilPeace2.pdf.

Hegselmann, R. & Krause, U. (2002). Opinion Dynamics and Bounded Confidence Models, Analysis and Simulation. *Journal of Artifical Societies and Social Simulation (JASSS)*, 5(3). http://jasss.soc.surrey.ac.uk/5/3/2.html.

isee systems (2009). Stella Software Website, http://www.iseesystems.com/softwares/Education/StellaSoftware.aspx.

Iyengar, R. & Monten, J. (2008). *Is There an "Emboldenment" Effect? Evidence from the Insurgency in Iraq*. NBER Working Paper No. W13839.

Kahl, C. H. (2007). COIN of the Realm: Is There a Future for Counterinsurgency? *Foreign Affairs*, 86(6), 169–176.

Kilcullen, D. J. (2006). Counterinsurgency Redux. *Survival*, 48(4), 111–130.

Kowalchuck, M., Singh, S. & Carley, K. (2004). *CORES – Complex Organizational Reasoning System, Report*. CMU-ISRI-04-131, Carnegie Mellon University, School of Computer Science.

Lawrence, T. E. (1935). *Seven Pillars of Wisdom: A Triumph*. Oxford: Alden Press.

Leites, N. & Wolf, C. (1970). *Rebellion and Authority: An Analytical Essay on Insurgent Conflict*. Arlington: RAND Corporation.

Lenin, V. I. (1975). Partisan Warfare. in Sarkesian, S. C. (Ed.) *Revolutionary Guerrilla Warfare* (pp. 187–203). Chicago: Precedent Publishing.

Leweling, T. A. & Sieber, O. (2006). Calibrating a Field-level, Systems Dynamics Model of Terrorism's Human Capital Subsystem: GSPC as Case Study. *Strategic Insights*, 5(8).

Marcella, G. (2003). The United States and Columbia: The Journey from Ambiguity to Strategic Clarity. *Strategic Studies Institute of the U.S. Army War College*. http://www.strategicstudiesinstitute.army.mil/pdffiles/PUB10.pdf.

Nagl, J. A. (2002). *Learning to Eat Soup with a Knife: Counterinsurgency Lessons from Malaya and Vietnam*. Chicago: University of Chicago Press.

O'Hanlon, M. E. & Campbell, J. H. (2007). *Iraq Index: Tracking Variables of Reconstruction & Security in Post-Saddam Iraq*. http://www.brookings.edu/iraqindex.

Parkman, J. (2005). Peace Support Operations Study. *MORS Workshop on Agent-Based Models and Othyer Analytical Tools in Support of Stability Operations*. McLean, Virginia.

Peters, R. (2006). The hearts-and-minds myth – Sorry, but winning means killing. *Armed Forces Journal*, 144, 34–38. http://www.armedforcesjournal.com/2006/09/1947271/.

Pierson, B. M. (2008). OSD/Joint Special Session: Irregular Warfare Activities in OSD and the Joint Staff. Proceedings of the *76th Annual Military Operations Research Society Symposium*. Alexandria, VA: MORS, 2008.

Quinlivan, J. T. (1995). Force Requirements in Stability Operations. *Parameters*, Winter 1995, 59–69. http://www.carlisle.army.mil/USAWC/Parameters/1995/quinliv.htm.

Robbins, M. (2005). *Investigating the Complexities of Nationbuilding: A Sub-National Regional Perspective*. Master Thesis, Air Force Institute of Technology. Wright-Patterson Air Force Base, Ohio.

Sentia Group (2008). *Implications of a U.S. Drawdown in Iraq*, Report, Sentia Group, Inc., http://www.sentiagroup.com/pdf/SentiaInsightMonthly-USDrawdownInIraq-July2008.pdf.

Silke, A. (1999). Rebel's Dilemma: The Changing Relationship Between the IRA, Sinn Féin and Paramilitary Vigilantism in Northern Ireland. *Terrorism and Political Violence*, 11(1), 55–93.

Singer, P.W. (2006). *Children at War*. University of California Press, p. 46.

Sterman, J. D. (2001). System Dynamics Modeling: Tools for Learning in a Complex World. *California Management Review*, 43(1), 8–25.

Tonge, J. (2006). *Northern Ireland*. University of Liverpool.

US Army (2006). Counterinsurgency Field Manual FM-3-24. http://www.fas.org/irp/doddir/army/fm3-24.pdf.

US Department of Defense (2007). *Joint Publication 1-02 Department of Defense Dictionary of Military and Associated Terms*, JP 1-02. http://www.dtic.mil/doctrine/jel/new_pubs/jp1_02.pdf.

Woodward, B. (2008). *The War Within: A Secret White House History 2006–2008*. New York, NY: Simon & Schuster.

Chapter 9
Crime and Corruption

Deborah Duong, Robert Turner, and Karl Selke

Like intergroup violence (Chap. 7) and insurgency (Chap. 8), crime and corruption are nearly inevitable companions of an international intervention. Both contribute to the reasons why the intervention occurs, and both may even grow and fester as side-effects of an intervention. Moreover, crime and corruption frequently serve as obstacles to a successful termination of an intervention.

High crime rates and frequent incidents of corruption are some of the main indicators and drivers of failed states, as well as some of the most important impediments to economic development (Frisch 1996, p. 68). A failed state cannot enforce laws against crime because the state itself is ridden with the crime of corruption, so much so that law enforcement is seen as unfair and illegitimate (The Fund for Peace 2008). Corruption is a particular type of crime that erodes the ability of the state to enforce the law or perform other functions. A widely cited definition of corruption is a "behavior which deviates from the formal duties of a public role because of private-regarding (personal, close family, private clique) pecuniary or status gains; or violates rules against the exercise of certain private-regarding influence." (Nye 1967). Because they are important drivers of state failure, both crime and corruption are among the most important phenomena to model for the purpose of international intervention.

1 Theories of Crime and Corruption

Most theories see crime and corruption as a breakdown of institutions. North (North 1990, p. 3) defines institutions as "the rules of the game in a society or, more formally, the humanly devised constraints that shape human interaction." Institutions "consist of both informal constraints (sanctions, taboos, customs, traditions, and codes of conduct) and formal rules (constitutions, laws, property rights)" (North 1991, p. 97).

D. Duong (✉)
NPS, 555 Dyer Road, Quarters "B" on Stone Road, Monterey, CA 93943, USA
e-mail: dduong@aciedge.com

A. Kott and G. Citrenbaum (eds.), *Estimating Impact*,
DOI 10.1007/978-1-4419-6235-5_9, © Springer Science+Business Media, LLC 2010

In the case of crime and corruption, the rules that are breaking down are laws, but in the case of corruption, traditional cultural patron–client relations are also breaking down (Smith 2007). Adam Smith (1994) saw social institutions as the "invisible hand" through which a miracle can occur: individuals acting purely in their own interest create a society that is good for the whole. If the emergence of good social institutions out of utility-maximizing individual acts is a natural process, then crime and corruption are the breakdown of that process. In crime and corruption, individuals seeking their own benefit create dysfunctional social patterns.

Crime and corruption are social forces and are often not volitional at an individual level. The worst critics of corrupt practices are often those who feel compelled to engage in them. There are coercive forces that drive people into crime and corruption. In some failing states, crime and corruption are the only way of doing business (Smith 2007). A good model of crime or corruption will take into account coercive social forces that draw individuals into vicious cycles of mutually harmful behaviors instead of the virtuous cycles of Adam Smith's free market. The purpose of such a model would be to detect and guide intervening actions at tipping points, points at which actions make a difference as to whether social institutions enter and leave such vicious cycles. If no action is taken at these tipping points, then future corrective action could be far more difficult or impossible.

Sociological theories of crime generally fall into three categories: theories of strain, theories of social learning, and theories of control (Agnew 2009). Theories of strain blame crime on personal stressors. Theories of social learning blame crime on social rewards from involvement with other criminals and look at crime more as an institution in conflict with other institutions rather than as individual deviance from institutions. In contrast, theories of control look at crime as natural and rewarding and try to explain the formation of institutions, such as religion, that control crime.

Theorists of corruption generally agree that corruption is a vicious cycle and an expression of the patron–client relationship. In patron–client relationships, a person with access to resources trades resources with kin and members of the community in exchange for their loyalty. According to Smith (2007), corruption is a result of globalization. In his anthropological study of corruption in Nigeria, Smith studied traditional patron–client relationships based on mutual obligations. Nigerians of all social strata make use of patron–client ties for access to resources but feel that the elites have come to betray the people. The integration of the patronage system with bureaucracy has produced a postcolonial state that facilitates corruption, the betrayal of patronage obligations.

2   Methods of Modeling Crime and Corruption

We begin by introducing briefly several modeling approaches that do not involve an explicit simulation, particularly rule-based systems, Bayesian networks, and game-theoretical approaches. Later, in this chapter, we will discuss simulation-based approaches.

2.1 Rule-Based Systems

Rule-based systems describe relations between variables that are Boolean (either true or false) in traditional systems or scalar (using degrees of truth and falsehood) in fuzzy systems. For example, Situngkir and Siagian (2003) use a fuzzy rule set to model how corruption causes inefficiency in nongovernmental organization (NGO) aid distribution and its effect on future aid. For example, one simple rule is, if an NGO receives a large amount of aid from a donor, then the NGO accomplishes a large amount of support activities to the population it serves. Situngkir and Siagian include one simple feedback loop. The feedback loop reflects the fact that if an NGO is effective in utilizing the donor's funds for the intended purposes, the donor is more likely to support the NGO in the future. The model points out that the development of standards for the evaluation of NGO programs can reduce corruption.

2.2 Bayesian Networks

A Bayesian network is a group of propositions connected by links, each of which describes the probability that one proposition is true given that a set of others are true. These do not involve degrees of truth or falsehood as fuzzy sets do. Fuzzy set advocates argue that fuzzy sets are more general than Bayesian networks and subsume them (Kosko 1994). Bayesian networks can be created manually or learned automatically from a given set of data. In modeling crime, Bayesian networks are often used to find patterns in crimes for forensic purposes. Baumgartner et al. (2005) presents a Bayesian network model of offender behavior for the purposes of criminal profiling. Their network links the action of the offender on the scene of the crime to his psychological profile for the purposes of predicting the likely suspects when another crime occurs with similar attributes.

Bayesian networks are descriptive rather than causal. They tell us the event that we may expect to observe, without explaining why the event occurs. Unlike a simulation, they do not describe the process that leads to the event. However, a Bayesian network can be an excellent complement to an agent simulation, which addresses causal mechanisms. For example, in an agent-based simulation reported in Duong (2009), a Bayesian network is used to generate a simulated agent's attributes by deriving the probability that an agent has an attribute given its other attributes. Then, the model simulates interactions of such agents in order to generate society-level patterns, which can be used to assess intervention policies.

2.3 Game-Theory Approaches

Rational-choice theory posits that humans are goal-driven and act to achieve their goals specifically to maximize their "utility," their measure of how well they have reached their goals. Game-theoretical approaches and neoclassical economic models of general equilibrium theory (Arrow 1951) are both based on the assumption of rational choice. Both approaches use mathematical techniques to find equilibria, points in the game at which no player can make a move that would improve the player's situation (utility value). Since agents in these models are rational, the agents gravitate toward these equilibria and stay there because no further move can improve their utility. The equilibria are thought to describe human behaviors such as whether a prisoner will testify against his accomplices in the prisoner's dilemma game (Axelrod 1984), or what the market prices of goods in the general equilibrium theory model will be.

Game-theoretical approaches analyze a criminal act in terms of the benefits and costs to each player in the act. For example, Eide (1999) uses a one-stage game to identify the conditions necessary for a behavior, such as crime and corruption, to occur by analyzing the cost and benefit of possible behaviors. Regression analysis of the effect of income on crime is used to support the rational-choice theory of crime in a game theory-based analysis.

Game theory is also used in the modeling of corruption. A common game-theoretical formulation for modeling of corruption involves a principal and an agent, in which the principal, seeking to maximize its utility, delegates decision-making power to an agent who may choose to maximize his own utility or that of a hidden principal (Farida and Ahmadi-Esfahani 2007). Corrupt acts are moves in the game.

2.4   Neoclassical Econometrics

Neoclassical econometrics is another tool based on rational-choice theory, suitable for modeling crime and corruption. Farida and Ahmadi-Esfahani (2007) present a study of the negative effect that corruption has on the production function important to economic growth, using a mathematical analysis within neoclassical theory called the Solow growth model. The Solow growth model includes several determinants of productivity such as capital and labor. Using the corruption index data (Transparency International 2009) and adding corruption to the productivity determinants, the study shows that economic data from Lebanon is consistent with a Solow model, and corruption acts as a detriment to production.

3   Simulation-Based Modeling of Crime and Corruption

Unlike the modeling techniques we discussed up to this point, simulation-based approaches are able to take into account greater complexities of interacting parts of social phenomena. In particular, fuzzy cognitive maps (FCM) and system-dynamics models are effective in describing complex systems, and agent-based models are well suited to modeling how systems become complex.

3.1 Fuzzy Cognitive Maps and System Dynamics

FCMs are fuzzy rule sets that incorporate representation of feedback loops. Feedback occurs when the output of a series of rules is input back into the same series of rules. The result is recomputed until it converges to either a steady state (called a fixed point in dynamical systems theory) or a repetitive state (called a limit cycle in dynamical systems theory). This state is then taken as an answer to the question the system was asked to compute. FCMs are well suited to modeling institutions such as commonly accepted forms of corruption, which a society learns when people perform acts that mutually reinforce each other.

Calais (2008) presents an FCM that models drug addiction, crime, economic productivity, international police interdictions, and America's image abroad (Fig. 1). In Calais's model, drug availability and drug usage are in a positive feedback loop. That means the more drugs are available, the more they are used, and the more they are used, the more they are available. There is also a positive feedback loop between American Image and tourism. Analysis of the model shows that international interdiction improves America's image abroad and economic productivity and decreases the prevalence of drug addiction. Calais also presents a guide for modeling crime with an FCM.

Like an FCM, a system-dynamics model describes relationships between variables but makes use of time-based differential equations to indicate the scalar value of a variable rather than Gaussian distributions to indicate "degree of membership" as in FCMs. Since feedback is involved, higher order effects can be observed. Dudley presents a system-dynamics model of corruption (Dudley 2006). The model (Fig. 2) includes positive feedback between corruption, bureaucracy, a weak legal system, lack of transparency, and resource rents (theft of resource revenues through corruption). Negative feedback occurs when more corruption leads to an improved legal system and decreased resource rents. In terms of individual

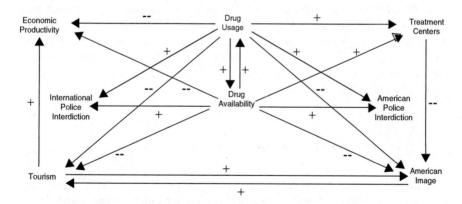

Fig. 1 Fuzzy cognitive map of the impact of drug addiction (figure reprinted with permission from Dr. William Allan Kritsonis, Editor-in-Chief, National FORUM Journals, Houston, Texas)

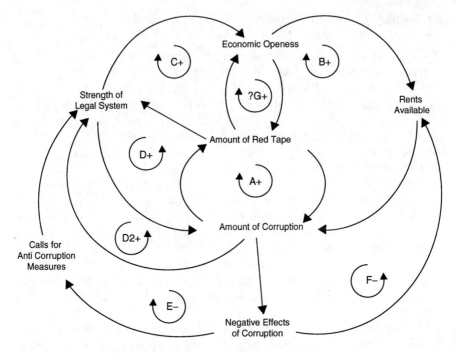

Fig. 2 A portion of Dudley's system-dynamics model of corruption

corrupt behaviors, the size of a bribe, the likelihood of payment, the value of service, and the effect of individual punishment are all factors in whether an individual takes a bribe. The need to keep a job, power, and loyalty can increase corrupt workplace behaviors. Analysis of the model leads to the conclusion that corruption is positively influenced by resource rents and negatively influenced by an improved legal system.

Both FCMs and system-dynamics models allow visualizations (e.g., Figs. 1 and 2) that appeal to nonspecialists. Practitioners often cite the appeal of presenting multiple factors of a system in a single visualization, which includes the direction in which the factors influence each other. Model users often value this visualization for encouraging insight into the system as much as they value the numerical answers obtained by these systems.

3.2 Agent-Based Simulations

Agent-based simulation can go a step farther by computing new social structures not previously identified in theory. FCMs and system dynamics are appropriate when the modeler knows all significant relations between entities. In contrast, agent-based simulation is suitable to those problems in which the modeler knows only a few relations and wishes to explore their implications. In effect, the implications are computed from these few relations as from first principles.

Agent-based models simulate the processes by which agents perceive their situation and make choices. Agents in such simulations come in two flavors: reactive and cognitive. Reactive agents have a few static rules that determine their behavior, with different macrolevel patterns emerging from different starting conditions. For example, an agent may have a static rule: avoid being close to another type of agent who is suspected of being likely to commit a crime. When the model simulates reactions of agents to each other, they may separate themselves from each other according to type, thus exhibiting a new macrolevel pattern not explicitly encoded in the model.

Unlike reactive agents who operate with a fixed set of predefined rules, cognitive agents can learn and change the rules by which they behave. Learning is important for the simulation of the emergence of institutions because it allows feedback from macro (society-level) rules down to micro (individual level) behaviors, a phenomenon known as "immergence." For example, a macrolevel rule could be the society's enforcement of a transparency program for reduction of corruption, while the microlevel rule could be the individual decision to avoid corrupt acts. Upper-to-lower feedback is essential for the emergence of new practices that are computed from the simulation's assumptions rather than being predetermined by the modeler beforehand (Andrighetto et al. 2007).

3.3   Reactive Agent Models

Reactive simulations, while less capable than cognitive-agent simulation, are adequate for testing a policy's effects with existing societal structures. For example, Dray et al. (2008) present a reactive agent model of drug enforcement policy, in which three law-enforcement strategies – standard patrol, hot-spot policing, and problem-oriented policing – are tested on a street-based drug market. Data from the urban environment of Melbourne, Australia, is used, and complex interactions between wholesalers, dealers, users, outreach workers, and police are modeled. Indicators include number of overdoses, fatalities, cash in dealers' hands, and numbers of committed crimes. The results show that problem-oriented policing is more effective in this environment than other strategies. Emergence of new structures is not required in a simulation in which the reactions of agents to policies are known and stay the same during the simulation.

In some models, reactive agents include limited elements of cognition, such as a simple memory based on past interactions. Makowsky (2006) presents a reactive agent model, CAMSIM, which uses a rational-choice approach to explain why people become criminals. In CAMSIM, agents have an age and choose a career based on maximum lifetime utility, from three possible careers – professional, labor, and crime. They infer the outcome of their own life from the lives of those around them. By simulation design, those around them are mainly their relatives.

The difference between the three possible careers is the amount of investment required, crime having a negative investment. Location and reproduction are also modeled. Changes in life expectancy matter to the career choice in this model. One conclusion of this model is that the effects of career choices extend over multiple generations. In effect, children "learn" from their ancestors' life experiences, even though the model does not include an explicit learning mechanism.

Another approximation of cognition occurs when agents operate within a genetic algorithm and learn new social structures as a group (Axelrod 1984). They do not learn through their individual experiences as an autonomous agent would. Instead, they learn through the experience of the "species," the group of agents within which they reproduce. Agents with better strategies reproduce in greater proportion so that the entire species evolves strategies that are more fit to their environment. In computational social science, these social strategies are mutually recognized rules of social interaction and social institutions (Axelrod 1984).

Much research in the field of computational social science models the social evolution of institutions by iterative game-playing and genetic algorithms. Axelrod's iterated prisoner's dilemma (IPD) was a pioneering study in which the strategy of cooperation emerged among agents even though they could have received an immediate benefit by cheating (Axelrod 1984). The IPD models the emergence of social behavior, which is relevant to the study of the breakdown of institutions through corruption and crime.

Situngkir (2003) applies a similar genetic algorithm and iterated game-theory approach to study corruption. The payoff matrix includes the cost of going to jail and the benefits of both corrupt and honest acts. As agents learn the best behaviors, they converge upon the strategy that is best for them given the strategies of other agents. Each agent reaches equilibrium and remains there because it can do nothing to better its situation. Situngkir shows that the behavior with the highest payoff is often corruption.

3.4 Cognitive Agent Models

In models that use cognitive agents, the agents learn how to perceive their environment and act upon the perceptions of their individual experiences. For example, Singh (2002) presents a cognitive agent model of urban crime patterns, in which agents with a common autonomous agent cognitive architecture called Belief, Desires, Intentions (BDI) use an artificial-intelligence technique called case-based reasoning. In BDI, agents deliberate over their beliefs and desires and commit to them as intentions. Using case-based reasoning, agents formulate a plan to achieve their goals by inferring from previous similar cases to which they have been individually exposed (Singh 2002). Singh's model includes variables of the law, the offender, the time, and the place. Criminal agents use their cognitive architectures to determine if a target of crime is a good target and to learn physical paths to their goals. The model yields a pattern of crime in a particular urban landscape.

4 Case Study: Cognitive Agents and Corruption

Nexus Network Learner (NNL), created with the Repast Simphony agent-based simulation tool kit (North et al. 2007), models the learning of social institutions of social network choice and role-based behaviors (Duong 2009).

NNL's model of corruption is based on Smith (2007). In the model, corruption is the result of conflict between the roles and role relations of the kin network and the bureaucratic network, two separate social structures with their own institutions forced into conflict by globalization. The model includes the kin network, the bureaucratic network, role behaviors that result in corruption, and the capacity of agents to learn new behaviors based on their cultural motivations.

The U.S. government used the NNL corruption model in the large simulation-based study described by Messer (2009). This study of hypothetical events in an African country examined the effects of international interventions on corruption, among other effects.

4.1 Overview of the Nexus Network Learner

The analyst initializes the NNL with data about individual behaviors and transactions, which are adjusted over time by the agents in the simulation, according to their goals. Agents use an artificial-intelligence technique to learn what traits to look for in the choice of network partners and in resource allocation behaviors. They base their choice on goals that are specific to their culture. Individual agents converge upon common practices and situations. When agents learn new behavior sets, a new social institution emerges.

Behaviors and goals that are input to the NNL corruption model include, for example, bribing, stealing, or whether to accept an offer of employment from an agent who has been rumored to steal from his employees. Behaviors such as stealing are input through a small rule set that implements a change in the flow of funds to role relations based on whether the agent or a network relation has learned to perform a behavior.

In sociology, the theoretical basis of NNL is in Symbolic Interactionism (Blumer 1986), in which roles and role relations are learned and created through the display and interpretation of signs (Duong and Grefenstette 2005). In the NNL corruption model, examples of roles include "Consumer," "Vendor," and "Maternal Uncle." An example of a role-relation rule is that the husband may choose up to three wives. The roles "Husband" and "Wife" belong to the Kin role network, while the roles "Vendor" and "Consumer" belong to the Trade role network. Examples of signs are social markers such as "Gender" and "Ethnicity."

NNL models the institution–individual linkage simultaneously with the individual–institution links. In this case, institutions are emergent social and legal norms that underlie collective activity and influence individual interaction. Figure 3 illustrates this process.

Fig. 3 Emergence with
cross-scale dynamics

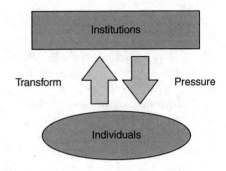

4.2 Networks of Agents

The NNL corruption model comprises three social networks: a network of bureaucratic
relations, a network of trade relations, and a network of family relations. Each of
these networks consists of a set of agents connected to other agents through a role
relation. Agents may have active roles, in which they have the job of initiating a
role relationship with a preferred partner, and passive roles, in which they may
accept the relationship. For every active role, there is one corresponding role, and
vice versa. All other roles are derived from these active–passive pairs.

The roles are described in an input file that includes a distribution for the typical
number of persons in each role relation (for example, a man has on average six
children and a standard deviation of 2). It also includes the demographic characteristics
of those eligible to choose role relations (such as a husband must be a male of
working age), the accounts that a role is responsible for (such as family support or
employee salaries), the flow of money to the accounts as expected by the proper
implementation of each role (for example, a person's dependents should get three-
fourths of his salary), and the conditional utility in a transaction (for example, if a
service providee has a plan to bribe and a service provider has a plan to accept a
bribe, the transaction has less direct utility). There are 65 different roles in the three
networks of the NNL corruption model, partly listed in Table 1.

The criteria for entering an active role are deterministic rules that are defined in the
role input file. In contrast, the criteria for choosing or accepting a role partner are
expressed in the probabilities of a Bayesian network. These probabilities may be
changed by learning; thus the name, "Nexus Network Learner." The NNL uses a
Bayesian network to characterize the demographic data of a country and to generate the
initial agent characteristics. The Bayesian network describes certain characteristics that
agents cannot change, for example, ethnicity or gender. It also describes other characteristics
that agents can change on an individual basis during the simulation, for example,
behavioral characteristics, such as bribing or stealing, and preferences for choices of
others in social networks (based on social markers or behavioral characteristics).

Finally, the Bayesian network describes demographic characteristics that
individual agents do not learn but are rather the output of the computations made

Table 1 Roles, network associations, and types (active, passive, derived) for the Nexus Network Learner corruption model

No.	Role name	Role network	Type	No.	Role name (continued)	Role network	Type
1	Wife (P)	Kin	p	33	Government purchaser	Trade	a
2	Child	Kin	p	34	Vendor	Trade	p
3	Government employee	Bureaucratic	p	35	Government vendor	Trade	p
4	Husband	Kin	a	36	Corporate receiver	Trade	a
5	Father	Kin	a	37	Head of government	Bureaucratic	d
6	Government employer	Bureaucratic	a	38	Government receiver	Bureaucratic	a
7	Retailer	Trade	p	39	Service provider	Bureaucratic	p
8	Employer	Trade	a	40	Home receiver	Kin	a
9	Sister	Kin	d	41	Head of corporation	Trade	d
10	Brother	Kin	d	42	Employee	Trade	p
11	Maternal aunt	Kin	d	43	Customer	Trade	a
12	Paternal uncle	Kin	d	44	Purchaser	Trade	a
13	Maternal cousin	Kin	d	45	Service providee	Bureaucratic	a
14	Paternal cousin	Kin	d	46	Head of household	Kin	d
15	Coworker	Bureaucratic	d	47	Provider	Kin	d
16	Mother	Kin	d	48	Corporate taxman	Bureaucratic	p
17	Sibling	Kin	d	49	Corporate taxpayer	Bureaucratic	a
18	Paternal grand parent	Kin	d	50	Income1	Trade	d
19	Maternal grand parent	Kin	d	51	Income2	Trade	d
20	Dependent	Kin	d	52	Income3	Trade	d
21	Spouse	Kin	d	53	Income4	Trade	d
22	Parent	Kin	d	54	Income5	Trade	d
23	Grade1	Bureaucratic	d	55	Income6	Trade	d
24	Grade2	Bureaucratic	d	56	Income7	Trade	d
25	Grade3	Bureaucratic	d	57	Income8	Trade	d
26	Grade4	Bureaucratic	d	58	Income9	Trade	d
27	Grade5	Bureaucratic	d	59	Income10	Trade	d
28	Grade6	Bureaucratic	d	60	Government pay distributer	Bureaucratic	d
29	Grade7	Bureaucratic	d	61	Pay distributer	Trade	d
30	Grade8	Bureaucratic	d	62	Government payee	Bureaucratic	d
31	Grade9	Bureaucratic	d	63	Household taxpayer	Bureaucratic	d
32	Grade10	Bureaucratic	d	64	Income taxman	Bureaucratic	p

during the simulation, such as unemployment statistics. The conditional probability that one agent property is related to another is used to generate agents with sets of properties similar to those of a population. The notional data could say, for example, that a working-age male of a given ethnic group is three times more likely to choose a person of his own ethnic group for his wife than he is likely to choose a wife of another ethnic group. Table 2 lists a set of nodes for the Bayesian network of the NNL corruption model, including their possible variables. Figure 4 illustrates a small portion of the Bayesian network.

In each of the three networks, there are eight types of corruption relations:

- Stealing/Trade Network (Scam)
- Bribing/Trade Network (Gratuity)
- Hiring Kin/Trade Network (Nepotism)
- Bribing to Be Hired/Trade Network (Misappropriation)
- Stealing/Government Network (Levy, Toll, Sidelining)
- Bribing/Government Network (Unwarranted Payment)
- Hiring Kin/Government Network (Nepotism)
- Bribing to Be Hired/Government Network (Misappropriation)

4.3 Agent Learning and Adaptation

In the simulation loop, agents perform two basic tasks. One is seeking and accepting role partners based on their traits and behavioral tendencies. The other is distributing money between financial accounts based on traits and behavioral tendencies of network partners. Agents also have the ability to observe and report behaviors based on their role, which may result in a penalty for some behaviors. Agents learn to keep the strategies that seem to increase the utility of their kin. Figure 5 illustrates the interaction between an individual's traits, his role interactions, and the institutions that result when these are combined with external government interventions such as penalties, foreign aid, or changes in resource pricing.

NNL's agents use genetic algorithms to learn and adapt to new role behaviors. Which behaviors are to be learned is an input to the simulation. NNL uses the genetic algorithm technique called the Bayesian Optimization Algorithm (BOA). Every agent includes an entire BOA that encodes a list of behaviors from which an agent may accept a subset. An agent tries each set of behaviors for a number of simulation cycles and then switches to another set. Table 3 illustrates the fit strategies of a single agent and how they change over time. The first seven learned behaviors are an agent's personal behaviors that determine the distribution of funds along networks. The last five behaviors determine the criteria for network choices. This particular agent learned behavior to employ his kin. The behaviors of accepting a bribe for employment, bribing for services, and stealing from a customer were tried and rejected early in the simulation run. The agent also learned not to choose employees who would offer him bribes but relearned the behavior later in the simulation run.

Table 2 Selectable characteristics of agents

No.	Agent properties	Options	Possible values/description
1	Role	4	Service providee, a service provider, employer, employee (can be many)
2	Hidden behavior	5	Steal from customer; Bribe for services; Accept bribe for services; Bribe employer; Accept bribe employer (can be many)
3	Know about behavior	2	Does or does not
4	Gender	2	Male or female
5	Ethnic preference	6	Four tribes, foreign, other (can be many) choice for spouse and employee
6	Corrupt	2	Is corrupt or is not corrupt
7	Ethnicity	6	Four tribes, foreign, other (can be many)
8	Zone	4	Region1, (can be many)
9	Age	3	Under 15, working age, over 60 (can be many)
10	Sector	3	Government, industry, agriculture (can be many)
11	Income	10	Low to high (can be many)
12	Reside (type of family)	3	Nuclear family, matrilocal, patrilocal
13	Wife age	3	Under 15, working age, over 60 (can be many)
14	Wife gender	2	Relative to the agent, if the agent is the wife then the selection is male
15	Wife ethnicity	6	Four tribes, foreign, other (can be many)
16	Child ethnicity	6	Four tribes, foreign, other (can be many) depends on the societal "Reside"
17	Child age	2	Working age, under 15
18	Employee income	10	Ten levels, could be many
19	Employee ethnicity	6	Four tribes, foreign, other (can be many)
20	Employee is kin	2 (Y/N)	Employer corruption
21	Accept bribe for services	2 (Y/N)	Employee corruption
22	Penalized	2 (Y/N)	Is or is not penalized
23	Employer steal from organization	2 (Y/N)	Employer corruption
24	Bribe employer	2 (Y/N)	Employee corruption
25	Bribe for services	2 (Y/N)	Employee corruption
26	Steal from customer	2 (Y/N)	Employee corruption
27	Steal from organization	2 (Y/N)	Employer corruption
28	Accept bribe employer	2 (Y/N)	Employer corruption
29	Rig election	2 (Y/N)	Government corruption
30	Commission for illicit services	2 (Y/N)	Government corruption
31	Unwarranted payment	2 (Y/N)	Government corruption
32	Gratuity	2 (Y/N)	Private sector corruption
33	Levies, tolls, sidelining	22 (Y/N)	Government corruption (could be many)
34	Misappropriation	2 (Y/N)	Government corruption
35	String pulling	2 (Y/N)	Employer corruption (employee is kin)
36	Productive	2	Is or is not productive

(continued)

Table 2 (continued)

No.	Agent properties	Options	Possible values/description
37	Employee productive	2	Is or is not productive (system related)
38	Scam	2 (Y/N)	Private sector corruption
39	Employed	2 (Y/N)	Government or private sector
40	Employee sector	3	Sector of employment (government, agriculture, or industry)
41	Employee: bribe employer	2 (Y/N)	Employee corruption (system related)
42	Service provider :steal from customer	2 (Y/N)	Private sector corruption (system related)
43	Service provider income	10	Low to high (can be many) used to relate corruption to income level
44	Taxman sector	1	Government
45	Service providee Bribe for services	2 (Y/N)	Government corruption
46	Factionalization	2 (Y/N)	More factionalization, less inter racial marriage and employment
47	Service provider age	3	Under 15, working age, over 60 (can be many)
48	Service provider employed	1	Employed (system related)
49	Taxman employed	1	Employed (system related)

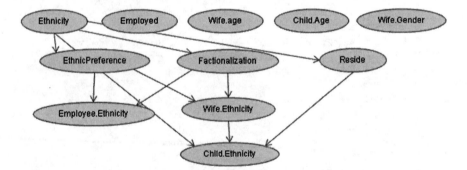

Fig. 4 A small portion of the Bayesian network that illustrates dependencies

The fitness of a strategy is measured with a utility, or "happiness," function. In the NNL corruption model, an agent's utility derives from the advantageous trade interactions of the dependent kin. The question as to which kin are dependent is defined in the role file. For example, agents of matrilocal tribes consider their mother's side to be their responsibility. Agents of patrilocal tribes consider relatives on their father's side to be their responsibility. Finally, modern urban neolocal families consider only their children to be their responsibility. However, agents are "happy" (in other words, they find utility) not when their kin receive funds but

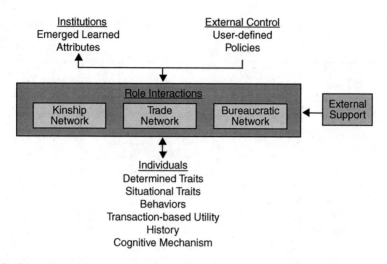

Fig. 5   Nexus Network Learner conceptual model

rather when they receive proper care, e.g., receive services and buy goods. Stealing and bribing can lessen the amount of happiness.

"Coevolution" occurs when two or more agents simultaneously learn from and adapt to each other. For example, one agent learns that choosing an employee who bribes his employer is an advantageous behavior. Simultaneously, another agent learns that offering bribes to his employer is advantageous. Such agents that learn from and adjust to each other create social structure: institutionally accepted corruption that exists throughout a society.

NNL agents learn to make fund allocation choices and network partner choices according to their individual incentives to support their kin. As they change each other's incentives, for example, by hiring employees who offer bribes, the choices they make become new social structures through the coevolutionary process. In some contexts, bribes flourish, and in others, they do not. For example, in the NNL corruption model, agents learn from their genetic algorithms the types of persons to include in their social network, based on criteria, including kinship, ethnicity, and bribing behavior. Their genetic algorithms also lead them to decide whether they divert funds across networks through bribing and stealing.

Because incentives are modeled as culturally-based, the effects of different interventions, such as increased penalties for stealing, foreign aid, and resource rents, can be studied in a particular cultural setting. Corruption is changed institutionally through synchronous changes in the habits of individuals, for example, groups of employees who decide not to accept employment from employers who steal, as well as groups of employees who tolerate abuse because no other employment is available. Agents are driven by new incentive structures that come both from intervening actions and from other agents' reactions to those actions.

For example, a run of the NNL corruption model that tested incarceration penalties for corrupt behavior displayed cyclical behavior at relatively low levels

Table 3 Example of an individual agent selection of top strategy over multiple learning cycles

Learning cycle	Utility result	1	2	3	4	5	6	7	8	9	10	11	12
1	66	N	Y	N	Y	N	N	N	N	N	Y	N	Y
2	85	N	N	N	N	N	N	N	Y	N	N	N	Y
3	633	N	N	N	N	Y	N	N	N	N	N	N	N
4	755	N	N	N	N	N	N	N	Y	N	Y	N	N
5	902	N	N	N	N	N	N	N	Y	Y	N	N	Y
6	925	N	N	N	N	N	Y	N	Y	N	N	N	Y
7	575	N	N	N	N	Y	N	N	N	N	Y	N	Y
8	748	N	N	N	N	N	Y	N	Y	N	N	N	Y
9	1,873	N	N	N	N	N	N	N	Y	N	N	N	Y
10	2,545	N	N	N	N	N	N	N	Y	N	N	Y	Y
11	105	N	N	N	N	N	N	N	Y	Y	Y	N	Y
12	1,743	N	N	N	N	N	N	N	Y	N	N	N	Y
13	2,747	N	N	N	N	N	N	N	Y	N	N	N	Y
14	2,803	N	N	N	N	N	N	N	Y	N	N	N	Y
64	18,275	N	N	N	N	N	N	N	Y	N	N	N	Y
70	20,459	N	N	N	N	N	N	N	Y	N	N	N	Y
74	23,011	N	N	N	N	N	N	N	Y	N	Y	N	Y
79	12,797	N	N	N	N	N	N	N	Y	N	Y	N	Y

Strategy component	Description
1	Bribe employer (Y/N)
2	Accept bribe employer (Y/N)
3	Steal from organization (Y/N)
4	Bribe for services (Y/N)
5	Accept bribe for services (Y/N)
6	Steal from customer (Y/N)
7	Factionalization (Y/N)
8	Employee: is kin (Y/N)
9	Employer: steal from organization (Y/N)
10	Employee: bribe employer (Y/N)
11	Service provider: steal from customer (Y/N)
12	Service providee: bribe for services (Y/N)

of bribing and stealing. In each cycle, bribing or stealing goes up, then the number of penalized agents goes up, and then bribery or stealing goes down again. In this example, agents reacted to each other in a path-dependent way typical of coevolutionary systems: at one point in the run, the cycle became very large so that a large proportion of agents learned to bribe. Bribing nearly became institutionalized; however, the penalty succeeded in damping the newly accepted behavior early on. The intervention was successful in keeping the bribing level constrained, but the social forces made the bribing persist, so much so that about a third of all agents in this 100-agent simulation were incarcerated at some point. Table 4 lists the properties of the penalized agents at a single time late in the simulation run.

Table 4 Properties of penalized agents

Properties for 31 "Penalized" agents	Values
Gender male/female	18/13
Ethnicity A/L/K/M/Frn/other	7/4/3/6/2/9
Region R1/R2/R3/R4	2/9/15/5
Age under15/working age/ over 64	9/22/0
Sector Govt/Ind	18/13
Family organization p/m/n	19/11/1
Income I1/I2/I3/I4/I5/I6/I7	0/2/3/6/12/7/1
Penalized	31
Employed (y/n)	29/2
Bribe employer (y/n)	0/31
Accept bribe employer (y/n)	0/31
Steal from organization (y/n)	0/31
Bribe for services (y/n)	24/7
Accept bribe for services (y/n)	20/11
Steal from customer (y/n)	0/31
Factionalization (y/n)	16/15
Employee: is kin (y/n)	4/27
Employer: steal from organization (y/n)	2/29
Employee: bribe employer (y/n)	0/31
Service provider: steal from customer (y/n)	1/30
Service providee: bribe for services (y/n)	20/11

4.4   Using the Nexus Network Learner

When using the NNL, the analyst or modeler typically follows these steps:

- Determine the social role networks that are relevant to the problem, and determine how the incentives for breaking the law can be represented by shifting resources from one network to the other; define rules that describe resource flows in transactions.
- Define roles with different powers of observing a law-breaking behavior, behaviors with different chances of being convicted in a justice system, and penalties with different incarceration lengths.
- Describe the demography of the population under study, accounting for demographic characteristics such as employment, education, and ethnicity.
- Assign notional but ultimately measurable probabilities to the Bayesian network, with the intent of replacing the notional data with real ones as the model matures.
- Execute the simulation; a single run of a model with 100 agents will take about 2 h on a typical laptop computer.
- Examine the output files, which are lists of the agents and their attributes, all defined in the Bayesian network. One output file lists the attributes that an agent actually displays in its simulated actions. Another output file lists the learned

strategies that encode the desired behaviors of each agent (but not necessarily the ones that the agent had an opportunity to perform) and how much an agent's kin actually benefits from the agent's behavior.

- Repeat the simulation multiple times to explore the possible outcomes; take note of the classes or characteristics of persons penalized in each simulation; consider the pattern of relations emerging between the intervention, such as a change in penalty policy, and the behavioral strategies that the agents evolve.

5 Practical Tips

- Recognize that computational modeling of crime and corruption is a young, immature field, and that current models are far from reliable. Make sure that model users understand this and that they apply modeling results with an appropriate degree of caution.
- Include in the modeling team a social scientist with a strong background in crime and corruption. This social scientist should collaborate directly and continuously with the modelers.
- When possible, use a recognized social theory of crime and corruption, or a consistent combination of several social theories. Often, a social theory postulates that certain foundational behaviors of individuals cause the emergence of social behavioral patterns. In such cases, first "hard-code" the underlying behaviors, and then develop the computational model until it demonstrates the emergence of theoretically asserted societal patterns.
- In modeling effects of interventions on crime and corruption, include representations of social institutions that control crime and corruption. Then, examine how interventions affect these institutions.
- Employ, when possible, well-accepted modeling tools. For instance: when implementing system-dynamics simulations, consider tools such as Vensim, Powersim, and iThink; for reactive-agent models, consider open-source tool kits such as NetLogo, Repast Simphony, MASON, or Swarm; and, for cognitive-agent models, consider Repast Simphony or MASON.
- Strive for a computer simulation that is causal in nature. Give preference to a model that involves few assumptions but demonstrates multiple real-world phenomena of crime and corruption.
- Maximize the number of variables in a model that are measurable in the real world. In the case of system dynamics models, for instance, recognize that stocks are more likely to reflect measurable quantities than flows, and attempt to maintain a ratio of stocks to flows of at least three to one.
- Usually, it is more efficient to begin the development of a model using a set of notional data. As the model matures and data requirements become better defined, notional data can be gradually replaced with real-world data.

6   Summary

Sociological theories of crime include: theories of strain blame crime on personal stressors; theories of social learning blame crime on its social rewards, and see crime more as an institution in conflict with other institutions rather than as individual deviance; and theories of control look at crime as natural and rewarding, and explore the formation of institutions that control crime. Theorists of corruption generally agree that corruption is an expression of the Patron–Client relationship in which a person with access to resources trades resources with kin and members of the community in exchange for loyalty. Some approaches to modeling crime and corruption do not involve an explicit simulation: rule based systems; Bayesian networks; game theoretic approaches, often based on rational choice theory; and Neoclassical Econometrics, a rational choice-based approach. Simulation-based approaches take into account greater complexities of interacting parts of social phenomena. These include fuzzy cognitive maps and fuzzy rule sets that may incorporate feedback; and agent-based simulation, which can go a step farther by computing new social structures not previously identified in theory. The latter include cognitive agent models, in which agents learn how to perceive their environment and act upon the perceptions of their individual experiences; and reactive agent simulation, which, while less capable than cognitive-agent simulation, is adequate for testing a policy's effects with existing societal structures. For example, NNL is a cognitive agent model based on the REPAST Simphony toolkit. NNL's Corruption Model structures corruption as arising from conflict between the roles and role relations of kin and bureaucratic networks. The NNL model includes three overlapping social networks each with roles: bureaucratic, trade, and kin or family networks. As agents make choices (e.g., whether to accept bribes), other agents with whom they interact observe the choices and draw conclusions about their utility. Different cultures are modeled by increasing the social or economic penalties attached to various behaviors.

7   Resources

The following is a list of links to data and software resources that aid in the study of crime and corruption, and the software that runs the models, mentioned in the text.
1. Open Source Software
   Weka Data Mining Software
   http://www.cs.waikato.ac.nz/ml/weka/
   Assortment of machine learning algorithms for analyzing data R
   http://www.r-project.org/
   Environment for statistical computing and graphics
   Sage
   http://www.sagemath.org/

Mathematics software system
Jess
http://www.jessrules.com/
Rule engine and scripting environment
FuzzyJess
http://ai.iit.nrc.ca/IR_public/fuzzy/fuzzyJToolkit.html
NetLogo
http://ccl.northwestern.edu/netlogo/
Environment for multiagent simulation
Repast Simphony
http://repast.sourceforge.net/
Environment for multiagent simulation
Mason
http://cs.gmu.edu/~eclab/projects/mason/
Discrete-event multiagent simulation library core in Java
Swarm
http://www.swarm.org/
Environment for multiagent simulation

2. Commercial Software
Vensim
http://www.vensim.com/
Environment for system dynamics modeling and simulation
IThink
http://www.iseesystems.com/
Environment for system dynamics modeling and simulation
Powersim
http://www.powersim.com/
Environment for developing many types of simulations

3. Corruption
Transparency International
http://www.transparency.org/
Datasets on numerous corruption indicators, such as the corruption perceptions index, the global
corruption index, and the bribe payers index
The Global Integrity Report
http://report.globalintegrity.org/
Resources and indicators on governance and corruption trends around the globe
Internet Center for Corruption Research
http://www.icgg.org/corruption.research.html
Links to academic research on corruption
World Bank Governance Data
www.worldbank.org/wbi/governance/data
Dataset on governance dimensions including *control of corruption*
Organized Crime and Corruption Bibliographic Database
http://www.osgoode.yorku.ca/NathansonBackUp/search.htm
Repository of articles and links concerning transnational corruption, human rights, and security
The Terrorism, Transnational Crime and Corruption Center
George Mason University
http://policy-traccc.gmu.edu/transcrime/corruption.shtml
Academic resources on the links among terrorism, transnational crime and corruption

4. Crime
Law Moose World Legal Resource Center

http://www.lawmoose.com/internetlawlib/1.htm
Legal reference materials
FBI Uniform Crime Statistics
http://www.fbi.gov/ucr/ucr.htm
US crime data and reports
National Criminal Justice Reference Service
http://www.ncjrs.gov/
Listings and/or repositories of justice and substance abuse information
National Archive of Criminal Justice Data
http://www.icpsr.umich.edu/NACJD/index.html
Listings and/or repositories of crime and justice data
Bureau of Justice Statistics
http://www.ojp.usdoj.gov/bjs/
Listings and/or repositories criminal justice statistics

References

Agnew, R. (2009) *Crime Causation: Sociological Theories – Strain Theory, Social Learning Theory, Control Theory*. http://law.jrank.org/pages/824/Crime-Causation-Sociological-Theories.html.

Andrighetto, G., Campennì, M., Conte, R. & Paolucci, M. (2007) On the Immergence of Norms: a Normative Agent Architecture. In *Proceedings of AAAI Symposium, Social and Organizational Aspects of Intelligence*. Washington DC.

Arrow, K. (1951) *Social Choice and Individual Values*. Wiley: New York.

Axelrod, R. (1984) *The Evolution of Cooperation*. New York: Basic Books.

Baumgartner, K.C., Ferrari, S. & Salfati, C.G. (2005) Bayesian network modeling of offender behavior for criminal profiling. In *Decision and Control, 2005 and 2005 European Control Conference. CDC-ECC '05. 44th IEEE Conference on*, pages 2702–2709.

Blumer, H. (1986) *Symbolic Interactionism: Perspective and Method*. Berkeley: University of California Press.

Calais, G. (2008) Fuzzy Cognitive Maps Theory: Implications for Interdisciplinary Reading: National Implications. *Focus on Colleges, Universities, and Schools*, 2(1).

Chen, Y. & Jeng, B. (2002) Yet Another Representation for System Dynamics Models, and its Advantages. In *Proceeding of the 20th International Conference of the System Dynamics Society*. Palermo, Italy.

Collier, P. (2007) *The Bottom Billion: Why the Poorest Countries are Failing and What Can Be Done About It*. Oxford: Oxford University Press.

Dray, A., Mazerolle, L., Perez, P. & Ritter, A. (2008) Drug Law Enforcement in an Agent-Based Model: Simulating the Disruption of Street-Level Markets. In L. Liu & J. Eck (Eds.) *Artificial Crime Analysis Systems: Using Computer Simulations and Geographic Information Systems*. New York: Information Science Reference.

Dudley, R. (2006) *The Rotten Mango: The Effect of Corruption on International Development Projects*. Beaverton.

Duong, D. (2009) Nexus Network Learner. *MORS Workshop on Irregular Warfare Analysis II*, MacDill AFB. http://www.mors.org/UserFiles/file/meetings/09iw/pres/WG2_Duong_D.pdf.

Duong, D. & Grefenstette, J. (2005) SISTER: A Symbolic Interactionist Simulation of Trade and Emergent Roles. *Journal of Artificial Societies and Social Simulation*, 8(1). http://jasss.soc.surrey.ac.uk/8/1/1.html.

Eide, E. (1999) *Economics of Criminal Behavior*, chapter 8100, pages 345–389. Edward Elgar and University of Ghent, http://encyclo.findlaw.com/8100book.pdf.

Epstein, J. & Axtell, R. (1996) *Growing Artificial Societies: Social Science from the Bottom Up.* Boston: MIT Press.

Farida, M. & Ahmadi-Esfahani, F. (2007) Modeling Corruption in a Cobb-Douglas Production Function Framework. For the *Australian Agricultural and Resource Economics Society 51st Annual Conference*, Queenstown, NZ. http://econpapers.repec.org/RePEc:ags:aare07:10400.

Frisch, D. (1996) The Effects of Corruption on Development. *The Courier, 158*, 68–70.

Humphreys, P. (2004) *Extending Ourselves: Computational Science, Empiricism, and the Scientific Method.* New York: Oxford University Press.

Kosko, B. (1994) *Fuzzy Thinking: The New Science of Fuzzy Logic.* New York: Flamingo.

Makowsky, M. (2006) An Agent-Based Model of Mortality Shocks, Intergenerational Effects, and Urban Crime. *Journal of Artificial Societies and Social Simulation*, 9(2).

Messer, K. (2009) The Africa Study. *HSCB Focus 2010 Conference.*

North, D. (1990) *Institutions, Institutional Change and Economic Performance.* Cambridge: Cambridge University Press.

North, D. (1991) Institutions. *Journal of Economic Perspectives, American Economic Association*, 6(1), 97–112.

North, D. (2005) *Understanding the Process of Economic Change.* Princeton: Princeton University Press.

North, M.J., Tatara, E., Collier, N.T. & Ozik, J. (2007) Visual Agent-based Model Development with Repast Simphony. *Proceedings of the Agent 2007 Conference on Complex Interaction and Social Emergence*, Argonne National Laboratory, Argonne, IL, USA.

Nye, J. (1967) Corruption and Political Development: A Cost-Benefit Analysis. *The American Political Science Review*, 61(2), 417–427.

Rao, M. & Georgeff, P. (1995) BDI-Agents: From Theory to Practice. *Proceedings of the First International Conference on Multiagent Systems* (ICMAS '95).

Schank, R. (1983) *Dynamic Memory: A Theory of Learning in Computers and People.* New York: Cambridge University Press.

Shao, J., Ivanov, P., Podobnik, P. & Stanley, E. (2007) Quantitative Relations between Corruption and Economic Factors. *The European Physical Journal B.* 56, 157–166.

Singh, K. (2002) *An Abstract Mathematical Framework for Semantic Modeling and Simulation of Urban Crime Patterns.* Masters Thesis, Simon Fraser University.

Situngkir, H. (2003) The Structural Dynamics of Corruption: Artificial Society Approach. *Journal of Social Complexity*, 1(3).

Situngkir, H. & Siagian, R. (2003) *NGOs and Foreign Donations.* Bandung: Bandung Fe Institute.

Smith, A. (1994) *The Wealth of Nations.* New York: Random House.

Smith, D. (2007) *A Culture of Corruption: Everyday Deception and Popular Discontent in Nigeria.* Princeton, NJ: Princeton University Press.

The Fund for Peace (2008) The Failed States Index, 2008. *Foreign Policy.*

Transparency International (2009) *2009 Global Corruption Barometer.* Transparency.org.

Witten, I. & Frank, E. (1999) *Data Mining: Practical Machine Learning Tools and Techniques with Java Implementations.* New York: Morgan Kaufman.

Chapter 10
Visualization and Comprehension

David Jonker and William Wright

The Achilles' heel of societal models, nearly universally, is their inability to convey their computational results to the human user. Data sets can be enormous and diverse, the uncertainties large and subtle, the dependencies complex and convoluted, and the products of the estimates often obscure and insubstantial and, therefore, difficult to convey and contrast. This chapter is about making the "invisible" visible.

Recognizing the diverse roles played by visualization, we divide them here into two categories. First is interactive data or information visualization, which is defined as the graphical expression of large amounts of data to enable efficient human perception and comprehension (Card et al. 1999). Second is visual analytics, which is the joining of interactive visualization with analytical reasoning and computational methods in order to derive and convey insight from massive, dynamic, ambiguous, and often conflicting data (Thomas and Cook 2005). This chapter pays equal attention to the two categories.

The chapter begins by introducing mental models and explaining why it is so important that they be made visible. Next, it turns to the challenges faced in understanding complex behavior, and indicates how effective visualization can help. The chapter then delves into the challenges associated with groups of diverse users, diverse model types, and disparate model authoring schemes, and again indicates the important role that visualization can play. Having uncovered the issues, the chapter presents a case study involving a large suite of PMESII models and discusses how (some of) the aforementioned challenges were successfully attacked. The chapter concludes with a set of tips for aspiring visualizers.

D. Jonker (✉)
Oculus Info Inc, 2 Berkeley Street, Suite 600, Toronto ON, M5A 4J5, Canada
e-mail: david.jonker@oculusinfo.com

A. Kott and G. Citrenbaum (eds.), *Estimating Impact*,
DOI 10.1007/978-1-4419-6235-5_10, © Springer Science+Business Media, LLC 2010

1   Models and Visualizations

1.1   The Importance of Mental Models

Analysts using PMESII models to gain insight into a complex geopolitical situation apply prior and tacit knowledge of experts to the task. In discussions with other experts, and with research, this "conceptual model" of the situation evolves. The visualization of these conceptual models aids the experts in sharing, testing, and evolving their concepts. In addition, the computer–human interface for computational PMESII models should be expressed in terms similar to these conceptual models, or linked to them, in order to maximize the usability of the PMESII tools in supporting and extending human expert knowledge. The effective communication of concepts requires that the conveyance aligns well with the receiving analyst's mental model.

The problem of mental model relation is a relatively new one. Computational methods and models in the realms of engineering, physics, and finance have seen few challenges in this area. Although they are applied to everything from weather forecasting to portfolio performance estimation, it has nearly always been possible to depict the outputs in terms of tangible, measurable observables, things that relate well with our mental models. The output of physical or financial models, for instance, is visualized using a simulated enactment of physical behavior, or tabulated spreadsheet charts. These are forms consistent with the real-world experiences of these phenomena and are, therefore, a natural representation for communication and comprehension.

The situation with social and political models is different. The subjects of primary modeling interest here are attributes such as satisfaction, attitudes, beliefs, goals, and cultural identity, things that cannot always be directly observed. Thus, there is no obvious or natural way of depicting them. Furthermore, the field is relatively young; therefore, there are few standards.

A case in point can be found in game-theoretical models. These have been used in economics, psychology, and sociology to model and understand attributes such as trust and reputation (Mui et al. 2001) and trust development over time (Axelrod 1987). However, few attempts have been made to visualize these interactions beyond the mathematical models themselves (Fig. 1). There are no standards, no precedents, and no common procedures for visualization of such models.

Another example is visual anthropology, in which emphasis is placed on creating a visual record of cultural interactions and leaving it to the analyst to interpret them (Collier and Collier 1986). No attempt is made at formulating an underlying visual expression. McCormick summarizes the state of affairs quite well with his observation that to visualize sociopolitical systems and their relationships to other social science domains such as economics and security requires "*a method for seeing the unseen*" (McCormick et al. 1987).

In the relatively short history of applying computational approaches toward the societal modeling, little emphasis has been placed on creating or adapting visual representations for such models. In fact, a survey of modeling and simulation applications reveals sets of time series graphs, often devoid of context, narrative, and summary,

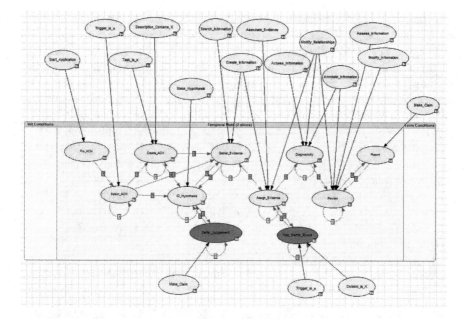

Fig. 1 Bayesian model visualization means little to the nonspecialist

which do not tend to map well to the users' conceptual constructs or *mental models* (Card et al. 1999; Sears and Jacko 2007).

To compensate for this gap, during model initiation and results analysis, human technical interpreters are often introduced between the model and the subject matter expert (SME). These interpreters view model results, translate them into summaries, and sometimes even draw preliminary conclusions. This is undesirable. The distancing of SMEs and decision-makers from the computational results not only impacts timelines and workloads, but also introduces the distinct possibility that the quality and accuracy of conclusions will be degraded. This diminishes the power of the models to explore issues and extend knowledge for the SME.

For elimination of this gap tool builders need to provide a means for easily interacting with and communicating information regarding a modeled situation directly to the decision-makers. The expert decision-makers need to be able to use the model tools to express the pertinent factors of the situation as they see them and then see the model results in the same terms. The modeling environment should be able to present the insights provided by the models, in forms consistent with domain SMEs' mental models.

Despite repeated appeals for support in this area, progress thus far has been relatively limited. Orford and coworkers (1999), for instance, provide a review of approaches and techniques across a variety of social science disciplines and find that adaptation of visualization techniques is for the most part limited to fields with strong ties to the physical sciences, for instance, geography.

However, progress is being made in the area known as *conceptual* social modeling. Conceptual models, similar to computational models, visually express aspects of a situation, but without the quantitative formulas and values required to simulate behavior under different conditions. As a result, conceptual modeling tools and techniques provide useful reference examples for visualizing a situation, albeit without going so far as to provide the ability to visualize change in that situation over time due to either the natural course of events or the effects of an intervention.

One such class of conceptual modeling tool is *link analysis*. Applications such as Analyst's Notebook™ (i2 Inc., Fig. 2) and VisuaLinks™ (Visual Analytics Inc.)[1] provide capabilities for building and maintaining diagrammatic visual representations that help acquire a rapid snapshot of key actors, interactions, and communications. These applications see heavy usage in the law enforcement, intelligence, and military communities, where they are employed to build visual maps of connections (e.g., transactions, phone calls, "is-related-to" relations, etc.) between various organizations, people, and concepts.

Link analysis tools tend to provide a meaningful snapshot of current reality, perceived or otherwise, but they do not generally provide capabilities for considering

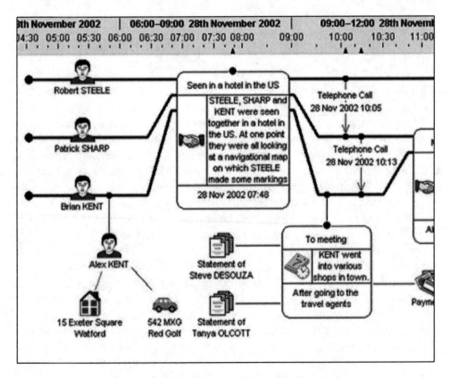

Fig. 2 Analyst's Notebook provides conceptual modeling using link analysis

---

[1]i2 Analyst's Notebook is a trademark of i2 Inc. VisualLinks is a trademark of Visual Analytics Inc.

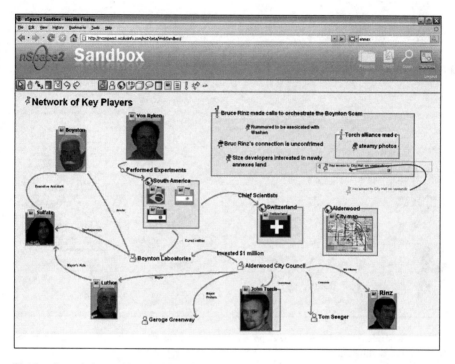

Fig. 3 nSpace helps rapid exploration of data

alternative or dynamic realities. An example of a tool that attempts to support the analysis of alternative social and political realities is nSpace (Fig. 3). nSpace includes integrated tools for system information gathering and analysis (Jonker et al. 2005; Wright et al. 2006) and the nSpace Sandbox (r) component[2] that allows analysts to conceptualize and evaluate alternative hypotheses. Links to original source evidence in the analysis provide a means to verify or re-evaluate evidence and assumptions.

A component missing from many conceptual modeling implementations is that of time. Few tools actually support analysis of *how a situation evolves over time.* Animation, while a seemingly logical solution, provides a poor means of visualizing change over time. Underlying this shortcoming are human limitations in the ability for visual memory. Human visual perception is strong; human visual memory is weak.

Human perception of change over time tends to be improved when change is displayed simultaneously instead of sequentially (Wickens and Hollands 2000; Parasuraman and Mouloua 1987). GeoTime Configurable Spaces TM[3] exploits the advantage of this phenomenon (Kapler et al. 2008). It extends established two-dimensional (2D) *X, Y* forms of expressing conceptual models by adding a third visual dimension for time in the *Z* dimension (Fig. 4). This makes it possible

---

[2] nSpace Sandbox (r) is a trademark of Oculus Info Inc.

[3] GeoTime Configurable Spaces TMis a trademark of Oculus Info Inc.

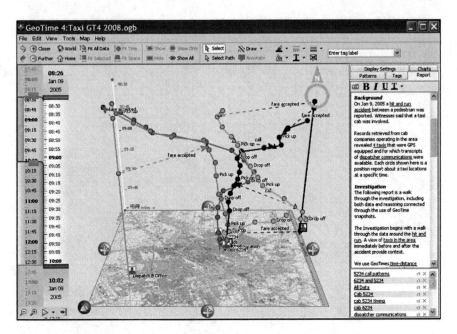

Fig. 4 GeoTime represents events within an X, Y, T coordinate space in which the X and Y planes represent geographic space, and the Z-axis represents temporal space

to visualize a social network of entities on a 2D plane, with communication and transaction events between those entities represented in time above that plane in 3D space (Fig. 5).

Another good example is the commercial game known as SimCity™ (Maxis, Electronic Arts). The player's objective in SimCity is to manage a city. Exogenous events occur throughout the game, some at random and some scheduled specifically to present challenges, and as the state of the city evolves, the player adjusts parameters in order to improve or stabilize the city's health. Health is gauged in terms of numerical metrics and is visualized in terms of graphics and tables. A neighborhood displayed in bright red, for example, indicates to the game player a current undesirable (as defined by the game) state in that region that may require further investigation. Events are presented to the user in a scrolling bulletin fashion as they occur and accumulate in list form for detailed examination. SimCity offers a diverse model of geopolitical growth over time and allows the user to explore many types of scenarios (Fig. 6). Since its purpose is to simulate experience in real time rather than to analyze it, the game neither attempts to address visualization of the causes and structural systems underlying behavior, nor does it generally include the dimension of time in its displays.

The bottom line is that while some relevant precedents can be found in conceptual modeling and in the realm of computer games, little has been done in the world of PMESII modeling to express models and simulation results visually in ways that are compatible with an expert's mental model. In our experience, when we ask an expert

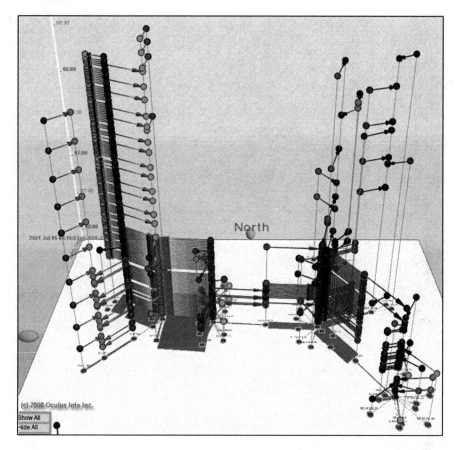

Fig. 5 Temporal view of activity in a social network in GeoTime. Note the summary network image on the ground plane, with constituent activity as events above in the time Z-axis

to describe visually their understanding of a complex societal situation, the result is likely to be a diagram of shapes, arrows, and words, and almost certainly not a series of charts and graphs. If the visual language used in conceptual models and in simulation games may serve as useful precedents in addressing this problem, the gap that must be bridged is the integration of time and causality into these vocabularies.

1.2 Explaining Complex Model Behavior

Another shortcoming in current visual vocabularies is the visual techniques for expressing model behavior. When analysts are unable to assure themselves that they understand the causal relationships and interactions associated with simulated events, they are unable to assign confidence in their observations and projections. A lack of insight into causes can also handicap analysts' ability to respond to

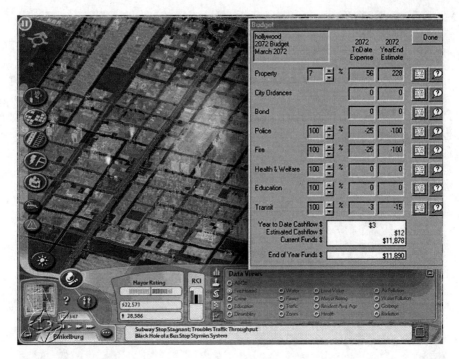

Fig. 6 SimCity 3000 allows fine control over a geo-political simulation, visually characterizing civic health on a map at a single point in time

projections with effective mitigating strategies. In turn, analysts are limited in their ability to support decision-makers.

In the case of societal behaviors, calibration is rarely possible. The factors involved in a situation are not only immensely complex but also ever-changing and often immeasurable. Even when measurement is physically or conceptually possible, it is often not viable due to political or cost considerations. As a result, societal models are not likely to become precisely tuned instruments of exact forecasting. They will remain, rather, repositories into which human experts may collectively "encode" their understanding of the society's behavioral structure and dynamics, albeit on a scale that is not limited by one person's mental capacity.

Accepting that PMESII simulation results will likely never be worthy of blind trust, the question remains: how does one go about assessing confidence in the results? How can a system help to build (or temper) confidence in observations that have not been anticipated through unaided human reasoning? One approach is to provide capabilities for analysts to explore a model until they are able to ascertain the cause of nonintuitive results. Once the causes of a computed result are understood, the analysts may decide whether it is the model that is representing a situation incorrectly or too simplistically, or whether it is *the analysts* who are reading a situation incorrectly or too simplistically. Either way, a basis is provided on which to judge the result and respond accordingly. In the former situation, the analysts may make note of future areas for improvement of the model and in the short term take steps to work around the shortcoming, while in the latter case, they may benefit from an improved situational understanding and adjust their strategies to make them more effective.

Insights into model behavior help to explain unexpected results and frame conclusions. Today, however, there are few robust methods for visualizing the causes behind model behavior. One of the more common strategies used is to display cause and effect chains using nodes and links. This technique is seen almost exclusively with causal models such as Bayesian networks. In other modeling paradigms, the causes of an effect are often more complex, with many factors contributing in various ways and to various degrees over time. In these cases, often no explanation is offered for behavior other than the one given by an expert intimately familiar with the model.

One technical development in societal modeling that holds promise for more intuitive visualization of the model structure and behavior is that of agent-based simulations. Visualization of the structure and behavior of agent-based simulations may intrinsically be more natural, due to their structural similarities to traditional social networks and the potential for reduction of behavior to the "decisions" of individual agents.

1.3   Accommodating Diverse Users

Analysis of intervention effects brings together people with knowledge of the diplomatic, military, aid, and nongovernmental organization (NGO) agencies involved, as well as those who have expertise in the humanities, economics, military, national policy, social sciences, and technical aspects of modeling. These stakeholders and analysts come with their own expectations, work styles, and doctrines. Consequently, designing the human–computer interface for a tool to be employed by such a broad user community can be a complex and inherently conflicted endeavor.

A key challenge for a user interface designed for this environment is to accommodate different methodologies and preferences within a collaborative process. Moreover, the process and the tools provided must meet the needs of professional social scientists and modelers as well as experienced and pragmatic leaders. These challenges have been explored in Computer Supported Cooperative Work (Neale et al. 2004), although it focuses on methods to overcome differences in group work in time and space rather than differences due to diversity of background, objectives, language, terminology, and experience.

Computer-supported collaboration can be distributed or co-located, synchronous or asynchronous, each with their own challenges. Asynchronous-distributed collaboration, for example, will have difficulties in communicating workflow and intent, while synchronous-co-located collaborators will encounter difficulties with shared screen space and interaction methods. Metrics have been proposed for evaluating these systems that combine coordination, communication, work coupling, and contextual factors into an activity awareness model (Fig. 7).

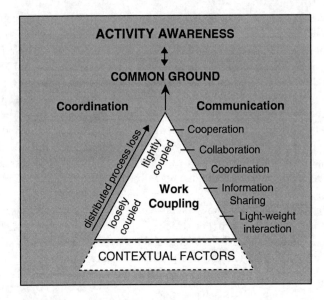

Fig. 7 Activity awareness model (Neale et al. 2004). Activity awareness is a key goal in collaborative environments

Although research from this field has been slow to reach mainstream analysis applications, SharePoint™ (a Microsoft product), InfoWorkSpace (an Ezenia product), wikis, and social-networking sites are excellent steps in the right direction. Consider, for instance, the InfoWorkSpace (IWS) tool, used extensively by the military. IWS provides shared white boards, bulletin boards, chat, and shared views, all based on a common physical office metaphor. Although this approach has become a de facto standard, Swanson et al. (2004) demonstrated that there were a number of areas in which IWS fell seriously short of the effectiveness of face-to-face collaboration.

Another example of a system for collaboration is the U.S. Army's Command Post of the Future (CPOF). This system provides distributed situation awareness and planning capabilities. CPOF provides an environment in which different disciplines and perspectives can work productively together. An essential reason for this success is CPOF's ability to visualize people's work in progress and to make this intermediate product easily available to others. CPOF work products include real-time situation monitoring on maps, analyses on maps, plans, and analysis using interactive charts and tables.

However, CPOF relies on a similarity of purpose and background in its community of users. Standard symbology and graphics, pervasive in the military, are an example of the common visual expressions used to support the exchange of information. Integrated PMESII models, on the contrary, must support political, social, economic, health professionals, justice, and law enforcement as well as military agencies– a truly diverse community. For this reason, PMESII visualization must be able to support different vocabularies, disciplines, and methodologies.

1.4   The Challenge of Heterogeneous Models

In modeling, effects of interventions, political, social, economic, and security aspects of a situation are intertwined. Therefore, diverse models from different disciplines must operate in an integrated fashion. However, modeling paradigms vary widely from domain to domain. Whereas an agent-based model may be the natural choice for modeling key actors in a situation, a system dynamics model may be the best choice for an economic model. And the diversity does not end there. Even within the same problem domain, there may be multiple instances of each type of model, applied to different subregions, subgroups, and time delineations. This diversity vastly complicates user interface design.

To discuss the resulting user interface challenges, it is necessary to first describe what model integration entails. If models of different types are to be integrated *generically*, such that one can affect another during simulation without possessing knowledge of the other's technical implementation, then a common technical language for expressing behavior must be established.

There are two common approaches to this problem. The first is to express and link behavior in the form of discrete and intermittent causal events. The second is to express and link behavior in the form of continuous scalar values over time. In the former option, noncausal models must be adapted to produce logical events at scalar thresholds, while in the latter, causal models must be adapted to produce scalar changes based on logical events. Note that an implication of either adaptation is that time between cause and effect must be resolved in some way for models which may not otherwise account for time.

Since scalar values offer a greater level of precision by representing any degree of change, at any point in time, it is often advisable to integrate models at this level so as not to handicap model classes that are able to work together at this level of granularity. Another potential advantage of this approach is that in a generic system of models, often less semantic interpretation is required for one model to "understand" the nature of a behavior produced by another model if the behavior is quantitative in nature. In addition, there is the assurance that adopting the scalar value approach does not preclude the translation or aggregation of scalar values to nonquantitative yet significant logical events for *user* consumption of simulated effects.

The homogenizing effect of a solution that dynamically integrates heterogeneous models into a single abstracted *supermodel* presents a significant challenge. Because the structure and properties of models vary so greatly, much texture and detail can be lost in the abstraction process.

Simply putting the onus on models to supply their own visualizations for display in the user interface is problematic for both users and model developers. The lack of consistency and visibility across models, as well as the reliance on modeling technology experts to supply visualization techniques and technology (an area of modeling that has traditionally been underdeveloped) often leads to poor results. Effective model integration at the end-user level requires a unified user interface.

Designing a unified user interface is inherently difficult. If the system is to be truly open to all models to be plugged in with minimal effort, and the variety of models is limitless, then there is typically little design that can be accomplished in advance for each model's integration. In many cases, given not much more than a large, abstract, unified set of labeled numbers that may change completely from situation to situation, the user must be able to construct meaningful visualizations on the fly; to reassemble the mental model as it were, with the supporting narrative that went into the development of these models.

Numerous tool kits offer capabilities to construct a "mash-up" of charts and graphs for dashboard-style analytics, including Cognos Visualizer™ and, more recently, the Google™ Visualization API. However, these are designed for technical staffs to use in creating data-driven report templates for end users, and not for end users to create supermodel inquiry and to monitor model execution results. Thus, there is a large capability gap in the extension of these techniques to nontechnical users. The needed capability, the ideal tool for constructing model visualization on the fly, would be one with the flexibility and ease of narrative and graphical expression provided by a story-based report-building tool (Eccles et al. 2007).

While postintegration visualization assembly can go a long way toward recapturing the theory and narrative aspects of a model for the benefit of other users (the "big picture"), the loss of visibility into other, more detailed aspects is not as easily regained in this way. If a user cannot audit the evidence behind a conclusion made in model development, it becomes difficult to assess a level of confidence in that model.

For this reason, a capability in the system whereby model assertions can be tagged in a consistent way with evidentiary document references, comments, dates, and sources can be useful in making this information generally available to the end user for the purposes of validation. Unfortunately, many of today's model editors do not provide capabilities for easily capturing this information such that it could be provided to a system. While not always practical or possible, the most efficient and thorough means of capturing this information for display would be to provide capabilities for authoring models in intuitive and visual ways from within the integrated system itself.

1.5 Model Authoring

Development of societal models involves knowledge of both the domain of interest and the appropriate modeling paradigms and systems. Currently, the latter requires trained technical expertise in specialized and relatively complex simulation tools. The parameters that govern a computational model and the tools used to configure them are often incomprehensible to the political or social science expert. As a result, they must rely on technical staff for this expertise and become separated from the construction of the models conceptualized by them. This is problematic. By divorcing experts from their models and their models' products, we introduce the possibility of invalid translation from theory to code and of misinterpretation of results from models.

Accordingly, a key challenge for user interface designers – one that is not even close to being met – is to develop methods that enable the SMEs to own and direct the model-authoring process. In particular, the SME must be able to author and interactively manipulate models in an intuitive, visual manner that closely aligns with his or her mental model.

In addition to these expert-driven "top-down" approaches, "bottom-up" machine learning approaches attempt to infer models automatically from raw data. Bayesian and neural networks can represent many complex systems, and dynamic versions of these algorithms can model processes over time (Antunes and Oliveira 2001). The difficulty of these approaches is often the opposite of that of expert-crafted models; the resulting models and predictions may not be transparent to the users of these models.

2   Technical Approaches: A Case Study

To illustrate some of the challenges and approaches involved in designing user interfaces for international intervention analysis, we now consider a specific case. The following study focuses on the user interface built for the COMPOEX system (Kott and Corpac 2007; Waltz 2008). Developed around the aforementioned nSpace framework, this visualization system was designed by the authors of this chapter.

2.1   Expressing Mental Models

A key objective of COMPOEX was to develop approaches for communicating the details of complex simulations to SMEs and decision-makers. It did not fully succeed in this endeavor, but it did make significant progress. One of these successes entailed the conceptualization and design of a set of constructs known as forms and panels. Forms are graphical building blocks such as event timelines, graph frameworks, node and link diagrams, geospatial maps, and flow diagrams. Panels are groups of forms populated with data and arranged so as to explain or summarize a situation. Panels are created via a drag-drop process and can be developed by SMEs without the support of software personnel or technicians. The process entails dragging elements and variables of interest into forms and arranging the forms into panels to fit problem-specific information needs. Using forms and panels, a user is able to rapidly build a live visual window into the modeled world and tailor it to the topics of interest. When the user then saves the assembled panel, the detailed *form* of presentation is maintained, but the data displayed in the panel remain live and updated as the model produces new data (Fig. 8).

To facilitate shared understanding when collaboratively conceptualizing or visualizing a problem, we developed a common visual language of expression. Modeled

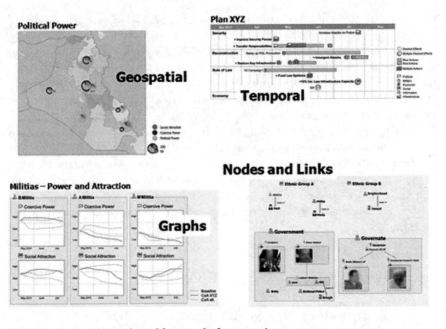

Fig. 8 Forms provide a variety of frameworks for expression

Fig. 9 Example elements of a common language of expression

entities such as people, places, and organizations can be rendered using a common symbology. To help users from a variety of backgrounds, we designed intuitive icons crafted specifically for rapid identification of key archetypes such as political, military, criminal, media, and social groups, among others (Fig. 9).

The common visual language developed for this system also includes time series state graphs and entity attribute thumbnails (Fig. 10). Time series graphs may be dragged into any of the visual frameworks in order to display the state of key named indicators. Entity attribute thumbnails permit a user to visualize common entity properties, such as sociopolitical power, across an entire panel, simply by dragging the property of interest into the panel's graphic legend. In this display option, small thumbnail charts appear to the left of each entity in either time series or scaled pie chart form, depending on the number of properties being visualized.

To accommodate the need to communicate information at higher levels of summation, and to navigate the information efficiently, varying levels of detail and

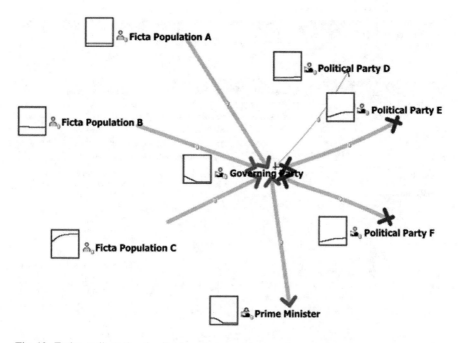

Fig. 10 Entity attribute thumbnails characterize power over time

aggregation are applied throughout the system. Outcomes of simulations are available for display at a summary level, and users are able to view lower levels of details for further information. For instance, a computational method of detecting significant effects from scalar data in a simulation result permits users to plot a summary of all effects as discrete event nodes on a timeline together with descriptive labels (Fig. 11). Hovering over an effect in the timeline provides further details in the form of a tooltip, and double-clicking the effect permits users to display a time series graph depicting the detailed behavior in relation to previous behaviors.

Effect summaries are also available in the aforementioned forms for users who prefer a geographic or conceptual context. In these forms, thumbnail pie charts (Fig. 12) are used to indicate the number and nature (beneficial, undesirable) of effects on each entity (a region or an industry, for example). Double-clicking the effect thumbnail displayed beside any of the entities in this context invokes a detailed list of individual effects. These techniques provide effective interactive methods for a user to visualize the bigger picture as well as to explore more detailed information.

Another objective of COMPOEX was to develop methods for assessing, capturing, and visualizing uncertainty within inputs and outputs. After experimenting with several approaches and finding them impractical, we evolved an approach whereby the user may choose to assert a hypothetical behavior that overrides the model-computed behavior for a certain subset of phenomena. The user then reruns the simulation to compare the detailed impact of modified assumptions. These assumptions

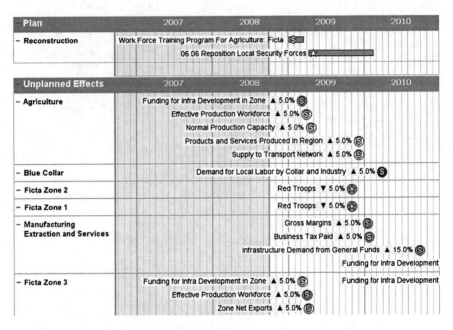

Fig. 11 The effect timeline summarizes actions and resultant effects over time

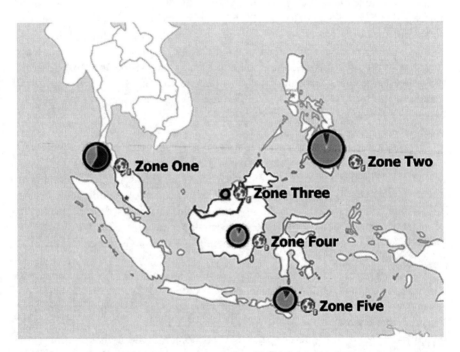

Fig. 12 Effect summaries on a map characterize localized impact of actions

are visually flagged with warnings. This technique proved to be an important capability, enabling an SME to set aside a disagreement with the model and continue to make effective use of the results.

The challenges in providing methods for the user to express *actions* that the system could simulate led to the development of a design principle referred to in this case study as "actions, effects, and desired effects anywhere." Using this principle, users can express desired effects and invoke actions from any context or view of the situation. For instance, when viewing a graph of a key economic indicator in a geographic context, a user can define a desired effect by simply drawing the desired change on the graph and moving on from there to a list of suggested actions for achieving that effect. This method significantly helped to ease the burden of communicating a conceptualized action to the modeling system.

Another important principle revolved around the need for planners to be able to compare and see change. Experience demonstrated that even subtle changes in the environment can be important at times, so a level of granularity is required whereby these changes can be detected and clearly displayed. Time-series graphs, with multiple instances and consistent scales, all displayed in context, became an important tool in the system for analyzing detailed differences. Several simulation runs (with different actions or assumptions) are overlaid on each graph for easy comparison (Fig. 13).

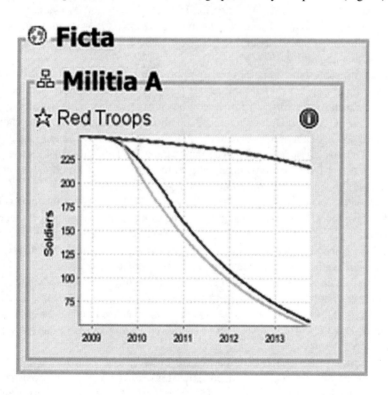

Fig. 13 Multiple results are shown together so differences can be seen clearly

For summaries of simulation results, change difference algorithms detect and highlight areas of significant change across the entire supermodel to guide detailed inspection by SMEs. Focusing on significant change helped improve the communication of effects of a plan simulation by eliminating information of lesser interest.

The provision of annotation capabilities satisfies another important principle. The users' ability to record their assumptions, thoughts, and observations as they worked has proved to be an important tool in framing and communicating a user's thinking and conclusions. This has enabled team members to brief decision-makers directly from the live application, with the ability to view lower levels of detail to answer questions.

2.2   Insights into Model Behavior

The models' inner logic should be transparent and comprehensible to a nonspecialist user. When something unexpected is observed in a model-computed result, it must be possible for analysts to rapidly view details and determine the sources of the surprise. Failure to provide such a capability can lead to misperceptions or the outright rejection of the model results. Since model authors will not always be available to explain the behavior of a model, analysts or decision-makers must be able to develop their trust in the model through effective interaction with the model and its visualization capability.

In our user interface, a causal investigation function provides cause-effect transparency via a view into model influences (Fig. 14). For any model behavior, this function displays downstream influenced behaviors (to the right). By dragging a behavior graph from the left or right into the middle, a user can follow the chain of influence downstream or upstream to "follow ripples in the pond" and to locate root causes. The display of both influence relationships and behaviors enables users not only to see that a relationship exists but also to observe the detailed nature of that relationship by comparing the pattern and degree of effects.

A second level of detail is provided in the system for investigating the logic behind model behavior. This involves providing a hyperlink above the variable of interest in the causal view that, when clicked, displays a detailed written document describing the theory behind the model responsible for that behavior. These documents are prepared by the model authors and include references used in their research.

These functions provide a useful first step but do not go far enough. The above system needs improvement in, for example, the ability to trace many steps of variable interactions and to distinguish causal from correlative relationships as easily as in conceptual models. Thus, more work is required in this area, and corresponding efforts are ongoing.

In addition to the generic approaches for investigating behavior applied to the collection of all models, specialized extensions are provided for a particularly important model type of interest, which display key structures and properties of that model. For the agent-based Power Structure model, which models the power and

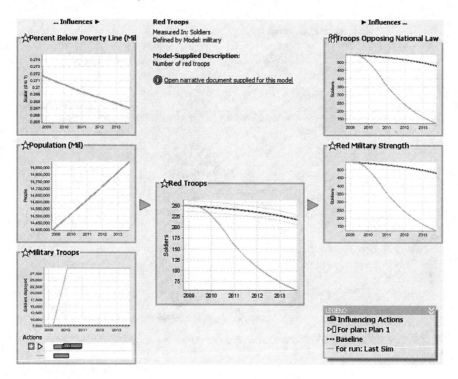

Fig. 14 Causal explanation provides insight into behavior

influence of key actors in the situation, the system shows relationships in the form of an interactive social network diagram. Positive or negative influence is indicated in this treatment using an arrowhead symbology (Fig. 15).

The ability to view actor properties in the application, such as their goals and role in the conflict, provides important narrative background information and clues as to their behavior.

2.3   User-Authored Narrative

The COMPOEX system relies on a modeling and simulation backplane that provides generic integration of heterogeneous models (Waltz 2008). At its core is a state vector consisting of scalar variables that models can both read from and write to at simulated time intervals. Models developed or gathered for a particular situation are integrated by plugging their inputs and outputs into the state vector. The homogenized state vector of potentially tens of thousands of variables, time series of named values produced by the collection of models, is available to the user interface that must use the data to produce meaningful visualizations.

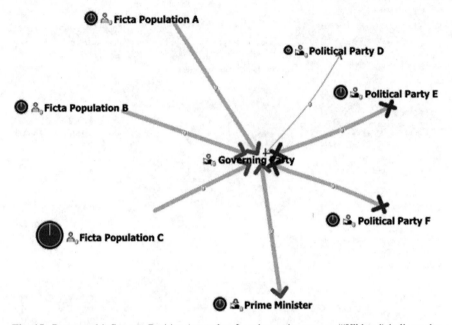

Fig. 15 Power and influence. Positive (*arrowhead*) and negative arrows ("X" head) indicate the type of relationship

One of the challenges of producing meaningful visualization in this context is the loss of texture and detail that can occur when integrating an abstract and diverse set of models. The problem is that while quantitative values produced by the model are readily available, the mental model or thinking that went into the design of the model is not. Compounding the challenge is the sheer volume of information available. The absence of data to express the conceptual aspects that go into the building of a model is not a problem unique to COMPOEX. Computational model interfaces today are almost universally highly technical; they are not designed for sense-making.

To overcome this shortcoming, in our interfaces, users are given the ability to assemble and organize panels with live data from simulations and add layers of meaning and narrative expression for the human user. Through arrangement and annotation with words, links, images, and other visual elements, a conceptual model can be expressed and shared, with live computational elements (Fig. 16).

2.4   Model Authoring by Decision-Makers

Model-authoring tools are most effective in the hands of the domain SME: the individual who possesses the detailed mental conception of the situation being modeled. While layering conceptual model aspects with computational data after the fact does help, the ideal solution would be to capture the mental model at the time when the computational model is being crafted. Unfortunately, this facility is absent from the vast majority of computational model-building tools, and the

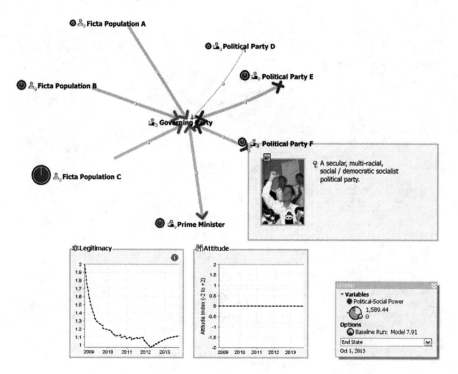

Fig. 16 Layers of narrative expression improve communication of information

degree of technical knowledge required to use these building tools presents a significant barrier for the SME.

Accordingly, a tool that proved particularly useful was the Power Structure model builder. SMEs and analysts (users without model-building skills) used this tool to create key actors, define their goals, and suggest how they influenced each other in the real-world environment being modeled. A targeted development effort produced both the principal constructs of the model and the user interface for the model in a way that closely fits the way an SME might think about these aspects of the situation.

The COMPOEX user interface provides visual exploration of the properties of the model, such as influence networks of actors, their roles, and their goals, in the larger context of the situation. In addition, users have editing capabilities that enable direct and intuitive model authoring in this context (Fig. 17), tied into the full suite of features provided for information gathering, research, and embedding of evidence and supporting narrative within model entities and relationships. By providing these capabilities, new levels of transparency and control enable SMEs to author models directly; however, much work is yet to be done.

In addition to the precedents it provides in addressing particular challenges, the COMPOEX case study serves to emphasize the importance of a number of key principles for the visualization of computational models in the social sciences. These are summarized in the following Practical Tips.

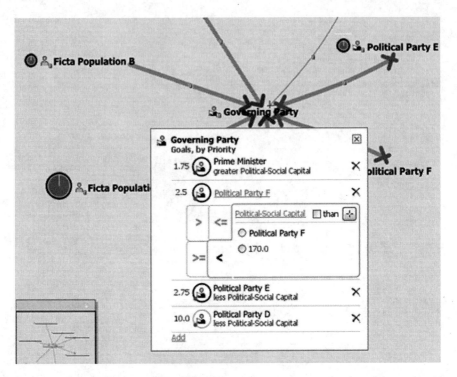

Fig. 17 Editing model of an actor's goals in the Sandbox

3 Practical Tips

- Look for ways to streamline the process of inputting intervention actions. Consider using system intelligence to translate or interpret the actions that a user wishes to take, or to suggest actions that might be appropriate for a desired effect.

- Provide end users who are not expert modelers with the means to author and manipulate models, using a language of expression that fits the users' mental model. Likewise, streamline model output by finding methods of expression that are natural to the way that domain experts conceptualize a situation.

- Use the simplest, most universal, and most accessible visual language. Account for the fact that methodologies and associated terminology can vary widely between user communities and change frequently.

- Provide means to present varying levels of detail and aggregation of computed data. Enable users to provide rich summary-level information to decision-makers, and, when appropriate, enable senior leaders to interact with the tools directly.

- Characterize intervention effects in summary form, indicating for instance the degree, potential desirability, scope, and nature of the effects, and enable analysts to perform comparisons without relying on visual memory.

- Recognize that time is a critically important dimension in analyzing PMESII effects. Thus, when possible, provide tools for interpreting the temporal sequence and spacing of the effects, and for examining short- and long-term effects.
- Consider permitting users to annotate computational results with descriptions, diagrams, and illustrations to communicate situations, actions, and anticipated effects.
- When possible, allow users to see the inner workings of a model by presenting the model's basic elements and the relationships among these elements.
- Develop capabilities that assist in understanding simulation outcomes – what chain of causes led to a particular effect – and in quickly diagnosing and understanding model operations.
- Expect and encourage a healthy level of skepticism from users and design computer-user interactions that are able to accommodate users' disagreements with the model.

4 Summary

Data visualization expresses data in concise and elemental graphical formats. Information visualization uses higher-level organizational structures in the graphic forms. Visual analytics combines visualization with analysis and further computation to derive meaning from large datasets. Typically, social science data has few features that can be depicted in physical form; consequently, Subject Matter Experts (SMEs) are often left to analyze and interpret a display of computational model output in complex, nonintuitive forms. This is inefficient and is a source of potential error. Conceptual models, which describe relationships qualitatively rather than quantitatively, have seen progress in visualization, e.g., link analysis and tools with additional dimensions, such as time. Several challenges are particularly strong in PMESII modeling visualization. Because PMESII models involve significant uncertainty, better visualization approaches are needed to depict uncertainty and causal relationships. Another challenge involves the need to provide sufficient accessibility and adaptability to accommodate a wide range of intervention partners and organizations, with standard symbols, vocabularies, and protocols. There is the need to integrate models from multiple social science domains operating as one system and user interface, and the need to capture and present the underlying theory and evidence behind the models. The COMPOEX user interface is a relevant case study. Key constructs and principles of this interface include forms (templates) for the display of model data including timelines, graph frameworks, link diagrams and geospatial maps; user-created panels that can combine and spatially arrange groups of forms to convey meaning; a common visual language, levels of summary and drill-down, tools for uncertainty and for detecting changes in outcomes. A causal investigation function is also provided, as well as capabilities for visual annotation, markup, and hyperlink references at both the summary and detail level within large and complex societal models.

5 Resources

VisualComplexity.com   http://www.visualcomplexity.com/vc/index.cfm?domain=Social%20
   Networks
Connectedness http://connectedness.blogspot.com/
Datawocky blog, 2008. Report on interview with Russel Norwig of Google research. http://anand.
   typepad.com/datawocky/2008/05/are-human-experts-less-prone-to-catastrophic-errors-than-
   machine-learned-models.html (as found on March 11, 2009)
Bertin, J. (2001). Matrix Theory of Graphics. *Information Design Journal*, 10(1), 5–19.
Bertin, J. (1983). *Semiology of Graphics, Diagrams, Networks, Maps*. University of Wisconsin
   Press.
Cohen, M.D., March, J.G. & Olsen, J.P. (1972). A Garbage Can Model of Organizational Choice.
   Administrative Science Quarterly, 17(1), 1–25.
Eick, S. & G. Wills, (1995). *High Interaction Graphics*, European Journal of Operational
   Research, 84 (445–459).
Epstein, J. M. (1999). Agent-Based Computational Models and Generative Social Science.
   Complexity, 4(5), 41–60.
Freeman, L. C. (2000). Visualizing Social Networks. *Journal of Social Structure*, 1(1).
Harris, R. L. (1996). *Information Graphics*. Management Graphics.
Hearst, M. (1999). *User Interfaces and Visualization*. In R. Baeza-Yates & B. Ribeiro-Neto, *Modern
   Information Retrieval* (Chapter 10). Addison-Wesley-Longman.
Herman, D. (1999). Spatial Cognition in Natural-Language Narratives, *Proceedings of the AAAI
   Fall Symposium on Narrative Intelligence*.
Mullet, K. & Sano, D. (1995). *Designing Visual Interfaces: Communication Oriented Techniques*.
   Mountain View, CA. SunSoft Press/Prentice Hall.
Larkin, J. & Simon, H. (1987). Why a Diagram is (Sometimes) Worth Ten Thousand Words.,
   Cognitive Science, 11(1), 65–99.
Scholtz, J. (2006). Beyond Usability: Evaluation Aspects of Visual Analytic Environments. In
   IEEE Symposium on Visual Analytics Science and Technology (pp. 145–150).
Tufte, E. R. (1990). *Envisioning Information*. Cheshire, CT: Graphics Press.
Tufte, E.R. (1983). *The Visual Display of Quantitative Information*. Cheshire, CT: Graphics Press.
Tufte, E. R. (1997). *Visual Explanations: Images and Quantities, Evidence and Narrative*.
   Cheshire, CT: Graphics Press.
W. S. Cleveland (1993). *Visualizing Data*. Hobart Press.
Ware, C. (2000). *Information Visualization – Perception for Design*. Morgan Kaufmann.
Wise, J. A., Thomas, J. J., Pennock, K., Lantrip, D., Pottier, M., Schur, A. & Crow, V. (1995).
   Visualizing the non-visual: spatial analysis and interaction with information from text docu-
   ments. *Information Visualization, IEEE Symposium on*, 0:51+.
Wood, D. (1992). *The Power of Maps*. New York: Guilford Press.

References

Antunes, C. & Oliveira, A. (2001). Temporal Data Mining: an Overview. Proceedings of *Workshop
   on Temporal Data Mining*. 1–13.
Axelrod, R. (1987). The Evolution of Strategies in the Iterated Prisoner's Dilemma. In Lawrence
   Davis (Ed.), *Genetic Algorithms and Simulated Annealing* (pp. 32–41). London: Pitman, and
   Los Altos, CA: Morgan Kaufman.
Card, S., MacKinlay, J. & Shneiderman, B. (1999). *Readings in Information Visualization – Using
   Vision to Think*. Morgan Kaufmann.

Collier, J. & Collier, M. (1986). *Visual Anthropology: Photography as a Research Method.* Albuquerque, NM: University of New Mexico press.

Eccles, R., Kapler, T., Harper, R. & Wright, W. (2007). Stories in GeoTime, *IEEE VAST 2007.*

Jonker, D., Wright, W., Schroh, D., Proulx, P. & Cort, B. (2005). *Information Triage with TRIST.* In *Proceedings of the International Conference on Intelligence Analysis.* Washington DC.

Kott, A. & Corpac, P. (2007). *COMPOEX Technology to Assist Leaders in Planning and Executing Campaigns in Complex Operational Environments*, Report No. A534184. Storming Media.

Kapler, T., Eccles, R., Harper, R. & Wright, W. (2008). Configurable spaces: Temporal Analysis in Diagrammatic Contexts. In *Visual Analytics Science and Technology, 2008. VAST '08. IEEE Symposium on Visual Analytics Science and Technology*, pages 43–50.

McCormick , B. H., Defanti, T. A. & Brown , M. D. (1987). Visualization in Scientific Computing – A Synopsis. *IEEE Computer Application Graphics*, 7(7), 61–70.

Mui, L., Mohtashemi, M., Ang, C., Szolovits, P. & Halberstadt, A. (2001). Ratings in Distributed Systems: A Bayesian Approach. In *Proceedings of the Workshop on Information Technologies and Systems* (WITS).

Neale, D. C., Carroll, J. M. & Rosson, M. B. (2004). Evaluating Computer-Supported Cooperative Work: Models and Frameworks. In *CSCW '04: Proceedings of the 2004 ACM Conference on Computer Supported Cooperative Work*, pages 112–121, New York, NY, USA. ACM Press.

Orford, S., Harris, R. & Dorling, D. (1999). Geography: Information Visualization in the Social Sciences. *Social Science Computer Review*, 17(3), 289–304.

Parasuraman, R. & Mouloua, M. (1987). Interaction of Signal Discriminability and Task Type in Vigilance Decrement. *Perception & Psychophysics*, 41 (1), 17–22.

Sears, A. & Jacko, J. (2007). *The Human Computer Interaction Handbook: Fundamentals, Evolving Technologies and Emerging Applications, 2nd Edition (Human Factors and Ergonomics).* CRC.

Swanson, K., Drury, J. & Lewis, R. (2004) A Study of Collaborative Work Practices in a Joint Military Setting. In *Proceedings of the International Command and Control Research and Technology Symposium.*

Thomas, J. & Cook, K. (Eds.) (2005). *Illuminating the Path: The R&D Agenda for Visual Analytics.* National Visualization and Analytics Center.

Waltz, E. (2008). Situation Analysis and Collaborative Planning for Complex Operations, *13th International Command and Control Research and Technology Symposia ICCRTS*, Seattle, WA.

Wickens, C. D. & Hollands, J. G. (2000). *Engineering Psychology and Human Performance (3rd ed.).* Upper Saddle River, NJ: Prentice Hall

Wright, W., Schroh, D., Proulx, P., Skaburskis, A. & Cort, B. (2006). The Sandbox for Analysis – Concepts and Methods, *ACM CHI*

Chapter 11
Verification and Validation

Dean Hartley and Stuart Starr

"All models are wrong; but some are useful." This well-known adage by the statistician, George Box, contains several important insights (Box 1979) – first, that a model is an abstraction of reality, and second, that the parts of a model which are important for its intended purpose should be emphasized, while those which are nonessential can be deemphasized or even left out. Consider, for example, "model" airplanes. Display models, such as those designed to be hung from ceilings, are concerned with proportions *vis a vis* size, material composition, or flying capability. If they look right, they make for satisfied users. Models intended for flying, on the other hand, sacrifice proportions and composition in order to produce a device that "performs." Since neither model reproduces reality, each is inherently "wrong"; however, each may meet its intended purpose quite well. The situation is similar with software models. A skilled modeler should focus on those aspects of reality that are germane to the issue of interest. If he or she succeeds, and the abstraction effectively addresses the factors relevant to the model's purpose, then it is possible that the model will be "useful."

The questions of adequate abstraction and of the model's usefulness are particularly challenging – and poorly explored at this time – for PMESII models (societal models). Like the challenge of visualization (Chap. 10), the challenge of validation handicaps the current generation of societal models. This is the topic of our chapter: having constructed a societal model, how do we know that the model is useful?

Verification, validation, and accreditation (VV&A) are processes that address the correctness of models. Just as models can have many forms, they can refer to different parts of the real world (e.g., hurricanes, flight characteristics of an airliner, price sensitivity of a commodity to scarcity, or the impact of international interventions). Further, models can have different purposes (e.g., prediction, training, equipment testing and evaluation, education, exploration of possibilities, or understanding).

D. Hartley (✉)
George Mason University School of Public Policy,
4400 University Drive,
Fairfax, VA 22030, USA
e-mail: dshartley3@comcast.net

A. Kott and G. Citrenbaum (eds.), *Estimating Impact*,
DOI 10.1007/978-1-4419-6235-5_11, © Springer Science+Business Media, LLC 2010

The VV&A processes apply to each combination of form, referent, and purpose; however, the details may differ from one combination to another.

1   The Basics of V&V

V&V (verification and validation) is the process of determining whether a model is useful. *Verification* is the process of determining that a model or simulation implementation accurately represents the developer's conceptual description and specification. Verification also evaluates the extent to which the model or simulation has been developed using sound and established software engineering techniques. Verification also applies to the data. *Validation* is the process of determining the degree to which a model or simulation is an accurate representation of the real world from the perspective of the intended uses of the model or simulation. It also applies to data. *Accreditation* (the 'A' in VV&A) is an official determination that a model is acceptable for a specific purpose. These definitions are taken from DoD (2007).

The IEEE standards are similar. Validation is the process of evaluating software during or at the end of the development process to determine whether it satisfies specified requirements. Verification is the process of evaluating software to determine whether the products of a given development phase satisfy the conditions imposed at the start of that phase (IEEE-STD-610). IEEE does not define accreditation, as this is a governmental term. The word "purpose" is present in the definition for a very good reason. A model that is good for one purpose need not be good for another.

Figure 1 illustrates the model-building process. We build a conceptual model based on the real world, keeping our model's purpose(s) in mind. Here, we run into our first practical problem: we do not really know what the real world is in the detail required for our model. We actually use a proxy for the real world, such as an established theory or our perception of the real world. We do this in all model building; however, the problem of knowing the real world is especially evident when we are dealing with societal models. Thus, we incur two risks for error in creating the conceptual model, one in which we can use the real world directly, and one in which we must employ a proxy. We incur a third risk when we create the coded model from the conceptual model. The risks numbered four through six in the figure relate to the collection of the data we must have to use the model. We may collect faulty data, or the data we collect may be inappropriate for the model. The final risk lies in the use of the model: the assumption that the model output tells us something "useful" about the real world.

Figure 2 illustrates the reversals of the arrows in Fig. 1. The reversed arrows are labeled as types of verification or validation, or both. Of particular interest is the fact that the model validation arrow connects the coded model to the proxy for the real world and that the connection going on to the real world does not actually reach that destination. This illustrates that, for societal models, validation will always be incomplete rather than definitive. As a matter of practice, verification for

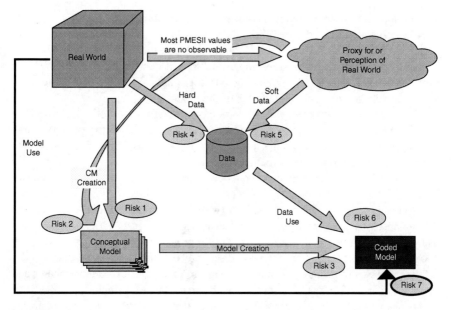

Fig. 1 Building and using a model

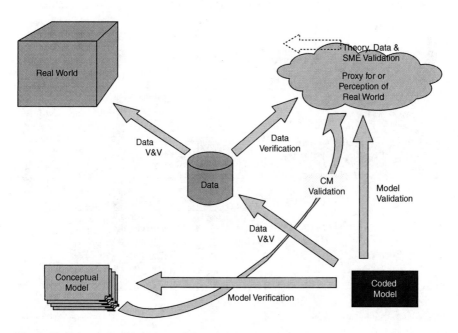

Fig. 2 Verifying and validating a model

any but the most simple of models will also be incomplete. The number of things to be checked will exceed our ability to check them.

Showing that a model is wrong is sometimes possible and useful. If it can be shown to be wrong, or wrong for some use, then it can be rejected for that use, avoiding Risk 7. However, convincing our users and ourselves that a model is approximately right justifies some uses of the model.

The literature on the theory of V&V is extensive. Two examples are especially useful. Balci (2001) provides an excellent brief description of the principles and techniques of V&V. He classifies more than 75 techniques into four categories: informal, static, dynamic, and formal. These techniques range from the informal *face validation*, in which domain experts judge whether the model seems to deliver believable results, to the formal *proof of correctness*, in which the model is proved to terminate and meet its specifications. In between, the techniques include the standard debugging, alpha testing and beta testing that are necessary facets of rigorous modeling and programming. Knepell and Arangno (1993) produced a very readable and useful book on the subject. They offer extensive discussions of techniques and provide guides for the processes of using the techniques to assess models. They describe phases and introduce characterization of attributes, which can lead to V&V metrics.

There are other useful sources. Davis (1992) introduces a categorical taxonomy for V&V techniques, which is expanded by Hartley (1997); see Fig. 3. Other taxonomies emphasize different aspects of V&V. For example, Giadrosich (1992) divides validation into structural (e.g., examinations of inputs, basic principles, and assumptions) and output validation. In another taxonomy, Henderson (1992) emphasizes the division of validation into five dimensions: application (e.g., training,

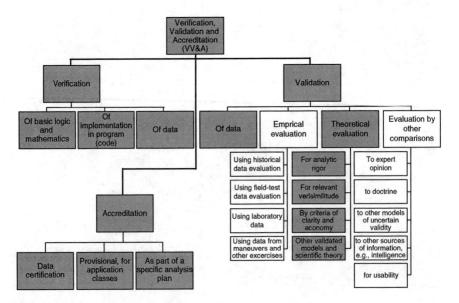

Fig. 3 Categorical taxonomy of V&V

analysis, prediction); truth basis (e.g., theory, other models, field trials, history); technique (e.g., statistical sampling, Delphi, graphics); composition (e.g., monolithic, amalgamated, decomposed ensemble); and depth (e.g., surface, measured, detailed).

The techniques of V&V are necessary but not sufficient for performing V&V in an effective and professional manner. That is, performing a random set of tests drawn from a list of V&V techniques will not necessarily lead to valuable information concerning the usefulness of a model. The V&V principles of Table 1, which are drawn from three sources, labeled Department of Defense Instruction (DoDI) 5000.61 (Youngblood 2004), DMSO Recommended Practices Guide (RPG) (DMSO 2007), and Balci (1997), provide some guidance that informs the choice of tests.

Table 1 Principles of V&V

Requirement and responsibility

V&V is required in all M&S [DoDI 5000.61]

M&S must be accredited as part of a VV&A process [DoDI 5000.61]

Organizations are responsible for VV&A [DoDI 5000.61]

   Universality

V&V of data must be an integral part of M&S V&V [DoDI 5000.61, DMSO RPG]

V&V must be incorporated in the M&S development/life-cycle [DoDI 5000.61, DMSO RPG, Balci]

Errors should be detected as early as possible in the M&S life cycle [Balci]

Intended use

Credibility can be claimed only for the intended use of the model or simulation and for the prescribed conditions under which the model or simulation has been tested [DMSO RPG, Balci]

A well-formulated problem is essential to the acceptability and accreditation of M&S results [DMSO RPG, Balci]

A simulation model is built with respect to the M&S objectives and its credibility is judged with respect to those objectives [Balci]

The whole M&S and its parts

V&V of each submodel or federate does not imply overall simulation or federation credibility and vice versa [DMSO RPG, Balci]

V&V of federations must address both technical and substantive interoperability issues [DoDI 5000.61]

VV&A process

VV&A must be planned [DMSO RPG, Balci]

V&V must be documented [DoDI 5000.61, DMSO RPG, Balci]

The level of V&V effort must be tied to criticality and risk [DoDI 5000.61]

Accreditation is not a binary choice [DMSO RPG, Balci]

VV&A is both an art and a science, requiring creativity and insight [DMSO RPG, Balci]

V&V requires some level of independence to minimize the effects of developer bias [DMSO RPG, Balci]

M&S use

There is no such thing as an absolutely valid model [DMSO RPG, Balci]

The success of any VV&A effort is directly affected by the analyst [DMSO RPG]

M&S validation does not guarantee the credibility and acceptability of analytical results derived from the use of simulation [DMSO RPG, Balci]

---

2   V&V Process Prescriptions

V&V has evolved from a best practice into a requirement in the last 30 years (Department of the Army 1999; Secretary of the Navy 1999; Youngblood 2004). The Australian Department of Defence V&V Guide (2005) illustrates the process of performing V&V on new or legacy models. The guide includes references to plans and reports (as well as the assessments), data, configuration management, management ownership, and the accreditation decision. The guide also discusses the difference between developer-team V&V and independent V&V (IV&V). The guide implies that there may be significant differences between V&V of single models and distributed or federated models by referring discussion of the latter situations to other documents. In addition, the guide devotes a chapter to the subject of the proper scope and cost of V&V efforts and also discusses the lack of V&V standards for general simulations. The standards it quotes for linking simulations, Distributed Interactive Simulation (DIS) and High Level Architecture (HLA) (IEEE 1996, 1998, 2000a, b), are certainly necessary but not sufficient for characterizing the confidence that should be placed in models following any given V&V effort. Youngblood (2005) describes an effort to remedy this situation. This effort introduces a Validation Process Maturity Model (VPMM), analogous to the Capability Maturity Models for software development. An early version of the VPMM is discussed in Conwell et al. (2000).

A very complete V&V process prescription is found in Knepell and Arangno (1993). The book includes the technique descriptions mentioned earlier, a chapter on formal assessment, another on limited and maintenance assessments, and chapters on performing V&V with man-in-the-loop and hardware-in-the-loop models. The book is written for V&V of general simulations and does not follow the terminology used here exactly; however, the terminology differences are minor. Although it does not discuss accreditation, the lack is not critical. Because it covers general simulations, there are many areas that are not pertinent to societal models.

Other approaches have also been suggested. For example, Balci et al. (2002) describe an automated environment for documenting and scoring the results of what they call Credibility Assessments, which can be conducted under V&V process guidance. The basic idea is to score (0–100) a number of elements, such as the quality of the conceptual model, weight the parts, and combine the weighted sums of the parts into an overall credibility score. On the other hand, Sargent (2004) does not believe in the use of scoring models, citing misplaced confidence and hidden subjectivity as reasons, and prefers developer-alone V&V, developer and user V&V, or IV&V methods, depending on the situation.

Many, if not most, discussions of validation tacitly assume a predictive role for models. That is, given a set of conditions the model calculates a result that is a prediction of the state the world would have given those conditions. In this role, the criterion for validation would be a test that proves that the results of the model correspond to a real-world state under those conditions. Other validation techniques are substitutes that may be necessary because of our imperfect knowledge or inability to perform this particular test.

Kott (2008) discusses a view of validation that concerns the impact of the model on the user's cognitive process, irrespective of the replication of reality in the model. Chapter 2 of this book introduces a similar view. Kott describes a type of model use that is very different from prediction, a variant of the course of action analysis. In course of action analysis, the goal is to decide which of several possible strategies for dealing with a situation is most advantageous. The set of possible courses of action is very large, and the human dynamics are very complex. In this use, the goal is not to score the courses of action but to explore the possibilities and stimulate the mind of the decision-maker. The desirable qualities of the model are the inclusion of all relevant factors, transparency of process to permit the decision-maker to "argue" with the model, and rapid execution to enable broad exploration. For this use, Kott maintains that the model is valid if it produces a desired quantitatively measured impact on the user's cognitive process. Conceivably, an otherwise inappropriate model could be valid if it caused the user to think *productively*. Such a model would indeed be wrong but, as noted by Box, could be "useful."

3   V&V in Practice

There are few documented examples of rigorous applications of the V&V process for three reasons. First, there is a limited audience for the fact that model "X" completely passed, say, 57 of 65 tests and completely failed only one of the remaining eight tests. Second, sponsors are often unwilling to discuss the limitations of their model in a useful report. Third, V&V are challenging processes that frequently imply the expenditure of significant resources. Often, those with funding authority believe those same resources could be better used in increasing the functionality of the model.

Still, there are several examples of V&V in practice, including the following: invalidation of a model (Hartley 1975); the comparison of models to actual historical events (Vector Research 1981; Bauman 1995); and a multiyear V&V effort on a set of societal models (Hartley 2004, 2005a, b; Senko 2005). There has been validation work on algorithms used in models (Hartley 1991, 2001). In addition to V&V of models serving analysis purposes (Wait 2001; Chaturvedi 2003; Coast Guard 2006; McDonald 2006; Sheldon 2006; Davis 2006), there has been V&V work on training models (Hartley et al. 1989, 1990, 1991). Three V&V examples merit discussion: V&V of a model before it was coded; V&V of a very large complex model; and efforts at historical validation.

In the first example, in 1994, the U.S. Department of Energy's Oak Ridge National Laboratory was asked to perform IV&V of a model in the conceptual phase of development. At that stage, there was little or no code in existence (Hartley 1994). The model was at that time named the Future Theater Level Model (FTLM). The performers of the V&V could find no documentation describing how to perform V&V on a conceptual model. Thus, they had to invent the procedures. The team performed V&V in a manner analogous to the independent audit of a company's accounting records: the final report contained a statement of findings and document analyses.

Included were the detailed examinations of the FTLM design documents (at all stages of development), the FTLM requirements document, and selected documentation for similar models. The V&V team examined these documents to assess the FTLM as to its design stage, its purpose as an analytical combat model, and its capabilities as specified in the requirements document. The requirements document defined the attributes of the FTLM by listing deficiencies in available models that required improvements, stating current capabilities that should be included, enumerating constraints, and defining milestones. Thus, the requirements document provided the standards for verification. The validation standards consisted of the experience and knowledge of the personnel of the performing organization. This experience and knowledge was informed by the code and algorithms of accepted, similar models. However, none of these models had been truly validated, considerably reducing their value as validation standards. This effort showed that V&V of a conceptual model was possible and could be performed before the model was coded.

In the second example, Metz (2000) describes lessons learned in the V&V of the JWARS model, a very large combat simulation in which no human intervention was required during the running of the model on a computer. The JWARS V&V was designed to be a part of the life cycle of the model, with twelve planned iterations. Table 2 shows the V&V requirements. While these requirements are specific to this particular model, several lessons can be derived for V&V of any large model. First, there can be cascading items requiring V&V among themselves prior to looking at any code. Second, there can be several referents against which the code is tested for V&V. Finally, all of the V&V efforts are documented, and serve as inputs to the accreditation decision. The entire process should be thought out in advance and revisited periodically.

Metz divides the lessons learned into three categories: planning for V&V, V&V activities, and V&V reporting. These lessons are shown in Table 3. Our experience is that these lessons carry over to every V&V effort and are not particular to the JWARS V&V effort.

Table 2 JWARS V&V requirements

The real world (often through surrogates such as test range results and other models) together with the requirements for the JWARS model drive the development of the Conceptual Model of the Mission Space (CMMS), which requires validation.

The JWARS requirements also drive the definition of the Model Validation Criteria, which must be approved (validated) against the CMMS.

The High Level Design (HLD) is created and must be verified against the Model Validation Criteria.

The Detailed Design is created and must be verified against the High Level Design and the choices for algorithms must be validated.

The M&S Code is created and must be verified against the Detailed Design.

The M&S Implementation is created and the integration is tested and verified against the Detailed Design.

The Implementation Data and Application Software is run and the results must be validated and Interim V&V Reports are created.

The total package includes Certified Data (from Test & Evaluation), Validated Application Software, and Summary V&V Reports, which are sent forward for accreditation.

Table 3 JWARS V&V lessons learned

Planning for V&V

There will be conflicting directions and advice. The DMSO RPG [34] only provides a starting point.

There will be resistance to V&V, based on cost (both V&V labor and the required supporting labor by the model developer).

Get help from outside experts.

Plan to add value through risk reduction, and so stakeholders are on board. Make sure this contribution is visible and measurable.

Base the V&V plans on the development process, not some arbitrary "perfect" V&V plan

Check to see if the development process will provide the items needed for V&V. If not, make sure the developer knows what is missing and needed.

The plan will change because the development process will change. Examine the changed development process for impacts in producing the items needed for V&V.

Ask for SME resources for validation assistance and plan for only partial fulfillment of the requests. Involve the user community.

Plan the V&V activities to coordinate with the developer testing.

V&V activities

Do not become a developer; however, communicate any observed risks and possible solutions. This adds value to the V&V project...

Develop a database and record everything. Share the contents. Set the database up to produce rapid reports for all stakeholders.

There will be delays and resource limitations that affect the V&V project.

There will be sponsor and developer concerns about reporting apparently negative results. Make sure there is internal communication and prior permission before reporting externally.

The developer will be sensitive to the contents of any V&V results. Make sure that the developer has a chance to respond to them and admit any mistakes made by the V&V process.

There will be uneven participation in the validation review process.

Record everything.

Make the V&V process transparent. The V&V process should help build confidence, not destroy it (if at all possible)

Make the reports available in multiple media for future accreditors use.

V&V reporting

Base the reports on the DMSO RPG, but tailor it to the use at hand.

Design the reporting process for use in accreditation, not for ease in producing.

Schedule the reports to have maximum impact. The emphasis is on early reports.

Release reports for review as early as possible and accept inputs from all stakeholders.

Make sure the reports focus on the extent that the simulation meets the requirements, not on the simulation itself.

The third example concerns what would seem to be the best measure of validity of societal models, comparison with historical results. The precise results of performing a set of actions A within the context of a set of conditions C are generally unknowable, particularly in broad societal situations, given our current state of knowledge. That is, if our model says that R is the set of results, we do not know what the true results would be, and so we cannot tell whether the model is valid or not. Even if we have a historical situation {A, C, R}, we do not have sufficient insight to know whether R was fully determined by A and C, was part of a range of likely results, or was an anomaly. Unlike models of physics, we are not able to do independent experiments (e.g., varying A and C and observing R).

In the examples of V&V practice, comparisons with military combat history (Hartley 1975; Vector Research 1981) showed that the models in question (with the proper parameter values) could yield results very similar to historical results. Most combat models built in the 1960s and 1970s had input parameters with uncertain connections to observable real-world data. A priori determination of the correct parameter values is problematic. Thus, matching model results with historical facts did not prove validity. However, the process did demonstrate that the models *might* be valid, which was better than discovering that the models could not match historical results and thus could not be valid. In general, many validation tests have this same characteristic: A determination is made that the model has not been proved invalid for the given purpose, not that the model has been shown to be valid.

These examples serve to demonstrate that the practice of V&V of complex models, such as societal models, is far from an exact science. To maximize the value of V&V, the documented practice of V&V should form the basis for any prescriptions for performing V&V. Where the nature of the models might affect the nature of the V&V, the documented practices should include V&V of similar models.

4   Peculiarities of Societal Modeling

To the authors' knowledge, the first well-known societal model was World3 based on Jay Forrester's World2 model. Used in a study commissioned by the Club of Rome, it served as the basis for the book *The Limits to Growth* (Meadows et al. 1972). The main systems modeled were agriculture and food production, industry, population, nonrenewable resources, and pollution. The central concept was that with finite resources and constant increase in their use, the resources will run out eventually, and society will experience a disaster. The book generated considerable controversy and debate about the model, which can be considered a form of V&V. Aside from questions about this particular model, the debate raises some important issues with respect to societal modeling:

- Assumptions are important;
- Social flexibility is important;
- Governmental reactions are not fixed;
- Economics is a powerful driver of change.

Within the military modeling community, the seminal works on identifying the characteristics of societal modeling were Hartley (1996) and Staniec and Hartley (1999). These documents reported and analyzed the results of a series of workshops on Operations Other Than War (OOTWs), which elicited contributions from a large number of operations analysts and military operators having a wide variety of knowledge and experience in the conduct of OOTWs. As the name implies, OOTWs include all military operations that are not war or garrison duty. Examples include Humanitarian Assistance and Disaster Relief and Stability, Security, Transition and Reconstitution Operations. These operations are usually conducted by the military in support of the Department of State, the United Nations, or some other nonmilitary authority. Another notable document was also the result of a series of workshops around the same time, sponsored by the Naval War College, in which the goal was to define a set of factors and relationships that modeled a societal situation (Hayes and Sands 1997). The following set of questions captures the essence of the societal modeling problem (Clemence et al. 2007). Items 1 through 4 directly affect the validation of the conceptual model. Item 5 addresses the problem of validating the data, and item 6 addresses the problem of validating the coded model.

1. *What are the relevant variables? Which variables influence each variable?* For example, gross national product (GNP) and percentage of the population with college degrees are candidate variables. Are they relevant in a particular model? What other variables are needed to calculate them? Are they inputs? What other variables should be calculated using these variables?
2. *What is the functional relationship between influencing variables and influenced variable?* For example, GNP might be the simple sum of several other variables, and the general educational level of a country might involve some complex formula that includes the population with college degrees as one factor.
3. *Which relationships are deterministic and which are probabilistic? What are the distributions for the probabilistic relationships?*
4. *Which of these variables and relationships are invariant with scenario? How do the noninvariant factors vary with scenario variations?* In chemistry, it may be a fact that two chemicals will combine to form certain products (invariant with scenario); however, their rates of combination may change with temperature and pressure (noninvariant with scenario). When people speak of human nature, they often mean that some behaviors are invariant. What are these invariant behaviors?
5. *What are the proper data to use as inputs?* The assumptions behind the data should fit the model's assumptions. If the model assumes a datum that represents an annual average for some quantity (e.g., cost of a gallon of gas), the most recent price is not the proper value. Further, the data should not be biased by a data producer's political agenda.
6. *What do the "answers" mean?* Models produce outputs, not answers. The models are used by humans who desire answers to a question. The question is whether there is a relationship between the outputs and the answers and, if so, what is that relationship.

These questions are not unique to societal modeling, except, perhaps, the one about human nature. However, in societal modeling the number of uncertainties and their magnitudes are particularly high. Our understanding of reality as expressed by societal theory is relatively limited when compared to physical systems of inanimate objects. The magnitude of the problem for V&V for societal modeling calls for a careful evaluation of the standards that should be used in making an accreditation decision. When we speak of validation of societal modeling, we must emphasize coverage of the appropriate domains of events and variables. We can expect, at best, only general directional and magnitude correspondence between model output values and what we know about the reality.

5   The Practice of V&V in Societal Modeling

One of the more thoroughly documented V&V efforts concerned a societal modeling system, the Flexible Asymmetric Simulation Technologies (FAST) Toolbox. The FAST Toolbox comprises DIAMOND-US (the U.S. version of the United Kingdom's Diplomatic and Military Operations in a Nonwarfighting Domain – DIAMOND), the Joint Conflict and Tactical Simulation (JCATS), the Interim Semistatic Stability Model (ISSM), Pythagoras, the Unit Order of Battle Data Access Tool (UOB DAT), the Canadian Forces Landmine Database (CFLD), and various support software (Hartley 2004, 2005a, b; Hartley et al. 2006; Senko 2005). We describe it in some detail because it typifies the issues that arise in V&V of societal models. The lessons from this example will figure strongly in the prescriptions we offer in the next section.

In 2001, the U.S. Defense Modeling and Simulation Office (DMSO) decided to collect a set of tools that would be useful for conducting OOTW. As mentioned earlier, these operations often involve humanitarian and peacekeeping activities. It may be relatively easy to model individually the actions that take place in OOTWs. The trouble arises when trying to connect the actions. For example, if great quantities of food are carefully imported and unloaded at the seaports and airports of a country, do the people stop starving? The answer is that it depends. It depends on factors such as culture, infrastructure, economics, and other societal factors. DMSO set out to create a toolbox that would address this problem and documented the corresponding multiyear V&V effort (Hartley 2004, 2005a, b; Hartley et al. 2006; Senko 2005).

Each year of FAST development was focused on a new scenario (e.g., Kosovo and Iraq). The scenarios and the related V&V served to expand the understanding of the OOTW domains and PMESII functions that the toolbox could model adequately. The scenarios also unified the development and testing elements into coherent plausible situations that would support face validation. Here, face validation refers to the validation method in which domain experts judge the extent to which the results agree with their knowledge and experience.

The FAST team defined four testing categories: Developer, Alpha, Beta, and Final. Developer tests were performed on each tool by its developer as part of the development process. These were informal tests and were not recorded in the FAST test database. Alpha tests were formal tests conducted on each tool by its developer during a weeklong test period. Intertool interface tests were also conducted. The team defined these tests and their data elements prior to the test period. The goal of these tests was to identify areas that needed further development. Beta tests were formal tests conducted by the tool developers but testing another developer's tool.

Again, toolbox-as-a-whole tests were conducted. Most of these tests were similarly predefined. However, some "freeplay" tests, in which the testers were permitted to create their own tests, were included. The standards for the Beta tests were higher than the standards for the Alpha tests, as the goal of the Beta tests was to ensure readiness for final or acceptance testing. Final tests were defined as user-acceptance tests, with the user conducting predefined tests and freeplay – any use that the user deemed appropriate for the tools.

Testing periods for the formal tests lasted up to a week and included more than 300 tests, each lasting from a couple of minutes to an hour in duration. The number of tests that potentially could be performed was very extensive; however, the time available for testing was strictly limited. This imposed a need for prioritizing the tests. The team developed a trace-back matrix that connected categories of tests to justifications for the tests and used this to control the selection of tests. The following list of justifications was used:

- Do not break the tool: changes to the tools should maintain previously existing functions.
- The function was specified: the Concepts of Operations (CONOPS) document described how the model or system should be operated and what it should do.
- The function was implied: the function was implied by other specifications in the CONOPS.
- User request: the function was requested by users but not specifically included in the CONOPS.
- Practical run-time requirement: the function was implied by usage (e.g., function f is justified by one of the items above and function g is not; however, function g must be performed to set up the system to test function f).

Each year, the CONOPS was revised to add new goals for the FAST Toolbox. For example, in 1 year the following objectives were appropriate to the state of development:

- Improved functionality: functionality was maintained or improved.
- Integration with XML: data transfer using XML was successful.
- Use of Command, Control, Communications, Computers, and Intelligence (C4I) data: data from an external C4I system was successfully imported.
- OOTW use-case: the OOTW scenario was successfully implemented.

The assignment of individual tests to the functions allowed for meaningful interpretation of the test result scores. These scores were used to determine areas for further work.

Most of the validation tests addressed fitness of purpose. In this example, supporting analysis, decision-making, and ease of use (required by the CONOPS) were assessed by subject matter experts (SMEs) using face validation. However, additional validation tests consisted of sensitivity analyses of the models and comparisons of algorithms with those used in other simulations. The tests for supporting analysis and decision-making included categorical tests of coverage of DIME-PMESII functions (e.g., modeling the causes and effects of migration of displaced persons). Clearly, if a given DIME-PMESII functionality was not modeled, the overall validity for use in modeling OOTW would be reduced in some measure. Coverage does not guarantee validity; however, lack of coverage negatively affects it.

The V&V task within the FAST project produced complete documentation, describing the process, the tests, and the results, over the years of the project. The documentation type most frequently used in practice was the recording of "hot

Table 4 FAST toolbox V&V lessons learned

Planning for V&V

Planning is critical and takes time and effort, both from the V&V team and the model developers.

Scenarios are useful for the face validation efforts; however, additional tests will be needed to cover all needed functions.

Testing categories (e.g., Developer, Alpha, Beta, Final) are important for defining expectations and goals concerning the results.

The system structure (models, modules, computer interfaces, human interface procedures, data) and versions must be defined.

Functions and subfunctions must be defined.

Requirements must be defined.

Objectives (for each testing event) must be defined.

The test venue and procedures must be defined.

Plan the V&V activities to coordinate with the developer testing.

The actual tests must be connected to the system structure and the objectives and must test particular functions and subfunctions. Most will have to be defined and sequenced by the developers.

The system for controlling the testing, recording the results, and analyzing the results must be ready prior to testing.

V&V activities

Communicate any observed risks and possible solutions.

There will be sponsor and developer concerns about reporting apparently negative results. (See testing categories, above.)

Record everything.

Make the V&V process transparent. The V&V process should help build confidence, not destroy it (if at all possible).

Validation

Validation of societal models may be largely a matter of checking for coverage of required functionality.

Face validation of societal models depends on the quality of the Subject Matter Experts

V&V documentation

Design the reporting process for use in accreditation, not for ease in producing a report.

Daily hot washes and a final hot wash of the results are critical.

Release reports for review as early as possible and accept inputs from all stakeholders.

Make sure the reports focus on the extent that the simulation meets the requirements, not on the simulation itself.

Accreditation

Accreditation is not controlled by the V&V team.

Accreditation may be formal or informal.

washes": reviews that immediately followed the testing. The records of hot washes consisted of PowerPoint slides that tabulated the test results in terms of the functions versus tools. The hot washes provided the sponsor with the testing results immediately upon conclusion of the testing.

Table 4 displays a set of lessons learned from the V&V of the FAST Toolbox. Note the similarities with the JWARS lessons learned (Table 3): both emphasize planning, documentation, coordination with developers, and sponsor concerns. The lessons from the FAST Toolbox add validation lessons that are specific to societal modeling.

6   A Prescription for V&V for Societal Modeling

This section is based mainly on the V&V methodology Clemence et al. (2007) created for a societal model described in Waltz (2008). The authors were members of the team that developed the methodology.

We will assume here that the overall model or tool set to be validated consists of several individual models that can be composed into an acceptable societal model. However, in the case of a monolithic societal model, think of the various modules as taking the place of the "models" in our discussion. When we use the words "test" and "testing," we include (a) dynamic tests in which the model is executed on a computer and the outputs are compared against some desired result, and (b) static tests in which the model's code, conceptual model, or data are examined and compared against some desired standard. A particular test may generate verification information or validation information or both. Where a distinction is not important, we will use the words in an inclusive sense.

Figure 4 presents a process flow diagram for V&V. Each of the branches in the figure represents a type of V&V. The types are defined by the circumstances under which the V&V takes place. Developmental V&V includes the V&V of conceptual models and debugging of the coded model during the development of the model. Periodic V&V is a formal process of testing the models and the entire system on a periodic basis (e.g., once or twice a year) and should take place during the development process and during the life cycle of the model when it is being used. Triggered V&V is a formal process that occurs whenever there is a significant change, either to the code of the model or in its application, for example, when there is a shift in use from analysis to training, or within analysis to predicting results rather than exploring possible results. The fourth branch represents capturing the increased knowledge of the model that accrues with each use. This knowledge can consist of negative occurrences, such as the discovery of bugs, or positive occurrences, such as the fact that the model performed as expected or that it delivered useful insights. Recording these occurrences should be part of the ongoing process of V&V.

Except for the final branch, the process blocks are similar. The first block, such as "code changes," represents the events that lead to the testing and define the type of testing. The "define," "execute," and "evaluate" steps are straightforward but extremely important. From personal experience, they can be extremely difficult to accomplish. Everything conspires against simplicity. For example, the people who best know what needs testing in the periodic testing of the development cycle are the developers. However, they are too busy getting the last bugs out of the code at the end of the development cycle to spend time defining the tests. By definition, triggered tests must be done quickly. If time were not of the essence, the tests could wait until the next periodic testing cycle (Fig. 4).

A good V&V process requires disciplined record keeping. The bulk of the formal testing concerns verification of major functional performance, with the addition of a few detailed tests suggested by the developers because of their importance or uncertainty. Additional tests are conducted to address the validity of the model with respect to its uses. These can include correspondence to real-world events, correspondence to other models, SME face validation, and Kott's (2008) measure of cognitive impact, as appropriate. In all cases, any knowledge that is gained is transitory and is quickly lost unless complete records are kept.

The authors also found practical reasons for multiphased formal testing. Alpha testing, beta testing, and final (or acceptance) testing all have the same structure but have different purposes. The purpose of alpha testing is to discover problems. Thus, its metrics are interpreted differently from those of the other types of testing. Lower scores mean that problems were discovered. The purpose of beta testing is to determine whether the problems discovered in alpha testing were fixed and to ensure that these

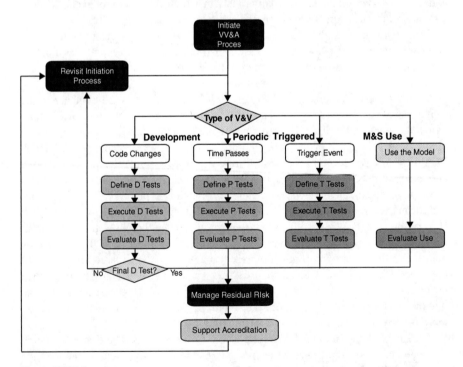

Fig. 4 V&V flowchart

fixes did not cause new problems. This also drives the definition of the beta-testing contents. The beta-testing standards should be significantly higher than the alpha-testing standards. The purpose of acceptance testing is to demonstrate to the sponsors and users of the system that the system does what it is supposed to do. The tests should include significant functional tests defined by the developers and free testing by the users.

Table 5 provides a notional illustration of the raw numbers for two cycles (alpha, beta, final) of periodic testing. The upper section gives the grand total results, the second section relates to the subset of tests that are concerned with the system as a whole, and the last two sections describe the results for the tests of the two models ("Name 1" and "Name 2") that make up this system. The "Total Tests" rows show a number of tests that were scheduled for each event, with the "Successful Tests" rows showing the numbers that were successfully completed. These numbers are useful in identifying trends.

Note the "New Tests" and "Repeated Tests" rows. In this example, the first alpha test event is the first formal test for the system so that all of the tests are "new." In the beta test events, some selected tests from the alpha tests should be repeated.

Table 5 Verification metrics example

Date	06/01/06	11/01/06	12/01/06	07/01/07	11/01/07	12/01/07
Event	Alpha test	Beta test	Final test	Alpha test	Beta test	Final test
Version number	1	1	1	2	2	2
Successful tests	81	106	109	99	116	119
Unresolved problems	6	3	0	3	3	0
Total tests	87	109	109	102	119	119
New tests	87	68	0	65	79	0
Repeated tests	0	41	109	37	40	119
System test						
Successful tests	20	25	25	25	30	31
Unresolved problems	1	0	0	2	1	0
Total tests	21	25	25	27	31	31
New tests	21	10	0	17	19	0
Repeated tests		15	25	10	12	31
Name 1						
Version number	2	2	2	3	3	3
Successful tests	30	40	42	35	41	42
Unresolved problems	3	2	0	1	1	0
Total tests	33	42	42	36	42	42
New tests	33	34	0	24	29	0
Repeated tests		8	42	12	13	42
Name 2						
Version number	3.1	3.1	3.1	3.7	3.7	3.7
Successful tests	31	41	42	39	45	46
Unresolved problems	2	1	0	0	1	0
Total tests	333	42	42	39	46	46
New tests	33	24		24	31	9
Repeated tests		18	42	15	15	46

These repeated tests will include those that failed and those that might be negatively impacted by imperfect fixes. Additionally, some new tests should be added. These new tests can be motivated by other areas that might be affected by fixes, tests of functionality that were not ready for alpha testing, or free-play testing. Do not introduce any new tests into the acceptance testing schedule. Note that some tests from the first cycle of testing are carried over into the next cycle's alpha tests (here labeled version 2). The point is that there is value in making sure that things that used to work still do.

The tests in Table 5 are mostly verification tests, although some are validation tests. There is some value in breaking down the failures to see if some part of the software development team is having more problems than other parts. However, the main goal is simply to drive the number of failures to zero.

With validation test failures, more than a description of the failure is required, and the metrics are not as simple. Not all validity problems will be solvable by code changes, and mitigations may be needed. For example, it might be appropriate to restrict the domain of applicability of the system, modify the usage process for the system, or modify the interpretation of the system outputs. The goal is to correct or mitigate all validity problems.

There are other useful metrics for validation of societal modeling besides the number of successes and failures of validation tests. One concept is the coverage of the societal domain. For example, if a societal model did not cover parts of the economic domain, its validity might be suspect, depending on the model's use. The elements of coverage include the determination of the pertinent issues, whether the model addresses a pertinent issue, how well an issue is addressed, and the question of metrics. Each element is discussed below. Because these pertinent issues will be coded as variables in models (perhaps complex variables), the term "variable," rather than "issue," will be used when the context is a model rather than the societal domain.

There are two basic methods for identifying the pertinent issues: bottom-up and top-down. The DIME-PMESII paradigm represents a top-down approach. First, the actions that might be applied in a situation requiring societal modeling are labeled as DIME actions, and then the descriptors of the situation are labeled as PMESII variables. For convenience, the categories can be broken into subcategories. On the other hand, one could begin by listing all of the activities ever undertaken within OOTWs, or any category of operations that might relate to societal modeling, and ask social scientists to provide lists of descriptors that might be appropriate for describing the state of the world. The interactions among the actions and descriptors must also be included in the list of pertinent issues. As a practical matter, neither the top-down nor the bottom-up approach works well alone. However, combining the two has proved productive. Ideally, a list of pertinent issues or variables should be complete and mutually exclusive. We do not have an ideal list; however, one such list is provided in Clemence et al. (2007).

Given a list of pertinent issues, determining whether an issue is addressed by a model is relatively straightforward. Determining how well the issue is addressed is more complicated. In validating the conceptual model, each issue that is addressed should be connected to the theory that is used in addressing the issue. The authors suggest rating the theories with the simple scale shown in Table 6. The second column explains the meaning of each rating. The third column identifies the expected result for a model that uses several theories. Because most social theories are defined within limited domains, any cross-domain connections must be supplied by the model designer, resulting in a reduced overall rating, despite the ratings of individual components. The shaded ratings represent those most likely to be used for societal modeling. In rating the coded model, the authors suggest a subjective assessment of the quality of the implementation of the theory into code with respect to the intended use of the model (a fraction between 0.0 and 1.0). Multiply the conceptual model rating by this assessment. This assessment should be produced by the most qualified individual available; however, given the lack of precision inherent in this process, the authors recommend simplicity over detail.

We are now ready to discuss the fourth and final element of coverage, model and system-level coverage metrics. The list of issues and variables, described above, yields hundreds of individual metrics. In the ideal list, each variable and thus each metric would be independent. Averaging the values of two metrics would make no more sense than averaging the x- and y-coordinates of a point would. We do not have an ideal list, and some of the issues are probably correlated; however, we do not know which ones. On the other hand, our taxonomy (derived from the top-down approach) gives us logical reasons for combining some issues. These combinations must be used carefully, as they tend to obscure

Table 6 Rating conceptual models

Label	Component	Ensemble
5	Expresses fully validated theory, e.g., Newtonian physics with caveats on operations near light speed or in regimes subject to quantum effects	Expresses multiple Level 5 theories with fully engineered interfaces, e.g., fly-out model of ground-to-air rocket involving chemical reactions of propellants, ballistics, air flow, electronics, etc.
4	Expresses well-researched theory involving considerable data checking and peer review, e.g., economic theory earning Nobel prize	Expresses multiple Level 4 or Level 5 theories with well-researched interfaces, e.g., economic model ensemble used by Federal Reserve in setting the U.S. interest and discount rates
3	Expresses theory supported by data and published in peer-reviewed literature	Expresses multiple Level 3, 4, or 5 theories with considerable peer-reviewed interfaces, e.g., some U.S. combat models
2	Expresses theory with rational basis, accepted by some experts as plausible (SWAG)	Expresses multiple Level 2–5 theories with plausible interfaces (SWAG)
1	Expresses a codified theory (WAG)	Expresses multiple Level 1–5 theories with codified interfaces (WAG)
0	Uncodified, mental model of uncertain consistency and completeness	Uncodified processes for connecting models of uncertain consistency and completeness

rather than illuminate problems. For example, in a particular list we identified, there are more than 50 variables that are classified as "Political." Any combination scheme that yields a single metric for "Political" maps a very large number of different individual variable metric values onto each possible Political-metric value.

To enhance the transparency of the model and system metrics, the authors advise the use of "spider" diagrams (Fig. 5). The spider diagrams support visualization of multiple dimensions in a single chart and support an overview and segmentation by each individual model. In this figure, the highest level of metrics are displayed for the system as a whole, for each model in the system, and for the group of inter-model connections that make up the "Connect" category in the system diagram. The distance from the center along an axis represents the value of the metric represented on the axis. The metric values are connected to permit easy comparison of the values on the axes.

There are two more levels of diagrams that are not shown. The first level contains diagrams for the system and each model describes results for only one of the categories (e.g., Political) and has subcategories as axes. The third level describes results for a subcategory and has variables as axes. Using these sets of

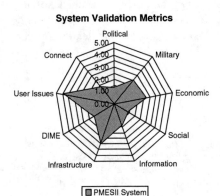

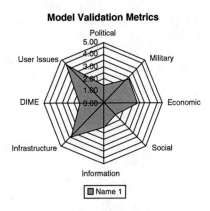

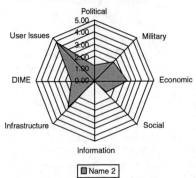

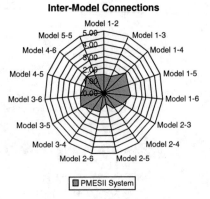

Fig. 5 Coded model validation display

diagrams provides the V&V team with insight into the model's strengths and weaknesses that emerge from the V&V process. Also not shown here are similar diagram sets for the conceptual model coverage validation metrics and data V&V metrics.

The final part of V&V is accreditation. The V&V team is not in charge of accreditation. It is responsible for gathering the information about the model, understanding it, and presenting it to the accreditation authority. The presentation should address the benefits the model will bring to the use for which it is proposed, the flaws in the model for that use, and the modifications to the use of the model envisioned to mitigate the flaws. The responsibility of the accrediting authority is to require that the presentation addresses these areas; it should understand the presentation, and make a decision. The decision can contain caveats, or restrictions, as the information about the model and the situation merit.

7   Entrenched and Risk-Based V&V

Consider V&V as part of the natural development cycle and continue V&V throughout the life of the model: *this is entrenched V&V*. Perform developmental testing while creating a model: this is internal V&V. If using a form of spiral development (Fig. 6), formal periodic tests will need to be performed: this may be internal or external V&V. When the version is ready for release, perform acceptance testing: this is generally external V&V and results in accreditation. Once a model is being used, certain events can trigger additional testing: a new model being brought into a system of models, a proposed change to an existing model (e.g., new submodel), or major changes to the system being modeled that require model changes. This triggered testing (V&V) should also result in a new accreditation decision. During use of the model, periodically perform supplemental tests to increase the understanding

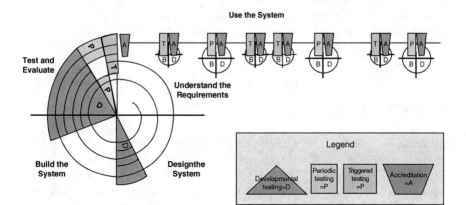

Fig. 6 Entrenched V&V

of the model. These supplemental tests include tests with new data sets or tests of functionality that were previously rated as lower priority.

While using the model, record successful results that support claims of the model's validity. It is too easy to view V&V as attempts to prove the model invalid instead of attempts to improve understanding of the strengths and weaknesses of the model so that it can be used better. If a model's weaknesses are understood, one has the opportunity to mitigate them. This is *risk-based V&V*.

Entrenched, risk-based V&V is only possible if V&V is viewed as a cumulative learning process. One must approach the process with discipline, deciding what the requirements are that should be met in each testing event, what the tests are and how they should be structured to determine if the requirements have been met, and record everything. The records of past testing are critical for cumulative learning. If each testing event is approached as if no testing has occurred in the past, one will be overwhelmed and disappointed.

Any social model large enough to address multiple DIME-PMESII elements will be too large for complete testing in any single event. Properly executed, entrenched V&V will allow one to be satisfied with incomplete testing because one will be adding to past knowledge and can expect to further extend one's knowledge of the model in the future. The risk-based V&V helps one to decide what areas need the most detailed testing.

Readers may notice that nothing is said here about any novel testing methods. Novel testing methods are of secondary importance to the more mundane task of disciplined, entrenched, risk-based, and recorded testing.

The final piece of advice concerns the impact of a model's purpose on its V&V. When setting requirements and defining tests, keep the model's purpose in mind. There is no point in requiring that a plastic scale model airplane passes a flying test! On the other hand, experience has shown that successful models are pressed into uses beyond their original purposes, often because there are no (known) alternatives. For this reason, it is a good idea to broaden the requirements somewhat so that information can be generated on a model's capabilities beyond those demanded by strict adherence to the stated purposes. On the other hand, if one is the accrediting authority, lean toward restricted accreditation and leave accreditation for broader uses for some later time when more knowledge of the model (entrenched V&V, again) has been gained.

8   Summary

Verification, Validation, and Accreditation is a process employed to determine if a model is capable of serving its intended purpose. Within this process, verification assesses how well a model's implementation matches the specifications, validation assesses the fidelity with which a model represents the real world, and accreditation officially specifies that a model is acceptable for use for a specific purpose. Fidelity is not necessarily the primary criteria of model utility; validity must be gauged with respect to the fidelity required for its intended use, not against a standard of absolute fidelity. Techniques of V&V range from the informal face validation, in which domain experts judge whether the model seems to deliver believable results, to the formal proof of correctness, in which the model is proved to terminate and meet its specifications. The lessons from the practice of V&V on complex societal models tell us that the selection and application of these techniques comprise only a small part of the V&V process. Planning, execution control, and documentation are the biggest factors in the success or failure of the V&V effort. In societal modeling, the number of uncertainties and their magnitudes are particularly high. Our understanding of reality as expressed by societal theory is relatively limited when compared to physical systems. The magnitude of the problem for V&V for societal modeling calls for a careful evaluation of the standards used for accreditation decision. Useful metrics for validation of societal modeling include successes and failures of tests and coverage of the societal domain. For example, if a societal model did not cover parts of the economic domain, its validity might be suspect, depending on the model's use. V&V that proceeds through the development cycle and continues throughout the life of the model is called the entrenched V&V. When a model is being used, certain events can trigger additional testing: a new model being brought into a system of models, a proposed change to an existing model, or major changes to the system being modeled that require model changes.

9   Resources

Balci, O. (2001) Verification, validation and testing of models. In S. I. Gass and C. M. Harris, (Eds.) *Encyclopedia of Operations Research and Management Science*. Boston, MA: Kluwer Academic Publishers.

Clemence, R. C., et al. (2007) *Verification, Validation, and Accreditation (VV&A)*. Evidence Based Research, Vienna, VA.

Hayes, B. C. and Sands, J. I. (1997) *Doing Windows: Non-Traditional Military Responses to Complex Emergencies*. Command & Control Research Program, Washington, DC.

Knepell, P. L. and Arangno, D. C. (1993) *Simulation Validation: A Confidence Assessment Methodology*. Los Alamitos, CA: IEEE Computer Society Press.

References

Balci, O. (1997) Verification, Validation and Accreditation of Simulation Models. In S. Andradottir, K. J. Healy, D. H. Withers, and B. L. Nelson (Eds.) *Proceedings of the 1997 Winter Simulation Conference* (pp. 135–141). Washington, DC.

Balci, O. (2001) Verification, Validation and Testing of Models. In S. I. Gass, and C. M. Harris (Eds.) *Encyclopedia of Operations Research and Management Science*. Kluwer Academic Publishers, Boston, MA.

Balci, O., et al. (2002) A Collaborative Evaluation Environment for Credibility Assessment of Modeling and Simulation Applications. In E. Yucesan, C. H. Chen, J. L. Snowdon, and J. M. Charnes (Eds.) *Proceedings of the 2002 Winter Simulation Conference*, (pp. 214–220). Washington, DC.

Bauman, W. (1995) *Ardennes Campaign Simulation (ARCAS)*, CAA-SR-95-8. US Army Concepts Analysis Agency, Bethesda, MD.

Box, G. E. P. (1979) Robustness in the Strategy of Scientific Model Building. In R.L. Launer, and G.N. Wilkinson (Eds.) *Robustness in Statistics*. Academic Press, New York.

Chaturvedi, A. R. (2003) *SEAS-UV 2004: The Theoretical Foundation of a Comprehensive PMESII/DIME Agent-Based Synthetic Environment*. Simulex, Inc.

Clemence, R. C., et al. (2007) *Verification, Validation, and Accreditation (VV&A)*. Evidence Based Research, Vienna, VA.

United States Coast Guard (2006) *Changes to Deepwater Plan Appear Sound, and Program Management Has Improved, but Continued Monitoring Is Warranted*. GAO, Washington, DC.

Conwell, C. L., et al. (2000) Capability Maturity Models Support of Modeling and Simulation Verification, Validation, and Accreditation. In J. A. Joines, R. R. Barton, K. Kang, and P. A. Fishwick (Eds.) *Proceedings of the 2000 Winter Simulation Conference* (pp. 819–828). Washington, DC.

Davis, P. K. (1992) Chapter VI – A Framework for Verification, Validation, and Accreditation. In A. E. Ritchie (Ed.) *Simulation Validation Workshop Proceedings (SIMVAL II)*. MORS, Alexandria, VA.

Davis, D. F. (2006) *Consolidated Report Consisting of three Research and Development Project Summary Reports*. Contract #W15P7T-06-T-P238. Peace Operations Policy Program. George Mason University, Arlington, VA.

Department of the Army, Pamphlet 5-11 (1999) *Verification, Validation, and Accreditation of Army Models and Simulations*. Washington, DC.

DMSO VV&A website (as of 2007) http://vva.dmso.mil/.

DoD Directive 5000.59 (2007) DoD Modeling and Simulation (M&S) Management. http://www.dtic.mil/whs/directives/corres/pdf/500059p.pdf.

Giadrosich, D. (1992) Chapter IV-A – Validating Models and Simulations. In A. E. Ritchie (Ed.) *Simulation Validation Workshop Proceedings (SIMVAL II)*. MORS, Alexandria, VA.

Hartley III, D. S. (1975) *An Examination of a Distribution of TAC CONTENDER Solutions*. National Military Command System Support Center, Washington, DC.

Hartley III, D. S., et al. (1989) *Sensitivity Analysis of the Joint Theater Level Simulation I K/DSRD-70*. Martin Marietta Energy Systems, Inc., Oak Ridge, TN.

Hartley III, D. S., Quillinan, J. D. and Kruse, K. L. (1990) *Verification and Validation of SIMNET-T*, KJDSRD-117. Martin Marietta Energy Systems, Inc., Oak Ridge, TN.

Hartley III, D. S. (1991) *Confirming the Lanchestrian Linear-Logarithmic Model of Attrition*, K/DSRD-263/R1. Martin Marietta Energy Systems, Inc., Oak Ridge, TN.

Hartley III, D. S., Radford, C. and Snyder, C. E. (1991) *Evaluation of the Advanced Battle Simulation in WARES 3-90*, K/DSRD-597. Martin Marietta Energy Systems, Inc., Oak Ridge, TN.

Hartley III, D. S., et al. (1994) *An Independent Verification And Validation of The Future Theater Level Model Conceptual Model*, K/DSRD-1637. Martin Marietta Energy Systems, Inc., Oak Ridge, TN.

Hartley III, D. S. (1996) *Operations Other Than War: Requirements for Analysis Tools Research Report*, K/DSRD-2098. Lockheed Martin Energy Systems, Inc., Oak Ridge, TN.

Hartley III, D. S. (1997) Verification & Validation in Military Simulations. In S. Andradóttir, K. J. Healy, D. H. Withers, and B. L. Nelson (Eds.) *Proceedings of the 1997 Winter Simulation Conference* (pp. 925–932). Washington, DC. http://www.informs-sim.org/wsc-97papers/0925.PDF.

Hartley III, D. S. (2001) *Predicting Combat Effects*. Linthicum, MD: INFORMS.

Hartley III, D. S. (2004) *FAST for the Warfighter: Test Strategy & Plan (Revision 3)* Dynamics Research Corporation, Vienna, VA.

Hartley III, D. S. (2005a) *MOOTW FAST Prototype Toolbox: FY04 Validation Strategy & Plan*. Dynamics Research Corporation, Vienna, VA.

Hartley III, D. S. (2005b). *MOOTW FAST Prototype Toolbox: FY05 Validation Strategy & Plan*. Dynamics Research Corporation, Vienna, VA.

Hartley III, D. S., Holdsworth, D. and Farrell, C. (2006) *OOTW FAST Prototype Toolbox: Analysis Process*. Dynamics Research Corporation, Orlando, FL.

Hayes, B. C. and Sands, J. I. (1997) *Doing Windows: Non-Traditional Military Responses to Complex Emergencies*. Command & Control Research Program, Washington, DC.

Henderson, D. (1992) Chapter IV-B – The Multidimensional Space of Validation. In A. E. Ritchie (Ed.) *Simulation Validation Workshop Proceedings (SIMVAL II)*. MORS, Alexandria, VA.

IEEE 610-1991 (1991) IEEE Standard Computer Dictionary. A Compilation of IEEE Standard Computer Glossaries. http://ieeexplore.ieee.org/xpl/tocresult.jsp?isnumber=4683.

IEEE 1278.1-1995 (1996) IEEE Standard for Distributed Interactive Simulation – Application Protocols. http://ieeexplore.ieee.org/xpl/tocresult.jsp?isNumber=10849.

IEEE 1278.1A-1998 (1998) IEEE Standard for Distributed Interactive Simulation – Application Protocols. http://usl.sis.pitt.edu/wjj/otbsaf/IEEE1278.1a-1998.pdf.

IEEE 1516.1-2000 (2000a) IEEE Standard for Modeling and Simulation (M&S) High Level Architecture (HLA) Federate Interface Specification. http://ieeexplore.ieee.org/xpl/standard-stoc.jsp?isnumber=19743.

IEEE Std 1516-2000 (2000b) IEEE Standard for Modeling and Simulation (M&S) High Level Architecture (HLA) – Framework and Rules. http://ieeexplore.ieee.org/xpl/standardstoc.jsp?isnumber=19334.

IEEE 1516.2-2000 (2001) IEEE Standard for Modeling and Simulation (M&S) High Level Architecture (HLA) – Object Model Template (OMT) specification. http://ieeexplore.ieee.org/xpl/freeabs_all.jsp?tp=&isnumber=19791&arnumber=915738&punumber=7319.

Kott, A. (2008) *Fiction vs. Reality: Validation of Decision-Support Models via Cognitive Impact*. Keynote speech at MODSIM World 2008 Conference, Sep. 15–18, Virginia Beach, VA.

Knepell, P. L. and Arangno, D. C. (1993) *Simulation Validation: A Confidence Assessment Methodology*. IEEE Computer Society Press, Los Alamitos, CA.

Meadows, D. H., Meadows, D. L., Randers, J. and Behrens, W. (1972) *The Limits to Growth*. Signet, New York.

McDonald, C. S. (2006) *Verification and Validation Report for the Synthetic Environment for Analysis and Simulation (SEAS)*. The Johns Hopkins University – Applied Physics Laboratory, Laurel, MD.

Metz, M. L. (2000) Joint Warfare System (JWARS) Verification and Validation Lessons Learned. In J. A. Joines, R. R. Barton, K. Kang, and P. A. Fishwick (Eds.) *Proceedings of the 2000 Winter Simulation Conference* (pp. 855–858). Washington, DC.

Sargent, R. G. (2004) Validation and Verification of Simulation Models. In R. G. Ingalls, M. D. Rossetti, J. S. Smith, and B. A. Peters (Eds.) *Proceedings of the 2004 Winter Simulation Conference*. Washington, DC.

Senko, R. M. (2005) *Flexible Asymmetric Simulation Technologies (FAST) for the Warfighter: FY05 Final Verification Test Report*. Dynamics Research Corporation, Vienna, VA.

Sheldon, B. (2006) Memorandum for the Record: V&V Report for the Synthetic Environment for Analysis and Simulation (SEAS). 27 October 2006.

Simulation Verification, Validation and Accreditation Guide (2005) Australian Defence Simulation Office, Department of Defence, Canberra, Australia.

Staniec, C. and Hartley III, D. S. (1997) *OOTW Analysis and Modeling Techniques (OOTWAMT) Workshop Proceedings*. MORS, Alexandria, VA.

Secretary of the Navy Instruction 5200.40 (1999) *Verification, Validation, and Accreditation (VV&A) of Models and Simulations*. Washington, DC.

Vector Research (1981) *Verification Analysis of VECTOR-2 with the 1973 Arab-Israeli War and Analysis of Related Military Balance Issues*. VRI-NGIC-1 FR 81-1. VRI., Ann Arbor, MI.

Wait, P. (2001) *Project Protection Comes at a Price*. Washington Technology. http://washington-technology.com/articles/2001/08/24/project-protection-comes-at-a-price.aspx.

Waltz (2008)

Youngblood, S. (2004) *DoDI 5000.61 and the VV&A RPG*. Defense Modeling and Simulation Office, Washington, DC.

Youngblood, S. (2005) *VV&A Standards Initiatives*. Briefing at the NDIA M&S Committee Meeting. Washington, DC. http://www.ndia.org/Divisions/Divisions/SystemsEngineering/ Documents/Content/ContentGroups/Divisions1/Systems_Engineering/PDFs18/Modeling_ Committee_PDFs/VVA%20Youngblood%20to%20NDIA%20MS%20Committee%20-%20 07%20Feb%202007%20v2.pdf.

Chapter 12
Conclusions: Anticipation and Action

Alexander Kott and Stephen Morse

International interventions are potentially massive undertakings intended to create a variety of desirable conditions on the ground – political, security, economic, social, etc. – in order to bring lasting peace and stability. Such endeavors are multifaceted in nature, and making progress is invariably an uncertain and vexing process. Clearly, no book can fully address this vast topic.

Moreover, interventions present unique challenges to those who would seek to model them to support better planning and decision making. Although this book looks specifically at computational methods for anticipating the real-world outcomes of an international intervention, the authors recognize that this is a young and rapidly evolving field of study in which many areas require further investigation and understanding.

Thus, important topics, such as dynamics of social structures, changes in cultural norms, effects on prior educational and social support systems, disruptions or improvements to infrastructure, and so on, are barely touched upon. Even if this were an extensively developed discipline, a computational model cannot be anything but a pale shadow of the infinitely rich, complex reality of a society in distress. The list of relevant and important issues and phenomena is truly endless.

Another important issue that we have only tangentially mentioned is the technique for integrating disparate models that belong to different disciplines and that often use highly dissimilar modeling paradigms. This is a largely unexplored area, and practitioners deal with it in a largely ad-hoc manner (Waltz 2008; Kott and Corpac 2007).

The inevitable incompleteness of any collection of models, along with poorly understood methods for combining heterogeneous models, leads to uncertainty regarding the reliability of computational tools. This uncertainty is exacerbated by difficulties in validation of such tools (Chap. 11). Throughout this book, the authors have stressed this uncertainty repeatedly and urged readers to use such tools with great caution, indicating that these tools should only be used as aids to human analysis and decision making.

A. Kott (✉)
Army Research Laboratory, 2800 Powder Mill Road, Adelphi, MD 20783, USA
e-mail: alexander.kott1@us.army.mil

A. Kott and G. Citrenbaum (eds.), *Estimating Impact*, 337
DOI 10.1007/978-1-4419-6235-5_12, © Springer Science+Business Media, LLC 2010

The authors believe that prudent, problem-specific methodologies are indispensable when using computational tools. Strictly speaking, it would be outside the scope of this book to outline such methodologies when using the modeling and estimating tools based on computational techniques discussed here.

However, given the practical orientation of the book, the question of how practitioners can make better use of computational tools cannot be ignored entirely. In particular, practitioners must wonder: how can we accommodate the uncertainty of a tool's results by applying human judgment appropriately?

1   Uses of Estimates

In two examples which we discuss below, planners and analysts used (or could have used) computational tools to obtain estimates of effects of various actions under consideration. They then considered these computational estimates to draw their own conclusions regarding the effects that would likely emerge from proposed actions taken by the international mission. The specific ways in which they could utilize these estimates deserve elaboration. Here we rely on observations made during extensive experiments with a suite of PMESII tools (Kott and Corpac 2007).

First, the most conventional use of computational estimates occurs when users largely agree with computer estimates – i.e., when they find the estimates to be consistent with their own intuition and expectations. In this case, the main value of computational tools is in producing a far greater degree of detail than a human analyst or expert can provide. The estimates of a PMESII tool can describe, e.g., the changes in hundreds of diverse political, economic, and social variables over many years, with a time-step of 1 week. Even a large team of expert analysts cannot produce such a detailed product (assuming that a decision requires this depth of analysis) in a practical amount of time. Still, users may find a need to adjust some of the estimates, or replace part of them with expert judgments.

Second, even if users disagree with computer estimates, a model can still point out potential effects that the users have not appreciated beforehand. Consider that a PMESII tool often produces estimates that describe changes in hundreds of variables. Even highly experienced analysts are unlikely to consider all of these variables. After examining the estimates, however, users often find important effects – i.e., changes in PMESII variables – that they had not considered previously but now find important enough to investigate further, although they may disagree strongly with the direction and magnitude of the computer-estimated effect.

Third, a computational tool is an important mechanism to elucidate and examine assumptions, which otherwise remain hidden or untested. As users explore or explain the reasons for their disagreement with computer-generated estimates, they often verbalize heretofore unspoken assumptions that should have been recorded or perhaps revisited. Occasionally, users are unaware of implicit assumptions which the process of interacting with a computer tool helps to reveal. When this learning process occurs among a team of analysts, some analysts may be surprised by the assumptions made by other analysts. Thus, computer-generated estimates can serve as a helpful learning mechanism to draw out unrecognized assumptions and biases.

Fourth, in a similar fashion, a computer tool helps users by providing them with an alternative or opposing opinion. By examining computational results and formulating the reasons why they disagree with them, users can arrive at clearer, more logical, and compelling explanations of their own positions. In effect, the computational models serve here as a useful intellectual punching bag with which users can learn and fortify their arguments.

2 Analogy: Model Predictive Control

We should also ask a broader question: how should one choose an action (or decide to do nothing) in an uncertain world in which the computational means for anticipating the action's impact are imperfect?

This very question is a subject of rigorous and extensive research in the discipline of control theory. Some key challenges and solutions developed in this discipline can give us useful insights in answering the broader question mentioned above.

A simple example of a control system is described in Fig. 1. On the left side is a depiction of an aircraft control system. An automatic system controller continuously senses the behavior of the aircraft (the "plant," in control-theoretical parlance); and based on an analysis of the sensor readings, it then issues control signals to the aircraft to produce the desired behavior. For an airplane, for example, the sensors tell the controller about such things as air speed, pressure, and wind; then the controller issues signals that automatically adjust the flaps and power to keep the plane on course (the course having been selected by the pilot).

In the context of an international invention shown on the right side of Fig. 1, the "plant" is the troubled country experiencing a crisis and in need of international

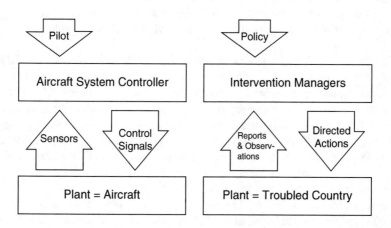

Fig. 1 An analogy between an aircraft control system and an intervention management system

assistance. The control signals are the directions issued by the intervention managers (so-called for the sake of generality) of the international mission regarding actions to be taken on the ground. These might include directions for enforcing a cease-fire, providing food aid, or arresting leaders of a criminal organization. Instead of obtaining sensor information from an automatic controller of an aircraft, intervention managers make observations and get reports of the situation within the country, such as reports of unwanted firefights, trends of malnutrition, or incidents perpetrated by a criminal organization.

Just as an automatic control system *continuously* monitors the state of the aircraft in flight and adjusts accordingly, so also should the intervention managers perform a periodic review of the progress achieved so as to exploit opportunities flexibly and respond rapidly to unwanted events within the country.

In the discipline of control theory, it is typically assumed that a *model of the system* to be controlled is available. The model can relate controllable inputs and the state of the current system so as to determine likely outcomes; that is, forecast how the current state of the system would change in response to a variety of user-selectable actions. Given a set of desired outcomes – to keep an aircraft stable and on course, for example – the controller uses the outcomes derived from the model to select from among available actions those that would best advance the desired goals.

But what if the model is imperfect? How would an intervention manager choose the most effective actions to make things better if the available model could not accurately represent the actual behavior of the society?

Control theorists have studied how the controller should behave if there is a disparity between the system model, on the one hand, and the actual system state and behavior, on the other. That is, if the model predicts one type of response, but the system behaves in some other way, the potential exists for erroneous (and perhaps catastrophic) outcomes to emerge. A control system should be able to perform within specified constraints, even when certain types of model disparities are present.

One method that the control theory provides for complex control problems is what is referred to as *model predictive control* (MPC). MPC is particularly appropriate when one is concerned about significant disparities between the model and the actual "plant," as is inevitably the case when modeling an intervention in a country in crisis.

The key steps of the MPC approach are to (a) perform predictive modeling of the plant for a relatively long period of time from the current moment forward; (b) select control signals that optimize long-term performance of the plant; (c) repeat the process of modeling (starting with an assessment of the new, most recent observations of the current situation), after a short period of executing the control signals; and (d) adjust the control signals (Mayne et al. 2000). In this manner, by frequent remodeling followed by timely adjustment of the control signals, the approach of MPC reduces the erroneous effects associated with the model's limitations in depicting reality in an uncertain world.

Let us elaborate on the MPC approach using the terminology associated with managing an international intervention. At a suitable time point during an intervention, the intervention managers assess reports on the current situation, reconsider their current strategy, and then revisit their plan of action, which covers a period of time called the planning horizon, say, 18 months forward. Then, they use the available PMESII models to assess the effects of their plan for the duration of the planning horizon. They will probably repeat the planning–modeling process several times, until they arrive at a plan that seems to promise the best combination of effects. Having generated a revised plan of action, they issue guidance and orders to various field organizations that carry out the actions directed by the intervention managers.

As time passes and field organizations perform the specified actions, the intervention managers collect further observations regarding the progress achieved. When they detect an unexpected or unwelcome event – a major crisis, a serious deviation from the anticipated effects, or simply a predefined number of months pass – the intervention managers perform replanning (Fig. 2). Again, they adjust their plan based on the most recent situation reports, use the available PMESII models to estimate impacts 18 months forward, adjust their current plan to produce the best combination of effects from this time forward, and issue new guidance and orders to the field organizations.

Note that the assess-remodel-adjust plan cycle tends to unfold within a significantly shorter time period than the 18-month planning horizon. The planning horizon could, for example, be equal to six update cycles; that is, the plan is revised six times (say, every 3 months) before its horizon (18 months) elapses. For this to happen, the international mission should be organized and staffed to perform such recurring updates. Here, periodic control activities (that is, frequent assess-remodel-adjust plans) are not the exception but rather they are the expected, normal approach for managing an international intervention in complex and uncertain environments.

To be sure, the model predictive control concept cannot be applied to intervention management in a blind, mechanistic fashion. Human beings are not mere sensors and actuators, and real-world interventions are immensely difficult to manage, often involving tragic upheavals that require policy-makers and decision-makers to apply deep human insights, judgment, experience, and leadership. Indeed, there are many complications.

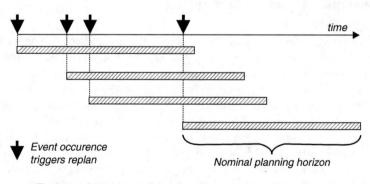

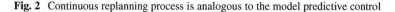

Fig. 2 Continuous replanning process is analogous to the model predictive control

Frequently, it is difficult to collect and interpret observations and reports regarding the country's situation at any given moment. Information is often incomplete, partly erroneous or intentionally distorted, contradictory, and subject to conflicting assessments. Deciding on the true meaning of the available information requires experience and mature judgment. Here, however, the use of computational models can help highlight the most consistent interpretations.

There are many real-world examples. Graham-Brown (1999) provides a cautionary note. In describing the experiences of international sanctions imposed on Iraq beginning in 1991, she illustrates the extreme difficulties that intervening organizations – governmental or nongovernmental – find in collecting reliable information on the effects of intervention actions and the multiple distortions of that information. One must not underestimate the challenges of objectively assessing conditions on the ground.

Further, as has already been emphasized, the PMESII models are inevitably imperfect in modeling a country in crisis, and they can produce results that can be potentially misleading in ways that may be unknown to users. Although the MPC paradigm serves to reduce the impact of such imperfections, users of models must apply common sense and interpret the computational results critically. When significant disparities occur between the model estimates and real-world results, the models may need adjustments and even major modifications.

The key point is that a mission's plan of action in an uncertain world is not immutable but only reflects current best knowledge. Consequently, it should be updated as new information on the situation becomes available. In an uncertain world, the MPC approach of periodic assessment and a rolling time horizon offers a structured way of organizing reporting systems, revisiting current strategies for change, and adjusting the mission's plan of action. This approach should be tailored to the specific requirements and constraints of a particular task and organization. We now describe two examples in which an approach conceptually similar to MPC is applicable.

3 Example: Advance Planning

In 2009, a U.S. government organization completed a detailed study that generated an advance plan for assisting a certain friendly country (referred to here as "Country-Y") that was experiencing a threat of insurgency. In performing this planning effort, the organizers of the study used a collection of PMESII models and developed a methodology that bears partial resemblance to the MPC paradigm. The following discussion uses only the publicly available information reported in Messer (2009).

The purpose of the study effort was to develop a preliminary, high-level intervention plan that involved assisting the government of Country-Y in defending itself against the insurgency while building an indigenous security capacity. The product of the study effort was to provide an analytical baseline for future additional planning, if any were to become necessary.

The study included a broad range of experts from several U.S. government agencies and one non-U.S. government organization. These experts represented many diverse areas: agriculture, justice, commerce, diplomacy and state relations, international development, international finance, intelligence gathering, and military services.

The study made use of numerous PMESII tools, including several discussed in this book: COMPOEX, Nexus, PSOM, Senturion, and others. To combine the use of PMESII models with expert judgments, the organizers of the study developed a human-in-the-loop war-gaming procedure they called the X-Game, which resembles the MPC approach in several aspects. Like a conventional war game, the X-Game involved several teams of human analysts, planners, and subject matter experts.

The participants were organized into five cells. The Blue Cell played the role of the intervening nations and officials of the government of Country-Y. This cell developed plans of action intended to accomplish the objectives of the international intervention. The Red Cell played the role of the various opposition parties, such as internal insurgent groups, that threatened the government of Country-Y as well as the international mission. This cell defined the objectives and plans of action of all such opposing parties. The Green Cell played the role of the various groups within the indigenous population and reflected their changing attitudes and reactions. The White Cell used the subject matter experts to assess the effects of all the actions on a broad range of PMESII-type metrics. In addition, a Modeling Team operated the PMESII computational models.

The overall planning horizon was 10 years. This long period was divided into phases. In each phase, the following process took place, approximately:

1. Taking into account all available information about the situation in Country-Y at the beginning of this phase, and the history of events up to that point, the Blue, Red, and Green Cells formulated their plans of action several years forward. They presented these plans to the Modeling Team and the White Cell.
2. The Modeling Team entered the plans into PMESII modeling tools and used the tools to generate estimates of PMESII effects as they unfolded to the end of the entire 10-year period. This step was also a good time to adjust the models if there was a major difference between expert assessments of effects of the previous phase and the corresponding computational estimates.
3. The Modeling Team presented the computed results to the White Cell.
4. The White Cell generated the assessment of how PMESII effects would unfold in Country-Y over time, taking into account the Blue and Red actions. In this assessment, the White Cell considered the predictions of the computational tools but did not necessarily follow these predictions.
5. The White Cell decided when a significant change occurred in the situation of Country-Y. This point in time became the beginning of the next phase, and the process iterated.

The organizers of the study reported achieving the required objectives of the study effort. To our knowledge, this was the first large-scale study to make systematic use of multiple PMESII computational tools for a practical planning purpose.

4   Example: Next State Planning

The second example is described by Kott et al. (2007) in what is now called the approach of Next State Planning for organizing planning efforts of an international mission. The paper emphasizes that PMESII computational tools would fit naturally into the Next State approach.

The international intervention in Kosovo in 1999 planted the seeds of Next State Planning. The combined UN-NATO mission (called UNMIK-KFOR) initiated this approach by preparing short-term plans to achieve specific near-term objectives on the ground in the next 3–5 months. This near-term approach was taken in addition to developing the mission's overall plan under a much longer time horizon of 3 years, the so-called End State Planning.

The method of Next State Planning derives from the lessons learned in the Kosovo intervention and generalizes some of the planning methods actually used there (Covey et al. 2005). Although mission managers in Kosovo did not use the PMESII modeling tools such as are discussed here, they do identify a beneficial role such tools could play if they were available at the time. Interestingly, while different from the X-Game, the Next State Planning approach also bears resemblance to the MPC paradigm in certain important aspects.

Although much smaller in size than later interventions, the UN-NATO intervention in Kosovo was no less extensive or complex. The UN-led civilian mission included 5,000 international police and 3,000 civil administrators, advisors, and trainers. Many humanitarian nongovernmental organizations sent thousands of relief workers and human rights investigators. Moreover, the NATO-led military command, KFOR, numbered 44,000 troops.

Based on the Kosovo experience, Covey et al. (2005) argue that planning of an intervention should occur on more than one temporal scale. At the most extensive, a 3-year mission plan for the entire intervention is needed to envision the emergence of a lasting political solution to the conflict. However, this long-range mission plan cannot attach great certainty to many key developments in the near term, and, of necessity, it defines a range of alternative approaches rather than a specific path. At this highest level of abstraction, the challenge is in ensuring that at least one of the paths will prove viable and effective once the details are understood.

Figure 3 illustrates how an international mission, operating in the presence of uncertainty, engages in a continuous process of moving in the direction of the desired end state, albeit in a stepwise process of achieving a series of "next states." This "next state thinking" involves continuously updating the mission's understanding of the situation and projecting it forward in time, and using this projection to guide near-term planning to achieve a desired "next state." The overall result is a jagged path of desired next states with a time horizon of 3–5 months that moves forward (seemingly haphazardly) but relentlessly in the direction of the desired end state over a longer time horizon of 3–5 years.

Consequently, to ensure that the path proceeds in the direction of the overall desired end state, a less-abstract layer of planning, or next state planning, is necessary to

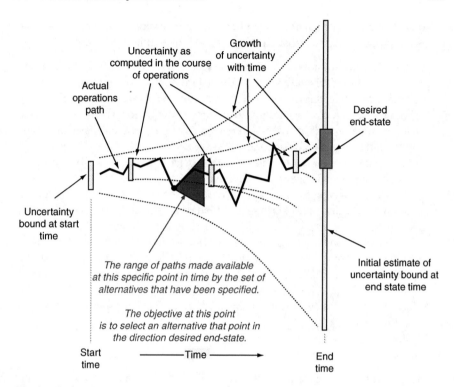

Fig. 3 High-level intervention plan faces a wide range of uncertainties

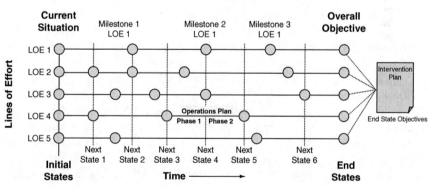

Fig. 4 The next states divide the overall intervention into short phases that present lower uncertainty range

generate a near-term, 3- to 5-month plan to achieve a specific intermediate next state of the intervention. Looking at the bigger picture, next states divide a long-term intervention into a sequence of well-defined, short-duration stepping stones. Figure 4 illustrates the relationship between the overall intervention plan and the next states.

At the beginning of each next state phase, the intervention managers perform the following process, adapted from Kott et al. (2007) in a simplified form.

1. Assess the current situation and its relation to the intended end state. Here, potentially, PMESII modeling tools help determine why the actions taken in the previous next state phase have not yielded the anticipated results. This may highlight an erroneous assumption. If there were major disparities between the model predictions and the real-world outcomes of the previous phase, this is the time to remodel specific phenomena.
2. Identify potentially feasible and desirable outcomes that describe favorable conditions of the next state (e.g., a national election).
3. Develop and analyze several alternative approaches to achieving the desired conditions of the next state. Form several planning teams and give each a different assumption. The teams generate alternative strategies and analyze their effects. Here, PMESII modeling tools can play an effective role: a planning team can use the tools to estimate and compare effects of alternative strategies, explore dependencies between various PMESII variables and their temporal dynamics, and test conceptual assumptions.
4. Select the best strategy, develop the next state plan, and perform risk and feasibility analyses.
5. Finalize.
6. Issue the action plan for execution.

Other authors emphasize the importance of periodic reassessment and replanning. For example, without offering a formal model or process for an interim-level replanning, Cuny and Hill (1999) provide recommendations for specific methods to assess periodically the state of famine in a region and specific changes in famine-response methods depending on the assessment.

The MPICE program (Dziedzic et al. 2008) has developed a broad-ranging recommendation for gathering a variety of in-depth PMESII metrics that can be useful in assessing the nature of the current next state. With a set of standardized methodologies and tools for such assessments, PMESII modeling techniques will acquire a reliable baseline of data.

The key idea, in both of our examples, is to generate a continuous cycle of anticipations and actions; in each cycle computational estimates of effects help intervention managers to determine appropriate actions, and then assessments of real-world outcomes guide the next increment of computational estimates. With a proper methodology, PMESII modeling tools can offer valuable insights and encourage learning, even if they will never produce fully accurate estimates useable in a customary, strictly predictive manner.

Having flourished only within the last decade, PMESII modeling approaches exhibit the limitations of a very young discipline. Yet the trend is unmistakable: these approaches are maturing, gaining popularity, and becoming indispensable tools for analysts, planners, and decision-makers in government and business.

References

Covey, J., Dziedzic, M. J. & Hawley, L. R. (2005). *The Quest for Viable Peace: International Intervention and Strategies for Conflict Transformation.* Washington DC: United States Institute for Peace.

Cuny, F. C. & Hill, R. B. (1999). *Famine, Conflict and Response: a Basic Guide.* West Hartford, Connecticut: Kumarian Press, pp. 117–126.

Dziedzic,M., Sotirin, B. & Agoglia, J. (2008). *Measuring Progress in Conflict Environment (MPICE).* Report ADA488249, US Army Corps of Engineers.

Graham-Brown, S. (1999). *Sanctioning Saddam: The Politics of Intervention in Iraq.* London: I.B. Tauris, pp. 315–319.

Kott, A. & Corpac, P. S. (2007). COMPOEX Technology to Assist Leaders in Planning and Executing Campaigns in Complex Operational Environments. *12th International Command and Control Research and Technology Symposium*, Newport, Rhode Island, June 19–21.

Kott, A., Corpac, P. S., Hawley, L., Brown, G. & Citrenbaum, G. (2007). Next State Planning: A "Whole of Government" Approach for Planning and Executing Operational Campaigns. *12th International Command and Control Research and Technology Symposium, Newport,* Rhode Island, June 19–21.

Mayne, D., Rawlings, J. B., Rao, C. V. & Scokaert, P. O. M. (2000). Constrained Model Predictive Control: Stability and Optimality. *Automatica,* 36(6), 789–814.

Messer, K. (2009). The Africa Study. Presented at the *HSCB Focus 2010 Conference*, August 5–7, 2009, Chantilly, VA.

Waltz, E. (2008). Situation Analysis and Collaborative Planning for Complex Operations. *13th ICCRTS*, Bellevue, WA, June 17–19.

Abbreviations and Acronyms

ABM	Agent-based modeling
ACTOR	Analyzing complex threats for operations and readiness
AST	Agenda-setting theory
BDI	Beliefs, desires, intentions
BOA	Bayesian optimization algorithm
CAST	Conflict assessment system tool
COIN	Counter insurgency
COMPOEX	Conflict modeling, planning, and outcomes experimentation
CONOPS	Concept of operations
CPOF	Command post of the future
CPT	Communication penetration theory
DARPA	Defense advanced research projects agency
DIAMOND-US	Diplomatic and military operations in a non-warfighting domain
DIMEFIL or DIME	Diplomatic, informational, military, economic, financial, intelligence and law-enforcement
DIS	Distributed interactive simulations
DMSO	Defense modeling and simulation office
DOD	Department of defense
ELT	Elaboration likelihood theory
FAST-	Flexible asymmetric simulation technologies
FCM	Fuzzy cognitive maps
FTLM	Future theater level model
GDP	Gross domestic product
HLA	High level architecture
HN	Host nation
HSCB	Human social and cultural behavior
IDP	Iterated prisoner's dilemma
IOS	Individual, organizational, and societal
ISSM	Interim semi-static stability model
IW	Irregular warfare
JFCOM	Joint forces command

JWARS	Joint warfare system
LOC	Lines of communication
MAS	Multiagent simulation
MC	Media channel
MIM	Media influence model
MPC	Model predictive control
MRM	Multiresolution model
NGO	Nongovernmental organization
NNL	Nexus network learner
ODE	Ordinary differential equations
OOTW	Operations other than war
OPT	Opinion leadership theory
PEBM	Public education and broadcasting model
PIE	Public information environment
PMESII	Political, military, economic, social, infrastructure, information
POFED	Politics, fertility, and economic development
PSOM	Peace support operation's model
PSTK	Power structure toolkit
PTT	Power transition theory
QVP	Quest for viable peace
RCM	Rational choice model
RPC	Relative political capacity
SCIPR	Simulation of cultural identity for prediction of reactions
SD	System dynamics
SEAS	Synthetic environment for analysis and simulation
SIT	Social influence theory
SJM	Social judgment model
SJT	Social judgment theory
SMCR	Source-message-channel-receiver
SME	Subject matter expert
SNR	Signal-to-noise ratio
SSM	State stability model
TFL	Transport for London
TOCU	Transport operational command unit
USG	United States government
VPMM	Validation process maturity model
V&V	Verification and validation
VV&A	Verification, validation and accreditation

Index